FOREWORD

The OECD/NEA Nuclear Science Committee set up a Working Party on Physics of Plutonium Recycling in June 1992 to deal with the status and trends of physics issues related to plutonium recycling with respect to both the back end of the fuel cycle and the optimal utilisation of plutonium. For completeness, issues related to the use of uranium coming from recycling are also addressed.

The Working Party met three times and the results of the studies carried out have been consolidated in the series of reports "Physics of Plutonium Recycling".

The series covers the following aspects:

- Volume I *Issues and Perspectives*;
- Volume II *Plutonium Recycling in Pressurized-water Reactors*;
- Volume III *Void Reactivity Effect in Pressurized-water Reactors*;
- Volume IV *Fast Plutonium-Burner Reactors: Beginning of Life*;
- Volume V *Plutonium Recycling in Fast Reactors*; and,
- Volume VI *Multiple Recycling in Advanced Pressurized-water Reactors*.

The present volume is the fifth in the series and describes the specific benchmark study concerned with multirecycle performance and toxicity behaviour of a metal-fuelled fast reactor with conversion ratios ranging from 0.5 to 1.0.

The opinions expressed in this report are those of the authors only and do not represent the position of any Member country or international organisation. This report is published on the responsibility of the Secretary-General of the OECD.

CONTENTS

SUMMARY

As part of a programme proposed by the OECD/NEA Working Party on Physics of Plutonium Recycling (WPPR) to evaluate different scenarios for the use of plutonium, fast reactor physics benchmarks were developed.

In this report, the multirecycle performance of the metal-fuelled benchmark is evaluated. Benchmark results assess the reactor performance and toxicity behaviour in a closed nuclear fuel cycle for a parametric variation of the conversion ratio between 0.5 and 1.0. Results indicate that a fast burner reactor closed fuel cycle can be utilised to significantly reduce the radiotoxicity originating in the LWR cycle which would otherwise be destined for burial.

CONTRIBUTORS AND BENCHMARK PARTICIPANTS

AUTHORS	*K. Grimm*	ANL	U.S.A.
	R. Hill	ANL	U.S.A.
	D. Wade	ANL	U.S.A.
PROBLEM SPECIFICATION	*D. Wade*	ANL	U.S.A.
DATA COMPILATION AND ANALYSIS	*K. Grimm*	ANL	U.S.A.
	R. Hill	ANL	U.S.A.
SECRETARIAT	*E. Sartori*	OECD/NEA	
TEXT PROCESSING AND OUTLAY	*E. Johnson*	ANL	U.S.A.
	P. Jewkes	OECD/NEA	
BENCHMARK PARTICIPANTS	*J. Da Silva*	CEA	France
	G. Rimpault	CEA	France
	T. Ikegami	PNC	Japan
	S. Ohki	PNC	Japan
	T. Yamamoto	PNC	Japan
	A. M. Tsibulia	IPPE	Russia
	P. Smith	AEA	U.K.
	K. Grimm	ANL	U.S.A.
	R. Hill	ANL	U.S.A.

Introduction

As part of a programme proposed by the Working Party on Physics of Plutonium Recycling (WPPR) to evaluate different scenarios for the use of plutonium, fast burner reactor benchmarks were developed; fuel cycle scenarios using either pyrometallurgical (metal fuel) or PUREX/TRUEX (oxide fuel) separation technologies were specified.

The results of simple, beginning-of-cycle physics benchmarks are discussed in Volume 4. There the geometry and composition were pre-specified, and the focus was on assessing the variability of physics predictions experienced among the participants owing to the non-traditional core leakage fraction and minor actinide content of the burner core designs.

In this report, the results of benchmarks which involve fast burner **recycle** are discussed for a metal-fuelled core. The fast burner cycle is thought of as being interposed between the LWR cycle and the geologic repository as a means to mitigate the amount of radioactive waste sent to the repository per unit of energy delivered by the symbiotic LWR/fast cycle, and the benchmarks were designed to elucidate the following issues as a function of the conversion ratio of the burner design:

1. What are the *rates of introduction and destruction* in the fast burner cycle of the *transuranic* (TRU) and *minor actinide* (MA) materials taken from LWR spent fuel – which otherwise would be destined for disposition in the repository?

2. What is the *rate of buildup and the composition of the waste stream* which is destined for ultimate disposal from the fast reactor fuel cycle? And what is the ***net** rate* – outgoing from the fast burner cycle minus incoming from the LWR cycle?

3. What are the *safety and operating characteristics* of the fast burner reactor power plants, i.e., do they suffer performance penalties as compared with traditional breeder designs?

Specifications

The metal-fuelled burner startup core and the once-through burner core models are described in Appendix B of Volume IV. The specification for multiple recycle burner core model is provided in Appendix B of this report.

Briefly, the benchmark design is a metal-fuelled burner core of 600 MWe (1575 MWth). As shown in Figure 1, the core region contains 420 driver assemblies and 30 control subassemblies surrounding a (37 assembly) central reflector/absorber island. The driver active core height is only 45 cm (17.7 in.), roughly half the height of conventional fast reactor designs. This pancaked, annular geometry greatly enhances neutron leakage giving a low conversion ratio of roughly 0.5. The axial design allocates a 15 cm reflector region directly below the core followed by a 30 cm shield region. The fuel pins extend above the active core region with a 70 cm fission gas plenum. Non-fuelled assemblies use a single composition for the entire axial height. The innermost three rows of the configuration shown in Figure 1 contain stainless steel assemblies, and the fourth row contains absorber (boron carbide) assemblies. Three rows of radial shielding surround the active core, a single row of steel and two rows of absorber.

The cycle length is one year at 85% capacity factor; one third of the core is refuelled annually. The spent fuel is cooled a year, reprocessed and refabricated during the second year, and reintroduced into the core at the start of the third year. Table 1 displays the fuel cycle parameters. In blending the refabricated assemblies, all recycled transuranic mass is utilised and a stream from processing spent LWR fuel provides the necessary transuranic makeup. The feedstream from the LWR cycle represents once through fuel with about three years of cooling prior to injection into the fast reactor closed fuel cycle, and a pyrometallurgical recycle technology to reduce LWR spent fuel and produce a fast reactor metallic feedstream containing all transuranics admixed together (Pu + Np + Am + Cm). For the purposes of the benchmark, the uranium recovered in the fast and the thermal reprocessing steps is set aside for future use, and makeup uranium for refabrication is assumed to be depleted tails from the enrichment plant.

This fuel cycle scenario was evaluated parametrically versus the fast burner transuranic conversion ratio:

- The reference core as described above having a transuranic mass conversion ratio near 0.5;
- A modification of the reference core (with an added lower axial blanket) having a conversion ratio near 0.75; and,
- A further altered core (longer lower axial blanket and radial blankets replacing radial reflectors) having a transuranic conversion ratio near 1.0.

In the 0.75 conversion ratio design, a 15 cm thick axial blanket is allocated below the fuelled region in each driver assembly; the radial configuration is identical to the reference core (see Figure 1). The use of a lower axial blanket is preferred to maintain the negative sodium void worth contribution associated with voiding of the upper plenum region. To obtain a conversion ratio of 1.0, the lower axial blanket thickness is increased to 45 cm, a single row of radial blanket is allocated, and internal blanket assemblies are substituted into the inner three rows of the central island of the reference configuration; the resulting annular configuration is shown in Figure 2.

For each of the three core geometries, the benchmark participants were to determine the blending ratio (recycle mass to makeup mass) such that upon multi recycle the end of cycle eigenvalue is unity. Reporting of mass flows, incore inventories, eigenvalue change versus burnup, and safety parameters was requested.

In addition to mass flow characteristics, the radiotoxicity of the fast burner fuel cycle inventories and makeup and discharged waste stream are evaluated in this benchmark. The mass flow results were converted to toxicity units using toxicity factors constructed using the methodology described by Bernard L. Cohen, [1] but using data from ICRP Publication 30, part 4, 1988 and BEIR III (1980). The benchmark toxicity data is shown in Table 2; the selected toxicity data includes most important heavy metal and some fission product isotopes. These isotopic toxicity factors quantify the fatal cancer doses per gram ingested orally. They denote the hazard of the material rather than the risk because they do not include account of any pathway attenuation processes, but simply assume total oral ingestion.

The three cases above address issues of the waste reduction potential of fast burner or fissile self-sufficient cores interposed between the LWR cycle and the geologic repository – after multiple recycle has caused their discharge compositions to relax to their equilibrium. A fourth case was specified to approximate the situation prior to reaching equilibrium: this case evaluates the performance of the reference burner (conversion ratio of 0.5) core design for a once-through fast burner cycle. Fresh assemblies are fuelled with a mixture of transuranics from LWR spent fuel (see Table 3) and depleted uranium; the enrichment requirement for an equilibrium three-batch cycle (equal parts blend of fresh, once-burnt, and twice-burnt material) is determined by each participant such that the end of cycle eigenvalue is unity. Again the mass and toxicity flows, eigenvalue changes upon burnup, and safety coefficients were reported by participants.

Results

Once-through results

The once-through fast burner case is the simpler of the recycle benchmarks and is discussed first. It utilises the reference (CR $\approx$ 0.5) core design. The heavy metal used to fabricate the fresh fuel is composed of recovered LWR transuranics (with isotopics specified in Table 3) and depleted uranium; all other material compositions are identical to the reference core configuration. The fresh fuel enrichment requirement must be calculated by each participant assuming a three-cycle residence time and end of cycle eigenvalue of unity.

This benchmark calculation was completed only by the Japanese [2] and United States [3] participants. Some of the safety parameters were calculated for the beginning-of-cycle metal benchmark (see Volume 4) in the European evaluation [4] and they are included in this discussion – although discrepancies can be expected because they are based on the prespecified material configuration of Volume 4.

The neutronic performance characteristics for the once-through benchmark problem are summarised in Table 4. A slight difference in the enrichment requirement predicted by the two participants is observed: 27.1% TRU/HM in the Japanese results as compared with 26.4% in the United States results. The predicted reactivity loss is 5.3% Δk in the U.S. results and 5.6% Δk in the Japanese results; this is roughly a \$1 difference for a burnup swing of about \$15. The delayed neutron fraction is 0.0035 for the JENDL-2, JENDL-3, and ENDF/B-5-based results. Despite differences in group constant generation schemes between the evaluations, the calculated Doppler coefficients agree well (ranging between -1.46×10^{-3}Tdk/dT and -1.72×10^{-3}Tdk/dT). The European and Japanese Doppler results are slightly (~5%) larger in magnitude than the United States value; and the JENDL-2 Doppler is noticeably (~10%) larger in magnitude than the JENDL-3 result. Fairly good agreement is observed between the Japanese and United States evaluations of the sodium void worth. The void worth in the active core region is \$1.91 (JENDL-2) and \$1.55 (JENDL-3) as compared with \$1.65 (United States); the calculated void worth is lower (\$0.66) in the European evaluation. The positive void worth in the core region is compensated by a large negative component for voiding the above-core plenum region; computed reactivity changes for this leakage effect range from -\$7.5 (United States) to -\$8.3 (JENDL-3). The resulting total void worth is significantly negative ranging from -\$5.8 to -\$6.7.

However, as shown in the United States evaluation [3] the void worth is near-zero if the flowing sodium only is voided.

The decay heat of the discharged fuel was evaluated for the once-through fuel cycle; results are summarised in Table 5. The fission products dominate the decay heat for the first year after discharge, and the fission product component is itself dominated by short-lived isotopes and the 30 year half-life isotopes Cs-137 and Sr-90. After a 100-year decay, the fission product component is significantly reduced and the heavy metal heating component dominates. The Japanese decay heat predictions are consistently higher than the United States results. The difference in total heating is 10% at discharge (26.8 MW in the Japanese evaluation as compared with 24.3 MW in the United States evaluation); at 1-year cooling the Japanese results are 0.25 MW and the United States results are 0.23 MW, a 9% difference. The discrepancy stays near 10% for cooling times out to 10,000 years.

Overall, the US and Japanese participants solutions of the once-through benchmark shows acceptable agreement – consistent with an earlier intercomparison on a high leakage core layout [5].

Multiple recycle results

Results for the fast burner multirecycle benchmark were completed only in the United States submission. The recycle benchmark included three core designs with a parametric variation of the transuranic conversion ratio (values of roughly 0.5, 0.75, and 1.0). The fuel cycle assumptions are summarised in Table 1. The ex-core segment of the fuel cycle includes a one-year cooling interval followed by a six-month chemical separation period, followed by a six-month refabrication time. Thus, the total ex-core interval between discharge and re-insertion is two years. It is assumed that the pyroprocessing treatment returns 99.9% of the transuranics to the core and loses 0.1% to the waste stream. In addition, 5% of the rare-earth fission products are recycled and all other fission products go to waste. The uranium is separated and stored for later use.

The neutronic performance characteristics of the three recycle benchmarks are reported in detail [1] in reference [3]. The discussion here will focus on performance differences caused by the parametric variation of the conversion ratio and on the radiotoxicity evaluation.

The neutronic performance characteristics of the three recycle cases are compared with the once-through results (U.S. solution) in Table 6. The reference burner (CR=0.5) recycle case requires a higher enrichment (increase from 26 to 30% TRU/HM) as compared with the once-through case. As discussed in [6], this increase can be attributed to a build-up of Pu-240 in the recycle isotopics. In addition, the burnup swing is significantly lower (4.3%Δk as compared with 5.3%Δk) in the recycle case, because Pu-240 is a superior fertile material. The enrichment requirement of the CR=0.75 and CR=1.0 designs improves the internal conversion ratio, and reduces the burnup reactivity swing from 4.3%Δk in the CR=0.5 recycle case to roughly 3.5%Δk in the CR $\approx$ 1.0 core.

As expected, the CR=0.5 core consumes transuranics at the quickest rate [2] (255 kg/y); the consumption rate of the CR=0.75 design is roughly half this rate, and the breeder design has a net transuranic production of 10 kg/y. The TRU consumption must be compensated by the external feed;

---

1 The complete report of the U.S. benchmark solution [3] is provided in Appendix A in order to aid any future comparisons.

2 At 1575 MWth (600 MWe) and 85% capacity factor, the maximum theoretical rate of a CR=0 core is 490 kg/y based on one gram of TRU fissioned per MWth day.

therefore, the TRU consumption rate specifies the external feed mass flow rate of LWR transuranics. However, the majority of the TRU mass reloaded into the fast burner each cycle comes not from makeup but from fast recycle material since the average discharge burnup is only ~10% of the initial heavy metal. For example, in the CR=0.5 design, the discharge burnup of TRU isotopes is slightly greater than 10%, but still 86% of the TRU reload material comes from the recycled inventory. As shown in Table 6, the HM mass flow rate is significantly higher (increases from 5.833 to 15.499 between the CR=0.5 and CR=1.0 cases) when external fertile zones have been allocated (i.e., in the CR=0.75 and 1.0 cores).

Variations in the safety characteristics are observed for the parametric variation of the conversion ratio. Isotopic changes between the designs lead to only small variations in the delayed neutron fraction. However, the Doppler coefficient increases in magnitude from -1.35×10^{-3}Tdk/dT in the CR=0.5 design to -1.8×10^{-3}Tdk/dT in the CR=1.0 design; this change is attributed to the higher U-238 concentration in the reduced enrichment fuel and blanket zones. In a similar manner, the lower enrichment requirement for the once-through burner design leads to a significantly higher Doppler coefficient. The sodium void worth values given in Table 6 are for the driver regions alone; the blankets are assumed to remain flooded. Thus, the void worth is slightly reduced in the CR=0.75 and CR=1.0 designs as the blanket power fraction increases. A significant increase in the void worth (over 0.5%Δk/kk') is observed between the once-through and multirecycle results; this change is caused by the buildup of Pu-240 in the TRU inventory with recycle [6].

The mass flows for the recycle benchmarks were converted to toxicity flows using the conversion factors of Table 2, and the toxicity inventory and/or flow rate for several points in the fuel cycle are summarised in Table 7 (these toxicity values include only the isotopes for which toxicity data is listed in Table 2). Using the CR=0.5 recycle case to illustrate the toxicity flows, the feed stream has a toxicity of 3.51E7 which is totally dominated by the toxicity of the TRU component. Within the reactor, the BOEC inventory has a total toxicity of 2.25E9, and the EOEC inventory has a total toxicity of 2.30E9. The toxicity of the fission products component increases significantly during the cycle as more fission products (FP) are formed. However, TRU toxicity totally dominates the incore working inventory and it is relatively constant with burnup due to the competing effects of inventory reduction (burning) and a temporary mass increase of short-lived highly toxic isotopes (e.g., Cm-244). Roughly $^1/_3$ of the TRU inventory, and ½ of the FP inventory are discharged from the reactor each year. As shown in the last four rows of Table 7, the vast majority (99.9%) of the TRU toxicity, 6.32E8, is recycled back to the reactor; and only 6.32E5 is discharged to the waste. Thus, the short-lived FP component (1.48E7) dominates the waste toxicity exiting the fast burner cycle.

A comparison of the toxicity flow data for the various cases reveals several features:

1. The toxicity inventory in the reactor is 30 to 60 times higher (~1.0E9) than the yearly feed or waste toxicity, and is dominated by the TRU component. The in-core toxicity is significantly lower (factor of 5) in the CR=1.0 and once-through system. This change is attributed to differing isotopic fractions – which depend on the core conversion ratio. As discussed in [6], low conversion ratios lead to a buildup of higher actinides (with associated higher toxicity levels) because the blending ratio of LWR feedstream to fast reactor recycle TRU favours the LWR feedstream containing Pu-240 as the conversion ratio is reduced, and this gives a higher mass starting point (than U-238) for further mass increase upon parasitic neutron capture;

2. For a once-through fuel cycle, the toxicity in the waste stream is dominated by the TRU component even at discharge;

3. On the other hand, in the recycle cases the bulk of TRU toxicity is returned to the core, and the short-lived fission product dominate the waste stream toxicity. It is found that for burner designs, the toxicity flow to the waste stream is significantly lower than the toxicity introduced in the feed stream from the LWR cycle. The toxicity is reduced by roughly ½ the introduction rate each year by the CR=0.5 core;

4. In the CR=1.0 recycle design only uranium feed is required (i.e., no TRU), and the toxicity introduction rate is dramatically slower (only 2.18E3/y, even including the uranium daughters), Thus, for the CR=1.0 design, (unlike the burner cores) the toxicity flow to the waste is higher than the toxicity feed rate.

The toxicity of the feed stream to the fast burner cycle is dominated by the TRU component – primarily long-lived isotopes; whereas, its waste stream toxicity is dominated by the short-lived fission product component. Thus, the net toxicity balance (relative toxicity of the fast burner waste stream as compared with its feed stream) will be significantly altered by radioactive decay subsequent to discharge. The time-dependent toxicity of the feed and waste streams for decay times up to a million years are shown in Figures 3 to 5 for the recycle CR=0.5, 0.75, and 1.0 cases, respectively. These figures display significant features:

1. The toxicity of the short-lived fission product dominated waste stream decays by three orders of magnitude over the first 500 years. Over the same time period the toxicity of the TRU-dominated feed stream decays by one order of magnitude (see Figures 3 and 4);

2. Thus, the net result of interposing a fast burner cycle between the LWR cycle and the repository is a large reduction (nearly two orders of magnitude) in **long-term** toxicity;

3. For all cases, the waste stream toxicity falls below the feed stream toxicity at roughly 50,000 years – a net beneficial long-term flow of toxicity to the earth's crust. A factor of fifteen reduction of TRU loss rate to the waste stream (fifteen times less than the 0.1% used in this study) would be required for the favorable toxicity balance to prevail at the shorter (500 y) decay intervals. For the first 500 years, the toxicity will be dominated by short-lived fission products (see Point 1).

Summary and conclusions

Results for the once-through fast burner core utilising the reference configuration were provided by the Japanese [2] and United States [3] participants. A comparison of decay heat predictions revealed differences near 10% for cooling times out to 10,000 years. Except for that, good agreement between the calculated results was shown.

Results for the multiple recycle benchmark problem were provided only in the United States benchmark submission [3]. The results assess the reactor performance and toxicity behaviour in a closed nuclear fuel cycle for a parametric variation of the conversion ratio between 0.5 and 1.0 and facilitate an evaluation of the three issues related to the efficiency to waste management of interposing a fast burner fuel cycle between the LWR thermal cycle and a geologic repository:

1. The influence of recycle and conversion ratio on the reactor operating and safety characteristics was evaluated. Significant, but predictable changes in the enrichment requirement, burnup reactivity swing, Doppler coefficient, and sodium void worth were observed. It is shown elsewhere [7] that safety performance of burner cores is in an acceptable range;

2. The rate of transuranic destruction ranges from 255 kg/y in the CR=0.5 design to a TRU production rate of 10 kg/y in the CR=1.0 design. These rates specify the attainable destruction rates of TRU generated from LWR spent fuel processing;

3. For burner designs, the toxicity of the waste stream at discharge (which is dominated by short-lived fission products) is significantly lower than the toxicity of the feed stream of TRU from LWR spent fuel – by a factor of two for the CR=0.5 design. Thus, interposing a fast burner cycle between the LWR spent fuel and the repository does indeed reduce the flow of toxicity to the repository per unit energy of the symbiotic cycle;

4. The payoff to **long-term** toxicity is much more dramatic than that for short-term toxicity. Payoffs of several orders of magnitude occur after 500 years when all short-lived fission products have decayed to levels below that of long-lived fission products and residual actinides;

5. A factor of 15 reduction in TRU loss rate in the recycle step would cause the **net** flow of toxicity to the earth's crust to become negative after only a 500 year sequestration of the waste – i.e., more toxicity removed in uranium ore than is returned in long-lived fission products and transuranics from recycle losses.

References

[1] B. L. Cohen, "Effects of ICRP Publication 30 and the 1980 BEIR Report on Hazard Assessments of High-Level Waste," Health Physics, 42, 133 (1982).

[2] S. Ohki and T. Yamamoto, "PNC's Results on the Metal-Fueled Fast Reactor Benchmarks," Power Reactor and Nuclear Fuel Development Corporation, August 1994.

[3] K. N. Grimm, "Metallic-Fuel Benchmark Calculations," Argonne National Laboratory, October 1994.

[4] G. Rimpault, J. Dasilva, and P. Smith, "European Neutronic Calculations for the Metal Fueled Fast Reactor Benchmark," Cadarache, April 1994.

[5] R. N. Hill, M. Kawashima, K. Arie, and M. Suzuki, "Calculational Benchmark Comparisons for a Low Sodium Void Worth Actinide Burner Core Design," Proc. ANS Topical Meeting on Advances in Reactor Physics, Charleston, South Carolina, March 8-11, 1992, Vol. I, p. 313.

[6] R. N. Hill, D. C. Wade, J. R. Liaw, and E. K. Fujita, "Physics Studies of Weapons Plutonium Disposition in the IFR Closed Fuel Cycle," Proc. ANS Topical Meeting on Advances in Reactor Physics, Knoxville, Tennessee, April 11-15, 1994, Vol. I, p. 43.

[7] R. N. Hill, "LMR Design Concepts for Transuranic Management in Low Sodium Void Worth Cores," Proc. Int. Conf. on Fast Reactors and Related Fuel Cycles, Kyoto, Japan, October 28 - November 1, 1991, Vol. II, p. 19.1-1.

Table 1
Fuel cycle assumptions

<u>Reactor Segment of Cycle</u>		
Cycle Length	365 days	
Capacity Factor	85%	
Power Rating	1575 MW_{th}	
Core Driver Refuelling	$^1/_3$ per cycle	
Blanket Refuelling	¼ per cycle	
<u>Recycle Segment of Cycle</u>		
Cooling Interval	365 days	
Chemical Separation	done on day 1 of second year	
Blending & Fab.	done on day 184 of second year	
Reinsertion into reactor	done on day 1 of third year	
<u>Chemical Partitioning Factors</u>	<u>% to Product</u>	<u>% to Waste</u>
All TRU isotopes	99.9%	0.1%
Rare Earth Fission Products* *(excluding Y, Sm, and Eu)*	5%	95%
All Other Fission Products*	0%	100%

* Recommend for Benchmark purposes, recycle zero fission products and send all to waste.
U.S. solutions are provided for recommended and for fission product recycle cases in Appendix A.

Table 2
Radiotoxicity data
(CD = Cancer Dose Hazard)

ISOTOPE	TOXICITY FACTOR CD/Ci	HALF-LIFE YEARS	TOXICITY FACTOR CD/g
Actinides and Their Daughters			
Pb-210	455.0	22.3	3.48E4
Ra-223	15.6	0.03	7.99E5
Ra-226	36.3	1.60E3	3.59E1
Ac-227	1185.0	21.8	8.58E4
Th-229	127.3	7.3E3	2.72E1
Th-230	19.1	7.54E4	3.94E-1
Pa-231	372.0	3.28E4	1.76E-1
U-234	7.59	2.46E5	4.71E-2
U-235	7.23	7.04E8	1.56E-5
U-236	7.50	2.34E7	4.85E-4
U-238	6.97	4.47E9	2.34E-6
Np-237	197.2	2.14E6	1.39E-1
Pu-238	246.1	87.7	4.22E3
Pu-239	267.5	2.41E4	1.66E1
Pu-240	267.5	6.56E3	6.08E1
Pu-242	267.5	3.75E5	1.65E0
Am-241	272.9	433	9.36E2
Am-242m	267.5	141	2.80E4
Am-243	272.9	7.37E3	5.45E1
Cm-242	6.90	0.45	2.29E4
Cm-243	196.9	29.1	9.96E3
Cm-244	163.0	18.1	1.32E4
Cm-245	284.0	8.5E3	4.88E1
Cm-246	284.0	4.8E3	8.67E1
Short-Lived Fission Products			
Sr-90	16.7	29.1	2.28E3
Y-90	0.60	7.3E-3	3.26E5
Cs-137	5.77	30.2	4.99E2
Long-Lived Fission Products			
Tc-99	0.17	2.13E5	2.28E-3
I-129	64.8	1.57E7	1.15E-2
Zr-93	0.095	1.5E6	2.44E-4
Cs-135	0.84	2.3E6	9.68E-4
C-14	0.20	5.73E3	8.92E-1
Ni-59	0.08	7.6E4	6.38E-3
Ni-63	0.03	100	1.70E0
Sn-126	1.70	1.0E5	4.83E-2

Table 3
LWR transuranic isotopics

Isotopic values are the weight fraction of the individual isotope in the total transuranic mass.

ISOTOPE	LWR AT 3.17 YEARS COOLING
Np-237	5.40-2
Pu-236	1.12-7
Pu-238	1.01-2
Pu-239	0.508
Pu-240	0.199
Pu-241	0.134
Pu-242	3.88-2
Am-241	2.51-2
Am-242m	1.11-4
Am-243	2.48-2
Cm-242	9.73-6
Cm-243	7.86-5
Cm-244	5.52-3
Cm-245	5.08-4
Cm-246	6.31-5
MA/fiss. Pu	0.172
MA/Pu	0.124
Np-237/MA	0.490
Am-241/MA	0.228
Am-243/MA	0.225
Np-chain	0.213

MA = sum of minor actinides
fiss. Pu = Pu-239 + Pu-241
Np-chain = Np-237 + Am-241 + Pu-241

Table 4
Once-through core neutronic characteristics

	EUROPE	**JAPAN JENDL-2**	**JAPAN JENDL-3**	**UNITED STATES**
Enrichment, TRU/HM		26.85	27.11	26.37
BOL Eigenvalue		1.0560	1.0562	1.0557
EOL Eigenvalue		1.0001	1.0001	1.0003
Burnup Swing, %Δk		5.60	0.943	0.944
Transuranic Inventory Ratio (EOL/BOL)		3.50E-3	3.48E-3	3.51E-3
Delayed Neutron Fraction	-1.89E-3[a]	-1.72E-3	-1.53E-3	-1.46E-3[b]
Doppler Coefficient, Tdk/dT				
Sodium Void Worth, %Δk/kk'				
Active Core	0.23	0.67	0.54	0.58
Above-Core Plenum		-2.85	-2.88	-2.63
TOTAL		-2.18	-2.34	-2.05

[a] includes structural Doppler, which is roughly -0.34E-3 Tdk/dT.
[b] see [3] for discussion of Doppler components.

Table 5
Once-through core decay heat of discharged fuel
($^1/_3$ Core) in Watts

TIME	**JAPANESE - JENDL-3**			**UNITED STATES**		
	HM	*FP*	*TOTAL*	*HM*	*FP*	*TOTAL*
Discharge	1.51E6	2.53E7	2.68E7	1.01E6	2.33E7	2.43E7
1 hour	1.01E6	5.11E6	6.15E6	6.52E5	4.91E6	5.56E6
1 month	2.83E5	6.17E5	9.11E5	2.57E5	5.84E5	8.42E5
1 year	1.14E5	1.31E5	2.48E5	1.05E5	1.26E5	2.31E5
10 y	5.51E4	9.92E3	6.50E4	5.11E4	9.10E3	6.02E4
100 y	3.05E4	1.10E3	3.16E4	2.78E4	1.01E3	2.88E4
1,000 y	7.61E3	0.48	7.61E3	6.98E3	0.41	6.98E3
10,000 y	1.96E3	0.45	1.97E3	1.77E3	0.46	1.77E3

Table 6
Recycle benchmark neutronic performance characteristics comparison

	BURNER REFERENCE	BURNER ONCE-THROUGH	BURNER RECYCLE	PARTIAL BURNER	MARGINAL BREEDER
Conversion Ratio	0.45	0.46	0.45	0.67	1.04
Enrichment TRU/HM	25.8	26.4	30.2	26.9	22.5
Burnup Swing %Δk	-5.9	-5.3	-4.3	-3.6	-3.5
TRU Consumption kg/y	250	244	255	148	-10
Recycled TRU kg/y	-	-	1502	1419	1316
Blending Ratio (Recycle/Total TRU Feed)	-	0.00	0.86	0.91	1.00
HM Loading kg/y	-	5785	5833	8018	15,499
Delayed Neutron Fraction	-	3.51E-3	3.15E-3	3.25E-3	3.33E-3
Doppler Coefficient,[a] *Tdk/dT*	-	-1.46E-3	-1.35E-3	-1.51E-3	-1.80E-3
Sodium Void Worth,[b] *% Δk/kk'*	-				
Active Core	-	0.61	1.10	1.00	0.80
Above-Core Plenum	-	-0.80	-0.73	-0.73	-0.77
TOTAL	-	-0.19	0.37	0.27	0.03

a for increase in the fuel temperature (structural effect not included, see [3] for details).

b for voiding of flowing sodium in the drivers only.

Table 7
Comparison of toxicity flows
(all toxicities in Cancer Dose, CD)

	CR=0.5 ONCE-THROUGH	CR=0.5 RECYCLE	CR=0.75 RECYCLE	CR=1.0 RECYCLE
Feed Stream, CD/y				
Uranium	6.54E2	6.34E2	9.87E2	2.18E3
Transuranics	2.06E8	3.51E7	2.03E7	-
BOEC Inventory, CD				
Uranium	4.06E1	1.40E2	1.24E2	1.29E2
Transuranics	6.57E8	2.23E9	1.54E9	4.11E8
Short-lived FP	1.90E7	1.84E7	1.90E7	2.02E7
Long-lived FP	1.29E2	1.20E2	1.25E2	1.32E2
EOEC Inventory, CD				
Uranium	6.65E1	1.87E2	1.56E2	1.38E2
Transuranics	1.04E9	2.26E9	1.57E9	4.25E8
Short-lived FP	3.72E7	3.589E7	3.67E7	3.90E7
Long-lived FP	2.50E2	2.40E2	2.43E2	2.56E2
Recycled Feed, CD/y				
Transuranics	-	6.32E8	4.39E8	1.21E8
Waste Stream, CD/y				
Transuranics	3.22E8	6.32E5	4.39E5	1.21E5
Short-lived FP	1.53E7	1.48E7	1.53E7	1.62E7
Long-lived FP	1.03E2	1.00E2	1.02E2	1.07E2

[a] includes toxicity of equilibrium daughters (e.g., Pb-210), which are dumped in the mill tailings.

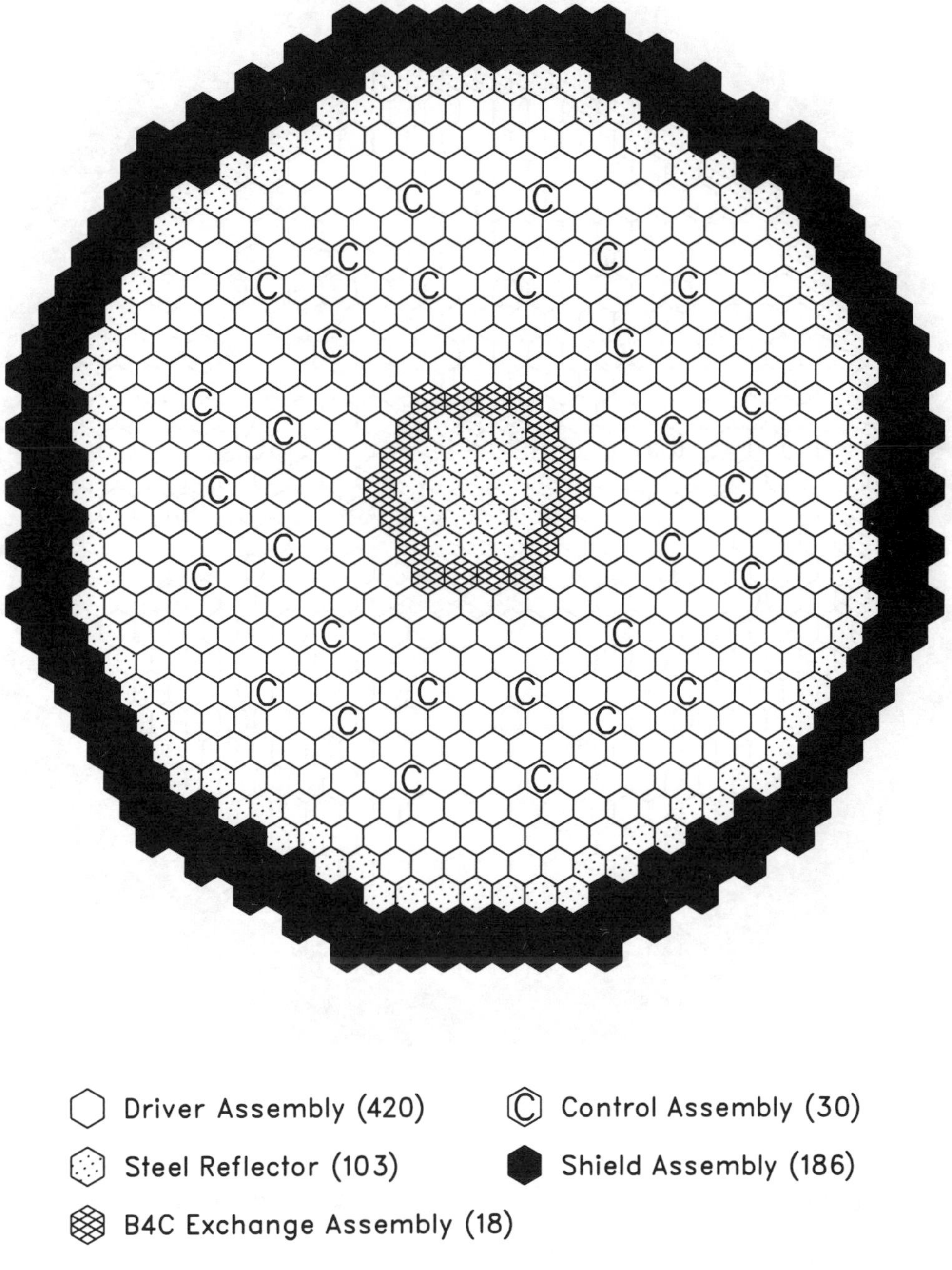

Figure 1 ***Benchmark reference core configuration***

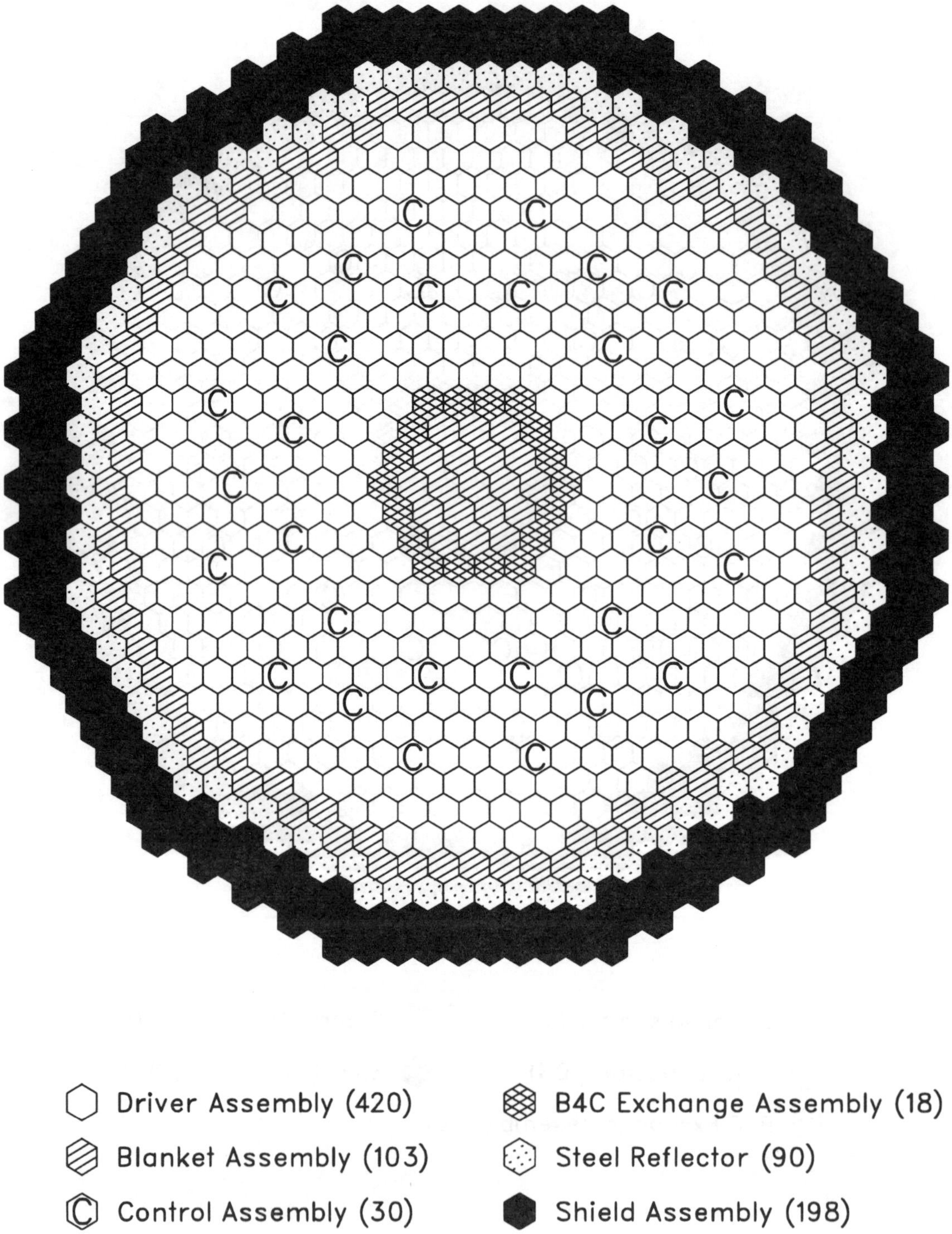

Figure 2
CR = 1.0
Converter core configuration

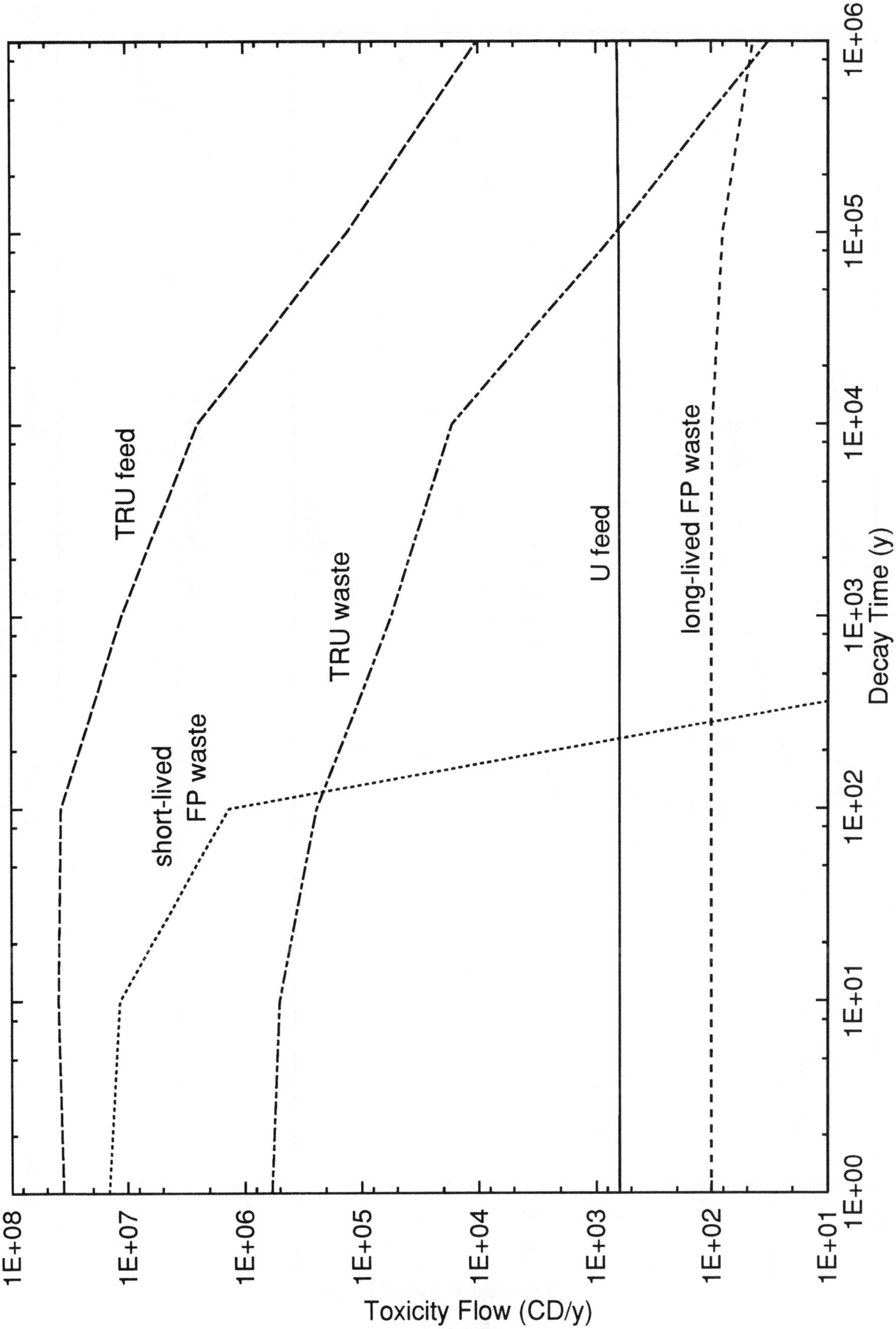

Figure 3
Toxicity components for CR = 0.5
Recycle case

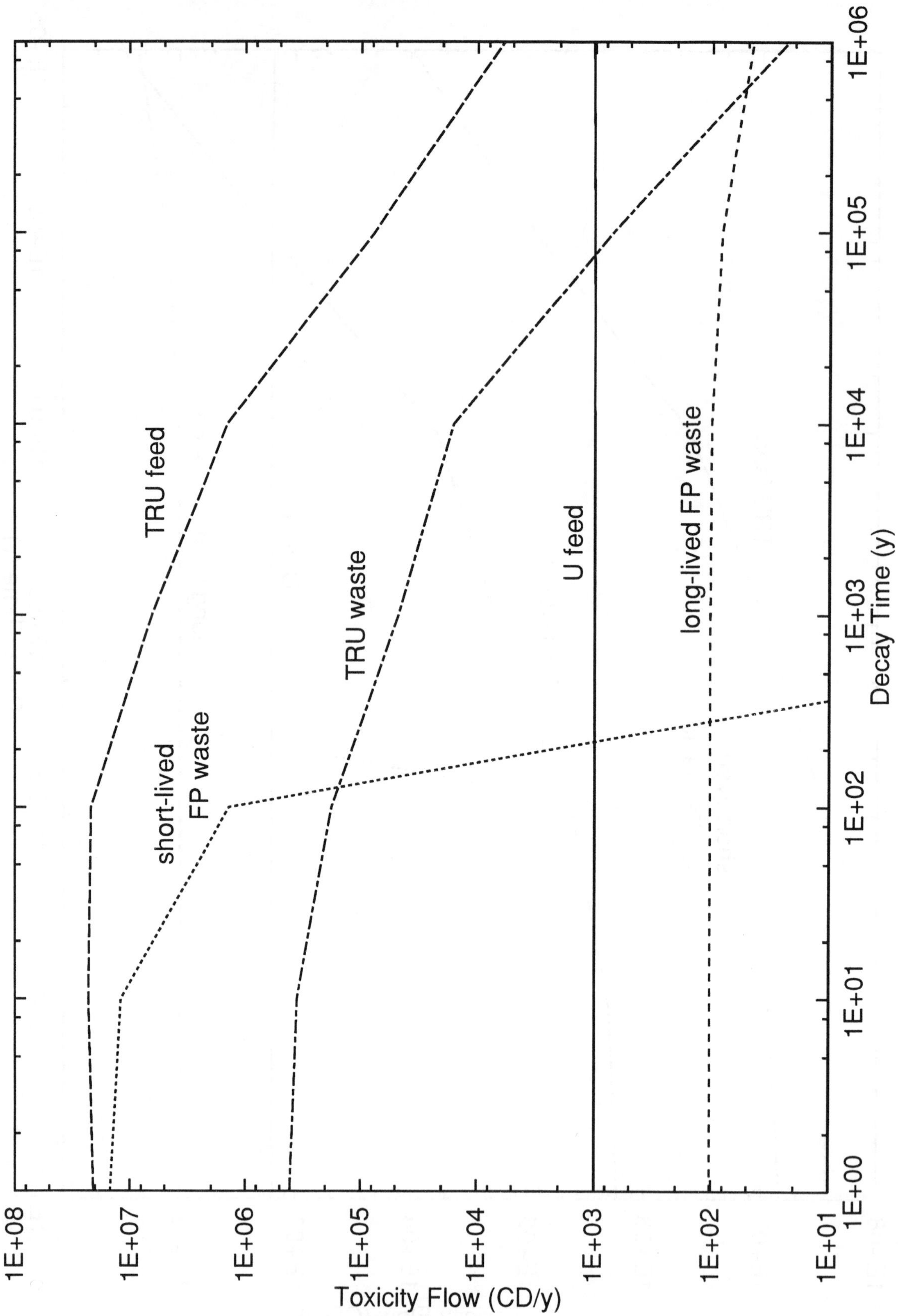

Figure 4
Toxicity components for CR = 0.75
Recycle case

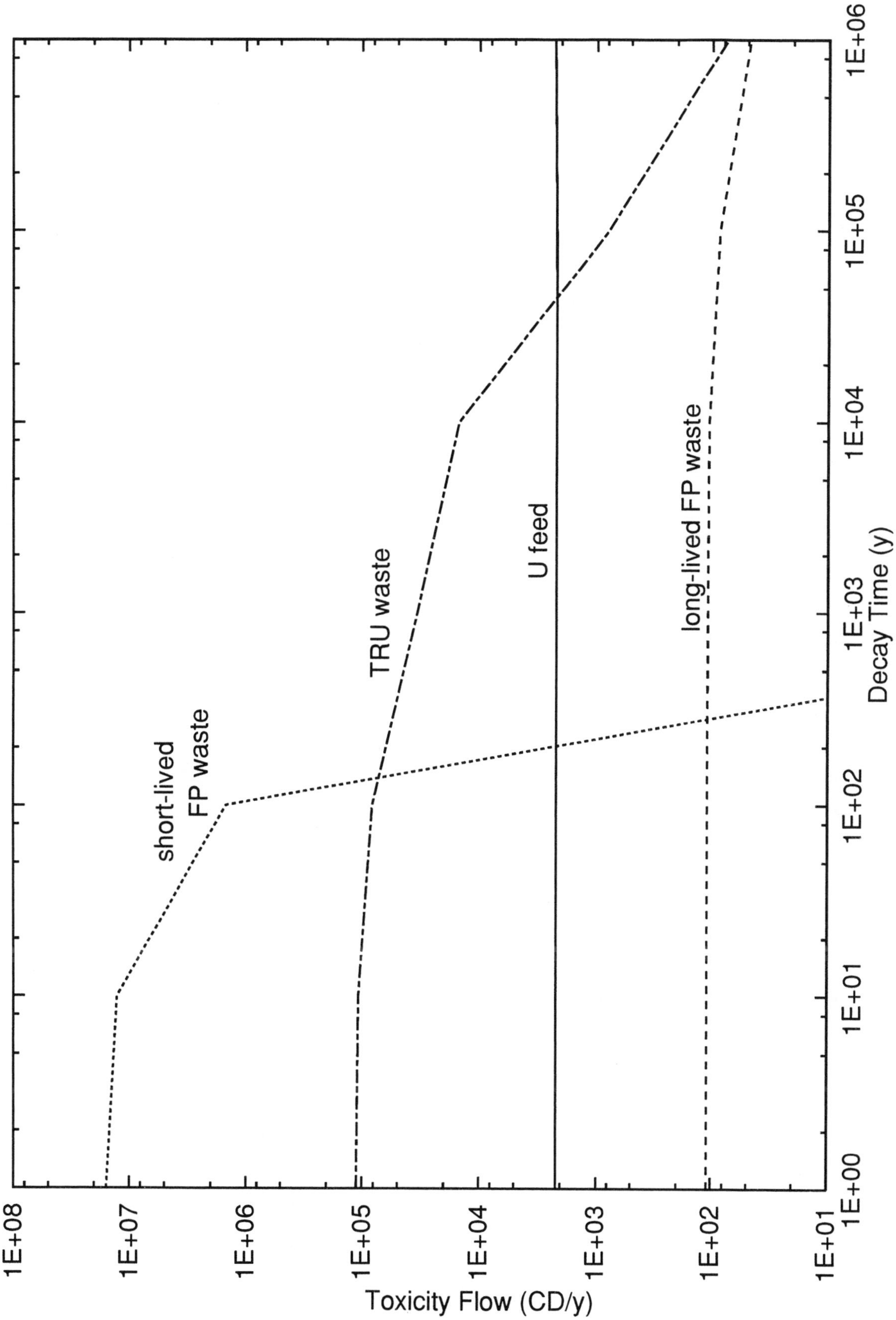

Figure 5

Toxicity components for CR = 1.0

Recycle case

Appendix A

U.S. metal-fuelled benchmark solution

Introduction

As part of the programme of the Working Party on the Physics of Plutonium Recycling, fast reactor core physics benchmarks are being calculated; fuel cycle scenarios utilising either PYRØ or PUREX/TRUEX separation technologies are being evaluated. In this appendix, the application of a metal-fuelled fast reactor with PYRØ recycle and transuranic (TRU) feed from LWR PYRØ-processing is investigated. The overall benchmark exercise was designed to evaluate the following issues:

1. What is the *rate of reduction of the TRU or minor actinide (MA) mass inventory* from LWR spent fuel processing (which would otherwise be destined for ultimate disposal)?

2. What is the *rate of buildup of the waste stream* which is now destined for ultimate disposal from the fast reactor fuel cycle?

3. What are the *safety and operating characteristics of the reactor power plant*?

Once the power rating of the fast reactor power plant (600 MWe in the case of this benchmark) and plant capacity factor (85% in this benchmark) are specified, the corresponding actinide destruction rate (at roughly 200 MeV/fission) is totally determined. Thus, the first issue (TRU destruction rate) is directly and exclusively dependent on the conversion ratio to which the burner reactor is designed; the lower the conversion ratio, the greater the rate of reduction of the LWR spent fuel TRU inventory. For this benchmark comparison, three core designs (conversion ratio near 0.5, 0.75, and 1.0) will be investigated.

The second issue (magnitude and composition of the waste stream) depends strongly on both the in-core transmutation characteristics and the ex-core chemical partitioning fractions between the recycle product and waste stream in the processing and refabrication steps. Predictions of the in-core material (and isotopic) distribution depends upon many factors such as irradiation time (average burnup), neutron cross-sections (as dictated by the neutron energy spectrum), nuclear interaction branching ratios, radioactive decay constants, and out-of-reactor cooling time. Chemical partitioning fractions for the PYRØ processing recycle are estimated to be 0.1% of the transuranics, 95% of the rare earth fission products, and 100% of the other fission products going to the waste stream.

The third set of issues (reactor operating and safety performance) is affected by both the transuranic management strategy and the operating characteristics (e.g., neutron energy spectrum) of the system. For systems designed to be net consumers of the TRU inventory, rapid reactivity losses are expected as the fissile inventory is reduced. The neutron energy spectrum depends strongly on both composition and leakage/absorption ratio so that it is controlled by:

- The material volume fractions of fuel, structure, and coolant;
- The fuel material (choice of MOX or metal fuel);
- The means to dispose of excess neutrons – leakage (loss of primarily high energy neutrons) or in-core parasitic absorption (with preferential loss of low energy neutrons); and,
- The blending ratio of external feed to recycle feed in the fresh fuel fabrication.

In this metal fuel benchmark problem, conventional fast reactor assembly designs are utilised, and the core has been designed to dispose of excess neutrons primarily by leakage (a pancaked geometry to enhance axial leakage).

The metal-fuelled fast reactor benchmark models are briefly described in Section 2 of this appendix; and the calculational methods utilised in this study are detailed in Section 3. The calculated benchmark results are summarised in Section 4. In Section 4.A, results for the startup core model are discussed; basic data, neutron balance, depletion methodology, and mass transition results are reported. In Section 4.B, results for a once-through fuel cycle are given; neutron balance, mass transition, safety parameter, and radioactive decay (and toxicity) results are reported. In Section 4.C, results for a recycle fuel cycle are given for three core configurations (conversion ratio near 0.5, 0.75, and 1.0); neutron balance, mass flow, mass increment, safety parameter, and radioactive decay (and toxicity) results are reported.

2. Benchmark models

A. *Burner core – Conversion Ratio near 0.5*

The geometric model for the 1575 MWth (600 MWe) benchmark burner reactor configuration is specified in Appendix B of Volume 4. Basically, the reactor core region contains 420 hexagonal driver subassemblies and 30 control subassemblies in an annular arrangement (subassembly pitch 15.617 cm) with an internal reflector and an external reflector plus shield (see Figure A.1). The driver active core height is only 45 cm. Below the active core in each driver subassembly is a 30-cm shield region and a 15-cm reflector region; a 70-cm plenum region is located above the driver fuel. The first three rows of the loading configuration contain steel shield subassemblies; the fourth row contains boron carbide shield subassemblies. Three rows of radial shielding surround the core (a single row of steel shield subassemblies followed by two rows of boron carbide shield subassemblies). A more detailed discussion (and description) of the development of this core configuration model can be found in [A.1].

B. *Converter core – Conversion Ratio near 0.75*

This configurations is essentially the same as the burner configuration; however an additional 15-cm blanket region is located below the fuelled section of each driver subassembly.

C. *Breeder core – Conversion Ratio near 1.0*

The geometric model for the 1575 MWth (600 MWe) benchmark breeder reactor configuration is specified in Appendix B of this Volume. Basically, the reactor core region contains 420 hexagonal driver subassemblies and 30 control subassemblies in an annular arrangement (subassembly pitch 15.617 cm) with an internal blanket and an external blanket plus shield (see Figure A.2). The driver active core height is only 45 cm. Below the active core in each driver subassembly is a 30-cm shield region, a 15-cm reflector region, and a 45-cm blanket region; a 70-cm plenum region is located above the driver fuel. This model contains internal and radial blankets with a 90-cm fuelled section. Below the fuelled section in each blanket is a 30-cm shield region and a 15-cm reflector region; a 70-cm plenum region is located above the blanket fuel. The first three rows of the loading configuration contain blanket subassemblies; the fourth row contains boron carbide shield subassemblies. Four rows of radial shielding surround the core (a single row of blanket subassemblies followed by a row of steel shield subassemblies and then followed by two rows of boron carbide shield subassemblies). A more detailed discussion (and description) of the development of this core configuration model can be found in [A.2].

3. Description of calculational methods

The cross-sections used in this analysis were generated specifically for this model from ENDF/B-V.2 datafiles using the MC^2-2/SDX code package [A.3, A.4]. Details of the group constant generation procedure are discussed later in Section 4.A.1. Both a 9-energy group set and a 21-energy group set of cross-sections were generated.

Depletion-related neutron flux calculations used the Hexagonal-Z sixth-core nodal diffusion theory option of the DIF3D code [A.5] and the 9-group cross-section set. The calculational mesh for the non-equilibrium problem, the once-through equilibrium problem, and the recycle equilibrium problem with the burner core utilised one planar node per assembly with 3 axial nodes in the lower shield region, 2 nodes in the lower reflector region, 5 nodes in the active core region, and 6 nodes in the plenum region. The calculational mesh for the converter core equilibrium problem utilised one planar node per assembly with 2 axial nodes in the lower shield region, 1 node in the lower reflector region, 1 node in the lower blanket region, 5 nodes in the active core region, and 6 nodes in the plenum region. The calculational mesh for the breeder core equilibrium problem utilised one planar node per assembly with 2 axial nodes in the lower shield region, 1 node in the lower reflector region, 4 nodes in the lower blanket region, 5 nodes in the active core region, and 6 nodes in the plenum region. *The blanket subassemblies in the breeder core equilibrium problem utilised one planar node per assembly with 2 axial nodes in the lower shield region, 1 node in the lower reflector region, 9 nodes in the blanket region, and 6 nodes in the plenum region.* In all flux calculations, the control rods were fully withdrawn to the top of the fuel region and remained fixed during the depletion step. Furthermore, all flux solutions were normalised such that the integrated energy deposition rate was 1575 MWth.

Isotopic depletion for all the problems were performed with the REBUS-3 code [A.6]. The regional depletion was calculated in five (equal length, 9 cm) axial zones within six radial zones based on a 310-day full-power burn cycle.

The safety coefficients for the equilibrium cycle once-through core configuration and the recycle configurations were calculated using perturbation theory. Forward and adjoint neutron fluxes, for the perturbation computations, were calculated using the triangular-Z sixth-core finite-difference diffusion theory option of the DIF3D code [A.4] and the 21-group cross-section set.

The depletion and transmutation of the stable actinides was modelled in the depletion calculation, however, an accurate evaluation of the global radioactivity and decay heating requires a detailed tracking of the numerous short-lived isotopes. This detailed analysis was performed with the ORIGEN code [A.7]. Heavy-element one-group effective cross-sections appropriate for each region of the model (i.e., collapsed using model regional fluxes) were used to replace values from the ORIGEN LMFBR library. The magnitude of the flux values used in the ORIGEN depletion calculation for each depletion region were obtained by averaging the beginning of equilibrium cycle (BOEC) and end of equilibrium cycle (EOEC) regional flux values. Fission product modelling in the decay heat calculations was based on the detailed ORIGEN fission product library **not** on the "lumped" fission products (described in Section 4.A.1) used in the depletion analysis.

4. Benchmark results

A. *Metal-fuelled burner startup core benchmark*

The first portion of the benchmark involved calculating end of life (EOL) masses (and other quantities), after a single 310-day depletion step, given beginning of life (BOL) compositions for a burner core configuration (conversion ratio near 0.5 - see Section 2.A). The atom densities used for the BOL configuration were given in Table B.1 of Appendix B.

A.1 Basic data reporting

As described in Section 3 the cross-sections were generated from ENDF/B-V.2 data files using the MC^2/SDX package. The ENDF/B-V.2 data was processed into "smooth" 2082-group cross-section and resonance parameter libraries for use in the MC^2-2 code; selected resonances (e.g., wide structural resonances) are processed into the multigroup data. Using the homogeneous core compositions given in Table B.1 of Appendix B, a 2082-group fundamental mode calculation was performed by MC^2-2 using continuous slowing-down theory and multigroup methods [A.3]. The 2082-group flux was used to collapse the "smooth" multigroup data to a 230-energy group structure.

The SDX code [A.4] was used to process the resonance data, correct for pin heterogeneity effects, and spatially collapse the cross-section data to a broad group energy structure. Resonance contributions were evaluated using the narrow resonance approximation at the 230-group level. *Note that broad resonances were previously evaluated in the MC^2 multigroup data.* One-dimensional unit cell calculations were performed for a typical fast reactor fuel pin lattice (pin outer diameter of 7.5 mm with a 1.2 pitch-to-diameter ratio) to generate 230-group heterogeneity correction factors. Finally, a one-dimensional diffusion calculation was performed to generate the 230-group collapsing spectra. The radial model utilised seven spatial zones (central reflector, central absorber, three concentric core rings, steel reflector, and shield); a single group and region independent buckling was varied to achieve criticality. Consistent SDX calculations were performed to generate both the nine and 21-energy group (shown in Table A.1) cross-section sets at the specified operating temperatures (850 K for fuel – including zirconium, 750 K for structural isotopes, and 700 K for coolant). In addition, a 21-group set was generated at elevated temperature (double the operating conditions) for use in the Doppler coefficient calculation; and a 21-group set was generated for sodium-out conditions for use in the sodium void calculation.

Ten lumped fission products are utilised representing the fission product distribution from U-235, U-238, Pu-239, Pu-240, and Pu-241 fission; each isotope had seperate rare earth and non-rare earth components. These lumps were collapsed using group constants for 180 fission product isotopes (generated from ENDF/B-V.2 data files using the MC^2/SDX package), fission yields, and a typical fast reactor discharge isotopic distribution. The assignment of each fissile isotope to one of these ten lumped fission products is shown in Table A.2.

The energy deposition rate was determined by the product of isotopic fission and capture reaction rates, calculated using the multigroup fluxes and cross-sections, and the group-independent isotopic energy deposition per fission and energy deposition per capture values given in Table A.2. Although not included in Table A.2, energy deposition per capture values for structural materials (roughly 8 MeV/capture) and coolant were included in the overall power normalisation. These energy deposition values implicitly assume that the gamma energy generated by fission and capture reactions is deposited at the site of the nuclear reaction; thus, this modelling does **not** account for gamma transport.

A.2 BOL neutron balance reporting

The BOL eigenvalue and eigenvalue convergence criteria are shown in Table A.3 for the startup core calculated using Hexagonal-Z nodal diffusion theory [A.8] and 9-energy groups. Shown in Table A.4 are the BOL groupwise fluxes and percentage of total flux within a group, for the central core region. *The central region of the core comprises axially integrated values within the middle rows of driver subassemblies.* These groupwise BOL fluxes, on a per lethargy unit basis, are shown in Figure A.3; the BOL and EOL spectra are virtually identical. The BOL k-infinity, defined as the regional fission production rate divided by the regional absorption rate, shown in Table A.3 was calculated for the above defined central core region. The core and model leakage/absorption ratios are shown in Table A.5; here quantities are summed over the fuelled portion of driver subassemblies, whereas model quantities are sums over the entire geometrical model. The large difference in core and model leakage fractions (61% of neutrons leak from the core, but only 3% leak from the model) is due to absorptions in boron-containing regions, especially the driver lower shield region. Core capture fractions (captures in material "*i*" divided by the absorptions in all materials) for heavy metal, structural, and coolant are also shown in Table A.5. Effective one-group microscopic capture, fission, ν*fission, and (n,2n) cross-sections using "midplane" central core region fluxes are shown in Table A.6 for all actinides, both BOL and EOL. *This "midplane" region is that node of the central core region that includes the core midplane; it is the third 9 cm axial region, out of five total core axial regions.*

A.3 Depletion methodology reporting

As mentioned previously in Section 3, the REBUS code was used to determine EOL atom densities. The solution process for a standard REBUS-3 burnup (density) iteration follows:

1. Calculates the *BOL fluxes* for each depletion region from the BOL atom densities (given values);
2. Calculates a *BOL burn* (gain and loss) *matrix* using the BOL fluxes;
3. Calculates *EOL densities* using an matrix exponential method given the BOL burn matrix and a user-defined burn time;

4. Calculates *EOL fluxes*, and then an *EOL burn matrix*, from the EOL densities;
5. *Averages the BOL and current EOL burn matrices* and repeats the iteration, calculating new EOL densities, fluxes and burn matrix. *Compares the two calculated EOL densities*;
   - If they are within an input convergence criterion, then the density iteration is assumed completed;
   - If the densities are not converged then the whole process is repeated from step 5;
6. This iteration procedure is repeated until the density convergence is within some user-defined value for all depletion zones or until a user-defined maximum number of iterations is reached. Details of the isotope transmutation chains used in the REBUS-3 depletion calculations are shown in Figure A.4 and Tables A.7-A.8. The TRU isotopes used in this calculation are shown in Table A.7. The reactions modelled in the depletion analysis were (n,fission), (n,gamma), (n,2n), alpha decay, $beta^+$ decay, and $beta^-$ decay. The energy normalisation method and "lumped" fission product model used by REBUS-3 was already discussed in Section 4.A.1.

A.4 BOL to EOL transition and EOL neutron balance reporting

Shown in Table A.9 are the BOL and EOL heavy isotope and fission product masses, along with the corresponding mass differences. The BOL and EOL eigenvalues and eigenvalue convergence criteria are shown in Table A.3. The burnup swing $[(k_{EOL} - k_{BOL}) / k_{BOL}\ k_{EOL}]$, from BOL to EOL, is also shown in Table A.3. The transuranic (TRU) breeding ratio, defined as the EOEC TRU mass divided by the BOEC TRU mass, and the TRU conversion ratio, defined as the production of "new" TRU isotopes divided by fissioned TRU isotopes, are shown in Table A.5. *Note that the TRU breeding ratio involves sums of **only** TRU isotopes.* The 0.94 "breeding ratio" indicates a 6% reduction in the TRU inventory during the startup cycle. The EOL neutron flux and spectrum for the central core region are shown with the corresponding BOL spectral results in Table A.4 and Figure A.3.

B. LMR once-through burner core benchmark

The second portion of the benchmark involved performing an equilibrium cycle calculation for a once-through burner core (see Section 2.A). Given are a $^1/_3$ core batch refuelling strategy, a specified time and energy extraction per burn cycle, and a specified composition of a TRU feedstream coming from LWR spent fuel processing. The fresh-fuelled enrichment is calculated such that the EOEC loading has a eigenvalue of 1.0.

B.1 BOEC neutron balance reporting

The equilibrium cycle fuel management scheme involved three radial zones each containing fresh, once-burnt, and twice-burnt subassemblies. Each driver subassembly will remain in its "zone" for three 310-day burn cycles. At the end of a 310-day cycle, one-third of each region, the subassemblies that have accumulated two burn cycles, will be replaced by fresh-fuelled subassemblies. "Batch-averaged" compositions (an equal mixture of fresh, once, and twice-burnt material) are used in the flux calculations. The REBUS-3 code [A.5] iterates the fresh fuel enrichment (in addition to the depletion iterations described in Section 4.A.3) to achieve EOEC criticality. The fresh-fuelled enrichment, calculated for this fuel-management scheme, is shown in Table A.10.

B.2 BOEC to EOEC transition and mass flow reporting

The BOEC and EOEC eigenvalues, eigenvalue convergence criteria, and burnup swing [(k_{EOEC} – k_{BOEC}) /k_{BOEC} k_{EOEC} at constant rod position] are shown in Table A.11. The TRU breeding ratio and TRU conversion ratio are shown in Table A.10 (see Section 4.A.4 for definitions). Once again, the 0.94 "breeding ratio" indicates a 6% reduction in the TRU inventory each cycle. The mass balance for BOEC and EOEC conditions, including mass differences, are shown in Table A.12. Also included as a footnote to this table is the total TRU (EOEC-BOEC) mass increment per unit energy production (-0.50 g/MWd). The heavy isotope mass flow for the whole core, including equilibrium loading, equilibrium discharge, and net mass gain, is shown in Table A.13. In Table A.13 the line labelled "Equilibrium Loading, kg/y" gives the feed rate to the equilibrium cycle, whereas the line labelled "Equilibrium Discharge, kg/y" gives the once-through cycle discharge rate. Also shown in this table is the yearly net change in mass for the heavy isotopes.

B.3 Safety parameter reporting

The system delayed neutron fraction (beta-effective), two Doppler coefficients and several sodium void coefficients are shown in Table A.14.

The Doppler coefficient for fuel (including Zr) and the Doppler coefficient for core structural material are both included in Table A.14. Doppler coefficient calculations used first-order perturbation theory (the adjoint flux was from the unperturbed model) where cross-sections evaluated at a nominal temperature are replaced by values generated at twice the nominal temperature. The calculated fuel (HM plus Zr) Doppler coefficient was modified by two scale factors, one factor (+4.4%) was used to model a more realistic temperature increase of 300 K (from 850 K to 1150 K) instead of the doubling of temperature used in the perturbation calculation while the other factor (+2.8%) accounted for using integral transport theory in the hyper-fine energy group calculations (RABANL approximation in the MC^2 code). *The first factor was generated by comparing first-order perturbation theory calculation results using 850 K, 1150 K, and 1700 K cross-sections without the RABANL approximation whereas the second factor was generated by comparing first-order perturbation theory calculation results using RABANL and non-RABANL cross-sections generated at 850 K and 1700 K. These scale factors were generated using simplified core models and cell-weighted, but not region-weighted, cross-sections.*

The components of sodium void for the total fuelled core region (drivers only), driver plenum region, and the sum of fuelled region plus plenum region are also shown in Table A.14 for the case where **all** the sodium was voided (sodium density set equal to zero) and the case where **only** flowing sodium inside the hex-can is voided. The sodium void calculations used exact perturbation theory (the adjoint flux was from the perturbed model) where cross-sections generated with a sodium-in model were replaced by cross-sections generated with a sodium-out model.

Both BOEC and EOEC decay heat level versus time for the drivers, calculated using the ORIGEN code, are shown in Table A.15 (see Section 3 for a discussion of the ORIGEN code). Also included in this table is the decomposition of the decay heating into light isotope, heavy isotope, and fission product components. The decay powers presented in Table A.15 are shown graphically in

Figures A.5-A.7. From both the Table A.15 and Figure A.5 it is seen that times less than ~60 years the EOEC decay power is greater then the BOEC value, whereas at times greater than ~60 years the decay powers are similar. This difference, at short decay times, is mainly due to the relatively large amounts of fission products in the EOEC configuration. At large decay times the BOEC decay heating is slightly higher than the EOEC values because the BOEC configuration has more actinide mass (alpha-decay).

B.4 Radioactivity and decay

As discussed in Section 3, the ORIGEN code was used to calculate radioactivity and decay. ORIGEN is a point depletion code which solves the equations of radioactive growth and decay, using the matrix exponential method, for large numbers of isotopes with arbitrary coupling. The ORIGEN code has three data libraries:

- A light element library,
- A fission-product library, and,
- A heavy element library.

The LMFBR portion of the light library includes 247 isotopes ranging from H-1 through W-167, whereas the LMFBR portion of the fission product library includes 821 isotopes ranging from Co-72 through Er-167. As described in Section 3, case-specific cross-section data was provided for actinides ranging from U-234 through Cm-246. The LMFBR portion of the heavy library includes 100 isotopes ranging from Tl-207 through Es-253. The decay heat is calculated using isotope disintegration rates and isotope-dependent decay-energy per disintegration values from the libraries.

The radioactivity (in Curies) as a function of time for each TRU isotope, for both fresh-fuel (feed) and discharged-fuel (3-burn cycle), are shown in Tables A.16 and A.17 respectively. Also included in the last line of these tables is the total TRU radioactivity versus time. Shown in Table A.18 is the isotopic radioactivity difference between discharged and fresh fuel. Also included in Table A.18, along with the total radioactivity difference, is the radioactivity difference per unit energy production. The summed TRU quantities for radioactivity, plus per unit energy production values, are also shown graphically in Figures A.8 and A.9. From Figure A.8 it is seen, for the times shown, that fresh fuel has a higher TRU radioactivity than discharged fuel; this follows since there are more TRU atoms, especially major isotopes of plutonium, in fresh fuel. *Although not shown in the tables or figures, at very short times discharged fuel has a higher TRU radioactivity than fresh fuel because discharge fuel has a large amount of quickly decaying Np-239.* As shown in Figure A.9, the total radioactivity difference per Megawattday (MWd) approaches zero, despite relative differences, because the radioactivity itself approaches zero.

Shown in Tables A.19-A.21 are the corresponding quantities for the toxicity hazard (in cancer deaths) of TRU isotopes. The factors used to calculate toxicity from radioactivity (in Curies) were given in Table 2, page 20; they are also shown in the last column of Tables A.19-A.21. The summed TRU quantities for toxicity, plus per unit energy production values, are also shown graphically in Figures A.10 and A.11. From Figure A.10 it is seen that at short times (less than approximately 50 years) the toxicity of discharged fuel is greater than the toxicity of fresh fuel; this is contrary to the results seen in Figure A.8 for radioactivity. This crossover results mainly from the effects of Pu-241 (half-life 14.4 years). *If the radioactivity resulting from Pu-241 decay is removed from the total radioactivity, the relative shape of the fresh and discharge radioactivities resembles the toxicity plot – Figure A.10.* At short times it is the major source of radioactivity whereas the curie-to-toxicity conversion factor for Pu-241, since it is a beta emitter and not an alpha emitter, is zero. This crossover in toxicity is also observed in the toxicity per unit energy figure (see Figure A.11).

C. LMR recycle benchmarks

The third portion of the benchmark involved performing an equilibrium cycle calculation with recycle for three core configurations:

- A burner core (conversion ratio ~0.5 – see Section 2.A),
- A converter core (conversion ratio ~0.75 – see Section 2.B), and,
- A breeder core (conversion factor ~1.0 – see Section 2.C).

Given are a $^{1}/_{3}$ core batch refuelling strategy, a specified time and energy extraction per burn cycle, and a specified composition of a make-up TRU feedstream coming from LWR spent fuel processing. The fresh-fuelled enrichment is calculated such that the EOEC loading has a eigenvalue of 1.0. When it is convenient, the tables showing equilibrium recycle parameters will repeat the equivalent parameter from the once-through equilibrium burner configuration.

C.1 Recycle BOEC neutron balance reporting

The equilibrium cycle fuel management scheme is the same scheme used in the once-through problem (see Section 4.B.1). Radioactive decay is modelled in all cooling, fabrication, and storage steps the same way it is modelled in the in-core steps (see Tables A.7 and A.8). "Batch-averaged" compositions (an equal mixture of fresh, once, and twice-burnt material) are used in the flux calculations. The EOEC isotopic densities are iterated until they are within some input convergence criteria.

The fresh-fuelled enrichments, calculated for this fuel-management scheme, are shown in Table A.22. Also shown in Table A.22 are the enrichments if no rare earth fission products are recycled, the blending ratio (TRU recycled mass/TRU total mass), the ratio of feed mass to recycle mass, and the TRU feed mass needed per year. The results shown in Table A.22 indicate that if the rare earths are not recycled the enrichment is reduced by ~0.5%. (Recall that in the standard recycle calculations 5% of the rare earth fission products are recycled.) The recycle stream isotopic mass fractions are shown for all recycle cases in Table A.23 whereas the isotopic masses from the LWR TRU feed are shown in Table A.24. The isotopic percentages are the same for the burner and converter configurations; only the amount of total mass is different. There is no feed mass for the breeder configuration because that particular configuration produces more than enough TRU mass. The corresponding fresh fuel isotopic mass fractions are shown in Table A.25; the fresh fuel is a mixture of recycled and make-up TRU with the blending ratios shown in Table A.22.

C.2 Recycle BOEC to EOEC transition and mass flow reporting

The BOEC and EOEC eigenvalues, eigenvalue convergence criteria, burnup swing [$(k_{EOEC} - k_{BOEC}) / k_{BOEC}\, k_{EOEC}$ at constant rod position], and TRU conversion ratio (see Section 4.A.4 for definition) are shown in Table A.26 for each recycle configuration.

C.3 Mass increments by heavy metal isotope

The mass balance for BOEC and EOEC conditions, including TRU totals and mass differences, are shown in Table A.27. Also included as a footnote to this table for each recycle configuration are the total TRU (EOEC-BOEC) mass increment per unit energy production.

C.4 Recycle safety parameter reporting

The system delayed neutron fraction (beta-effective), Doppler coefficient for core fuel only (including Zr) are shown in Table A.28 for all the recycle configurations. Doppler coefficient calculations used first-order perturbation theory (the adjoint flux was from the unperturbed model) where cross-sections evaluated at a nominal temperature are replaced by values generated at twice the nominal temperature. As with the once through fuel benchmark Doppler coefficients discussed previously (see section 4.B.3), the base coefficients were modified by factors to account for a smaller temperature change and using integral transport theory in hyper-fine energy group calculations. The components of sodium void for the total below core region (drivers only), fuelled core region (drivers only), driver plenum region, total driver region (below core, core region, and plenum region), and the sum of fuelled region plus plenum region are also shown in Table A.28 for the case where **only** flowing sodium inside the hex-can is voided. The sodium void calculations used exact perturbation theory (the adjoint flux was from the perturbed model) where cross-sections generated with a sodium-in model were replaced by cross sections generated with a sodium-out model.

Both BOEC and EOEC decay heat level versus time for all fuelled regions, calculated using the ORIGEN code, are shown for the recycle configurations in Tables A.29-A.31 (see Section 3 for a discussion of the ORIGEN code). Also included in these tables is the decomposition of the decay heating into light isotope, heavy isotope, and fission product components. These tables show the decay heating for **all** fuelled regions in the models; the regional components (core, axial blanket, etc.) of the all-region sums are shown in Tables A.32-A.37 for the converter and breeder configurations. *The burner configuration only has the one component – core.* The total equilibrium cycle decay heating for BOEC and EOEC conditions for each recycle-type are shown in Figure A.12 and A.13. The decay heating for the breeder cycle is lower than the other cycle-values because of the reduced higher actinide level in the breeder cycle. Figure A.14 shows the regional components of the total EOEC decay power; it indicates that the majority of the decay heat in the breeder cycle is due to the central (fuelled) section of the core region and not due to any blanket region. *Although not shown, the same statement can be made for the converter cycle.*

C.5 Recycle radioactivity and decay

The time and isotopic dependence of the all-region fresh-fuel (feed plus recycled fuel) radioactivity (in Curies), discharged-fuel (3-burn cycle) radioactivity, plus the radioactivity difference between discharged and fresh fuel are shown for each recycle configuration in Tables A.38-A.46. Also included in the last line of the fresh and discharged radioactivity tables is the total TRU radioactivity versus time. Furthermore, included in the radioactivity difference tables, along with the total radioactivity difference, is the radioactivity difference per unit energy production. As with the decay heating results given above, the components of the all-region sums are shown in Tables A.47-A.64. The radioactivity difference per unit energy production is plotted for the different cycle types in Figure A.15. The large differences seen in the figure result from the differing amounts LWR feed material (mainly Pu-241) in the fresh fuel (see Table A.22). Figure A.16 shows the regional components of the breeder core radioactivity difference per unit energy production; it indicates that the majority of the radioactivity difference in the breeder cycle is due to the central (fuelled) section of the core region and not due to any blanket region. *Although not shown, the same statement can be made for the converter cycle.*

Shown in Tables A.65-A.73 are the corresponding quantities for the toxicity hazard (in units of cancer deaths or cancer deaths per Megawattday) of TRU isotopes. The factors used to calculate toxicity from radioactivity (in Curies) are given in Table 2, page 20; they are also shown in the last column of the tables under the heading "FACTOR". As with the radioactivity results given above, the components of the all-region sums are shown in Tables A.74-A.91. The toxicity difference per unit energy production is plotted for the different cycle types in Figure A.17. The large difference seen in the figure for the once-through cycle results from the large amount of LWR feed material (mainly Pu-241) in the once-through fresh fuel cycle (Pu-241 beta decays to Am-241 which in turn alpha decays – see Figure A.4). Figure A.18 shows the regional components of the breeder core toxicity difference per unit energy production. This figure indicates that blanket regions contribute significantly to the discharge fuel versus fresh fuel toxicity difference in the breeder cycle; mainly due to the buildup of plutonium isotopes, especially Pu-239, in the depleted uranium blankets.

References

[A.1] R. N. Hill, M. Kawashima, K. Arie, and M. Suzuki, "Calculational Benchmark Comparisons for a Low Sodium Void Worth Actinide Burner Core Design," Proc. ANS Topical Meeting on Advances in Reactor Physics, Charleston, SC, U.S.A., March 1992.

[A.2] R. N. Hill, "LMR Design Concepts for Transuranic Management in Low Sodium Void Worth Cores," Intl. Conf. on Fast Reactors and Related Fuel Cycles, Kyoto, Japan, October 2- November 1, 1991, p. 19.1-1.

[A.3] H. Henryson II, B. J. Toppel and C. G. Stenberg, "MC^2-2: A Code to Calculate Fast Neutron Spectra and Multigroup Cross Sections," ANL-8144, Argonne National Laboratory (June 1976).

[A.4] W. M. Stacy Jr, et al., "A New Space-Dependent Fast-Neutron Multigroup Cross-Section Capability," Trans.Am.Nucl.Soc.,15, 292 (1972).

[A.5] K. L. Derstine, "DIF3D: A Code to Solve One-, Two-, and Three-Dimensional Finite-Difference Diffusion Theory Problems," ANL-82-64, Argonne National Laboratory (April 1984).

[A.6] B. J. Toppel, "A Users Guide to the REBUS-3 Fuel Cycle Analysis Capability," ANL-83-2, Argonne National Laboratory (March 1983).

[A.7] M. J. Bell, "ORIGEN – The ORNL Isotope Generation and Depletion Code," ORNL-4628, Oak Ridge National Laboratory (May 1973).

[A.8] R. D. Lawrence, "The DIF3D Nodal Neutronics Option for Two- and Three-Dimensional Diffusion Theory Calculations in Hexagonal Geometry," ANL-83-1, Argonne National Laboratory (March 1983).

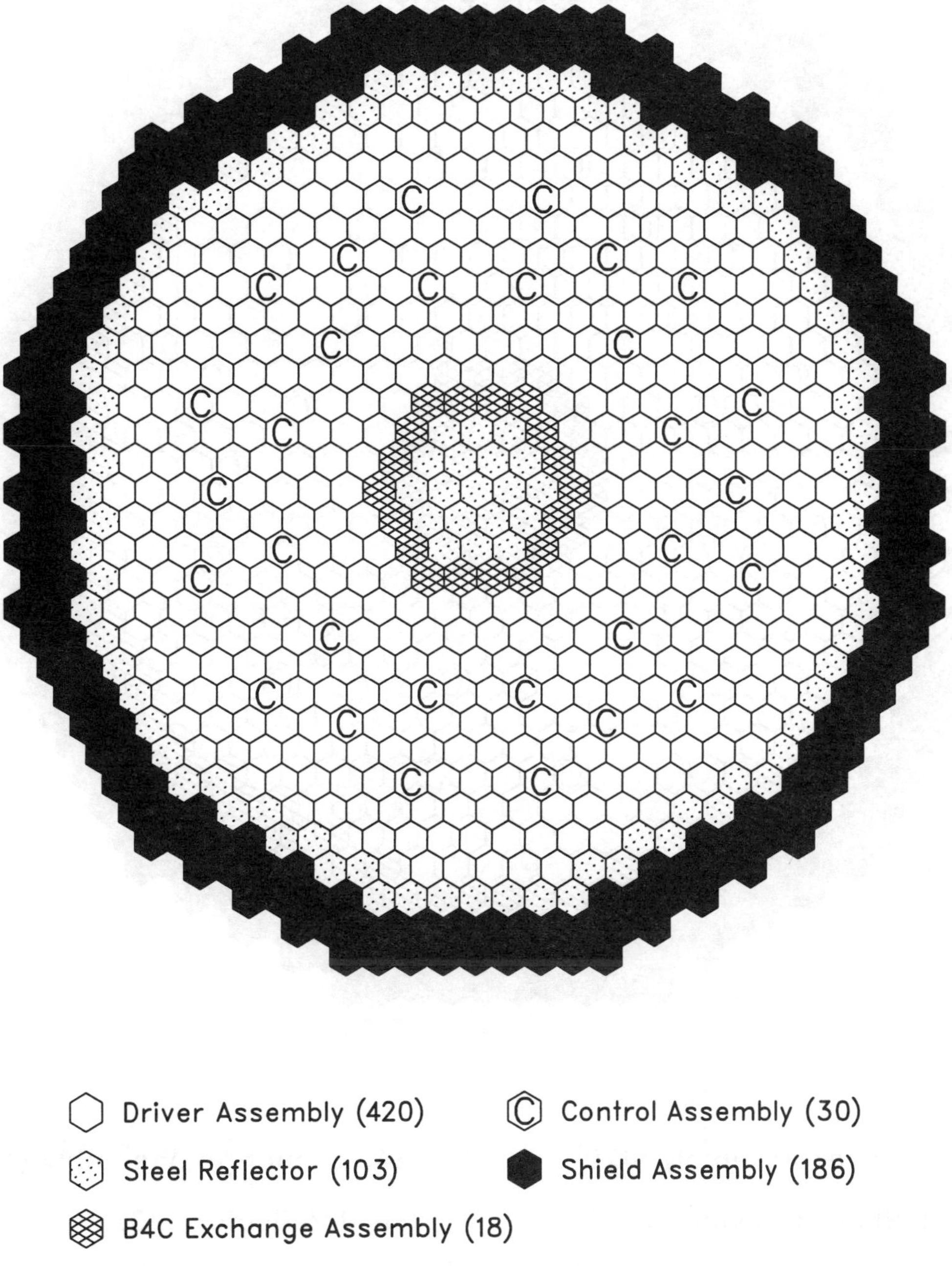

Figure A.1 ***Benchmark reference core configuration***

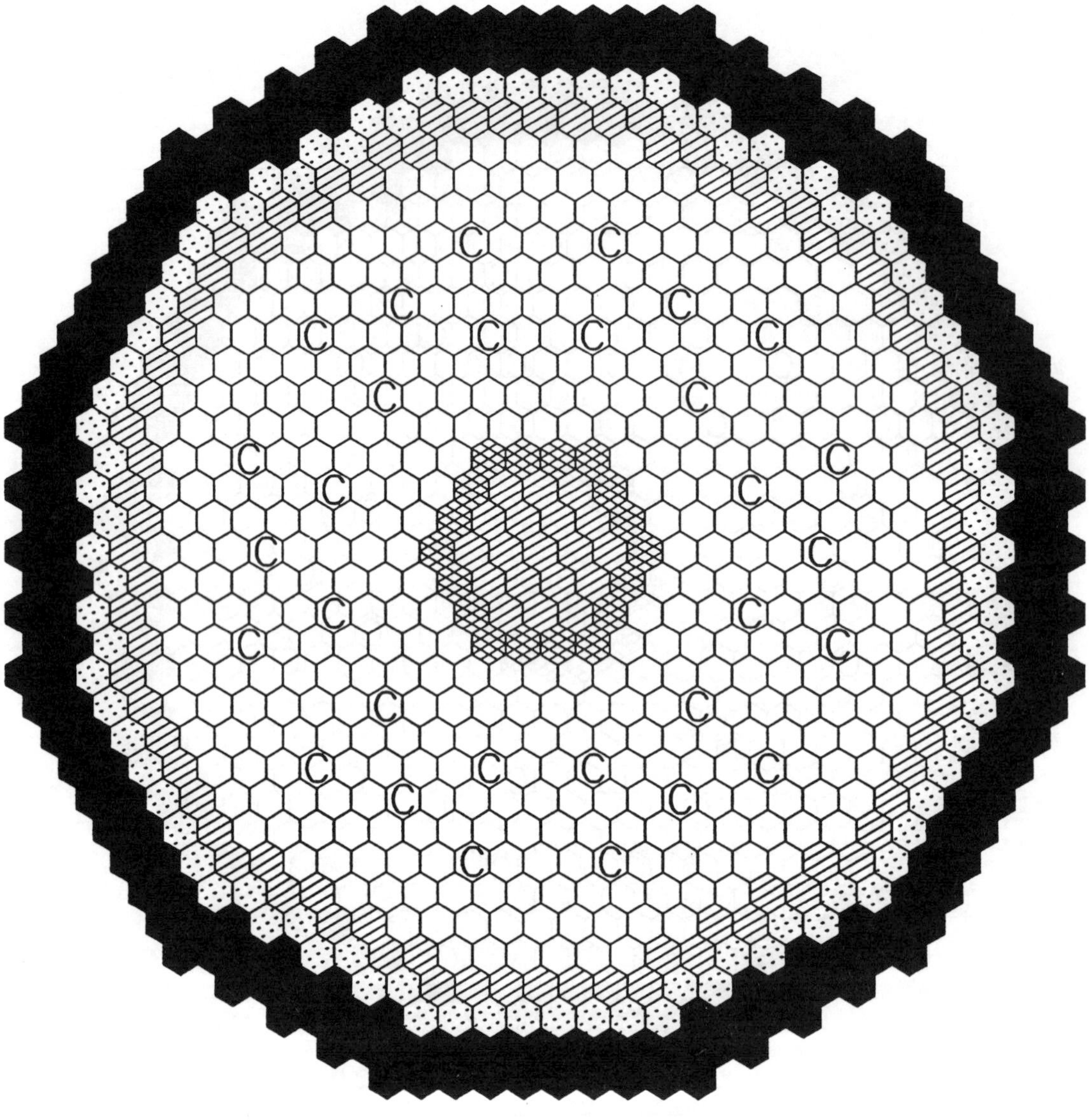

Driver Assembly (420)

Blanket Assembly (103)

Control Assembly (30)

B4C Exchange Assembly (18)

Steel Reflector (90)

Shield Assembly (198)

Figure A.2 ***CR = 1.0 Breeder core configuration***

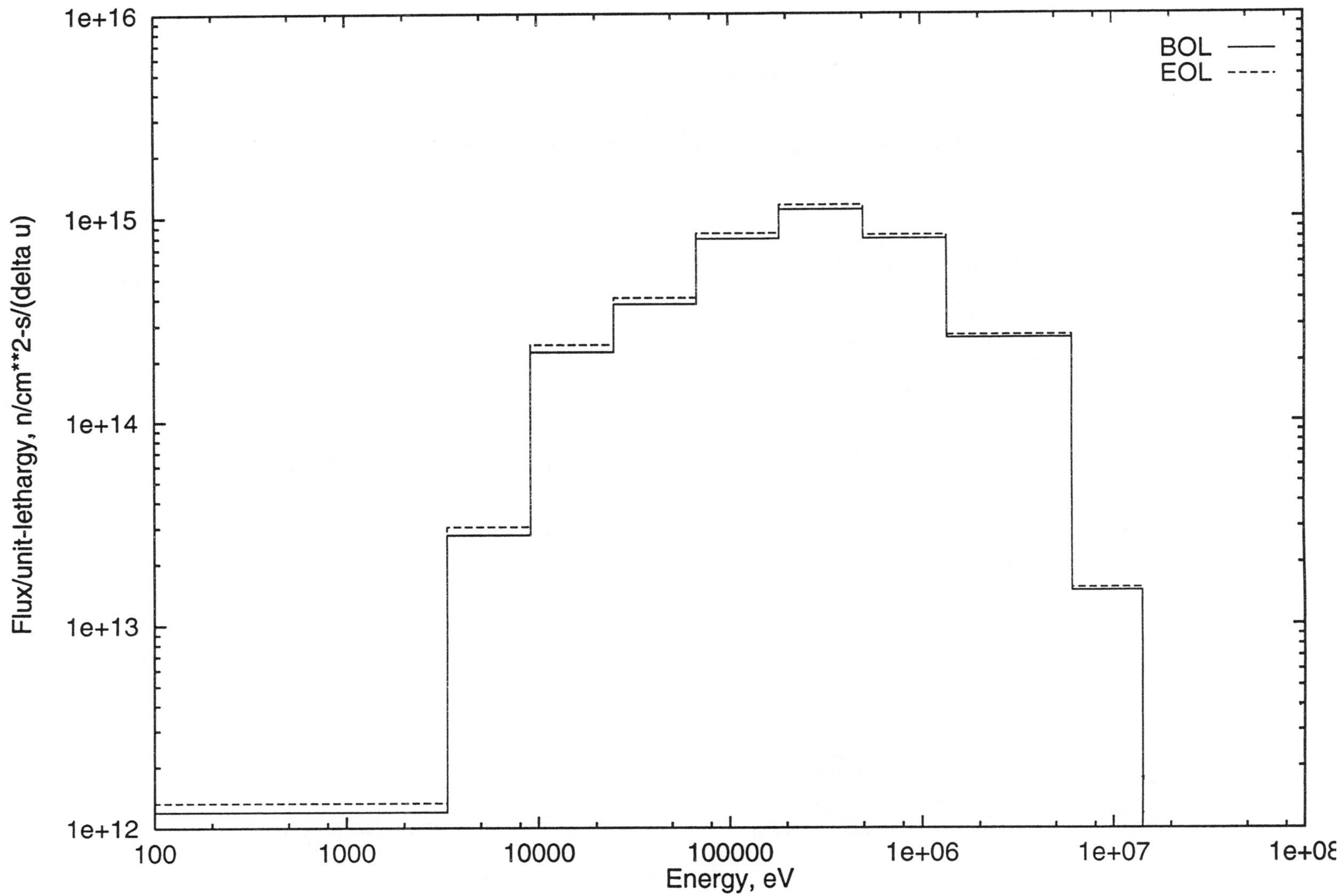

Figure A.3 ***BOL and EOL flux/unit lethargy***

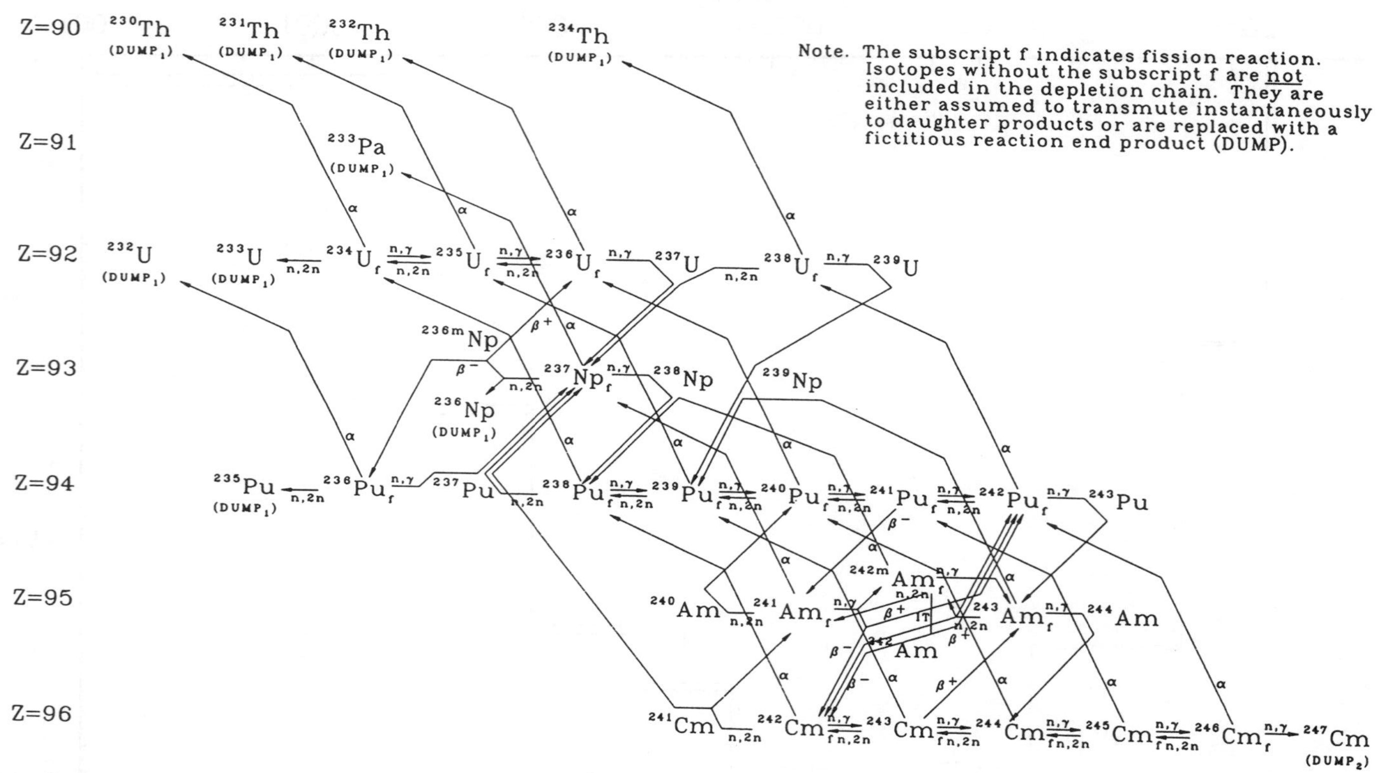

*Figure A.4 **Isotope transmutation chains used in REBUS-3 depletion calculations***

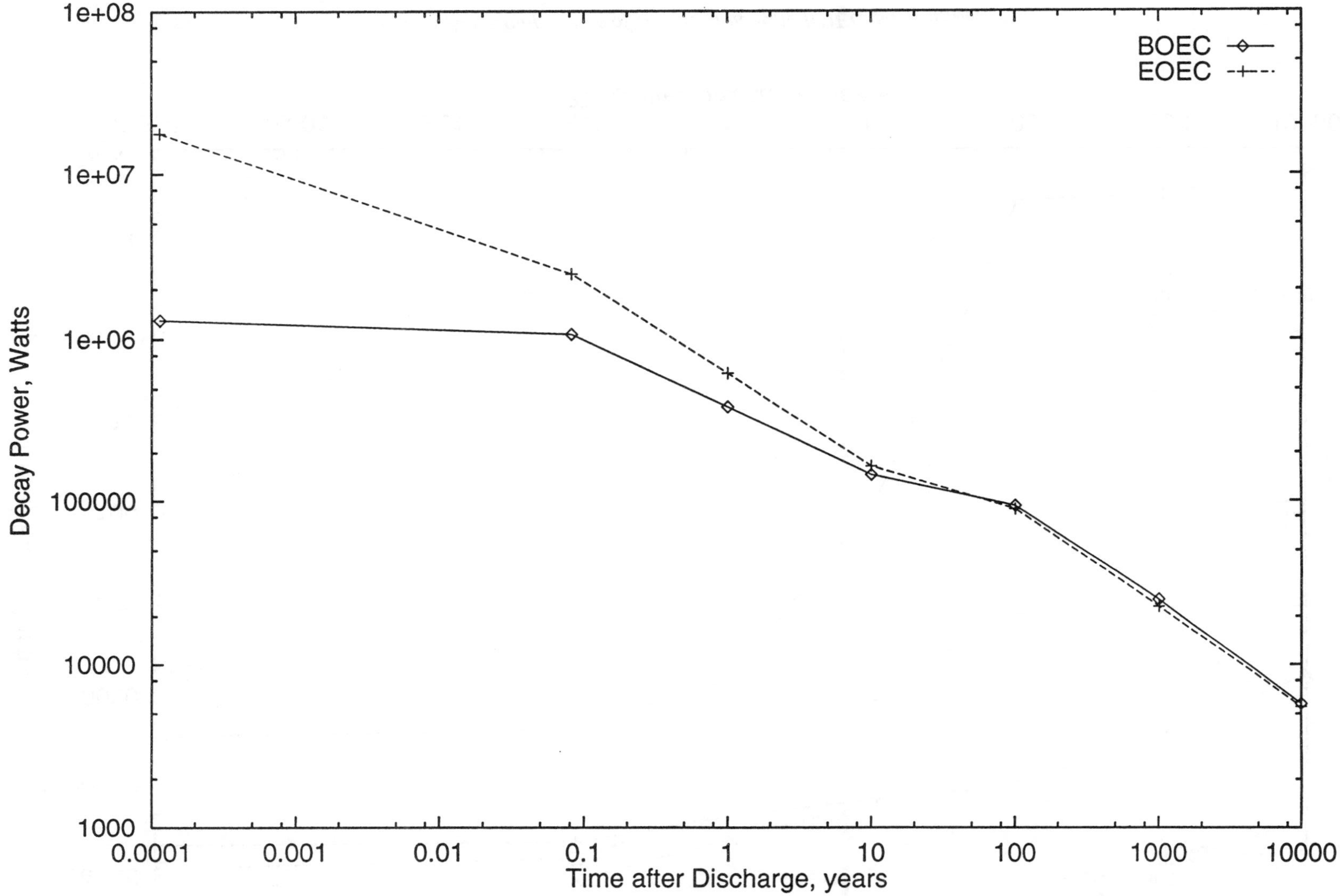

Figure A.5 ***Equilibrium cycle BOEC and EOEC decay power***

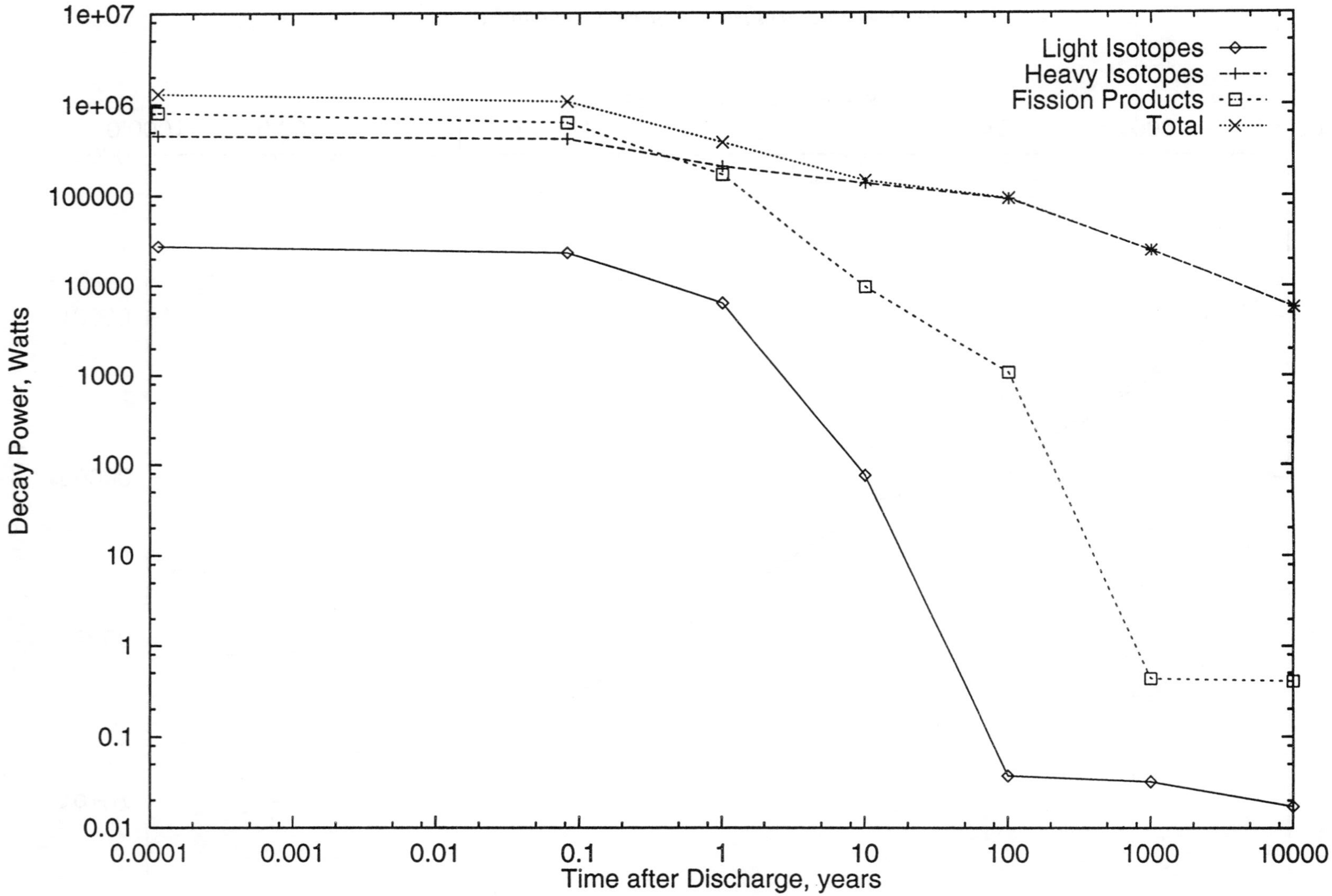

Figure A.6 ***Components of equilibrium cycle BOEC decay power***

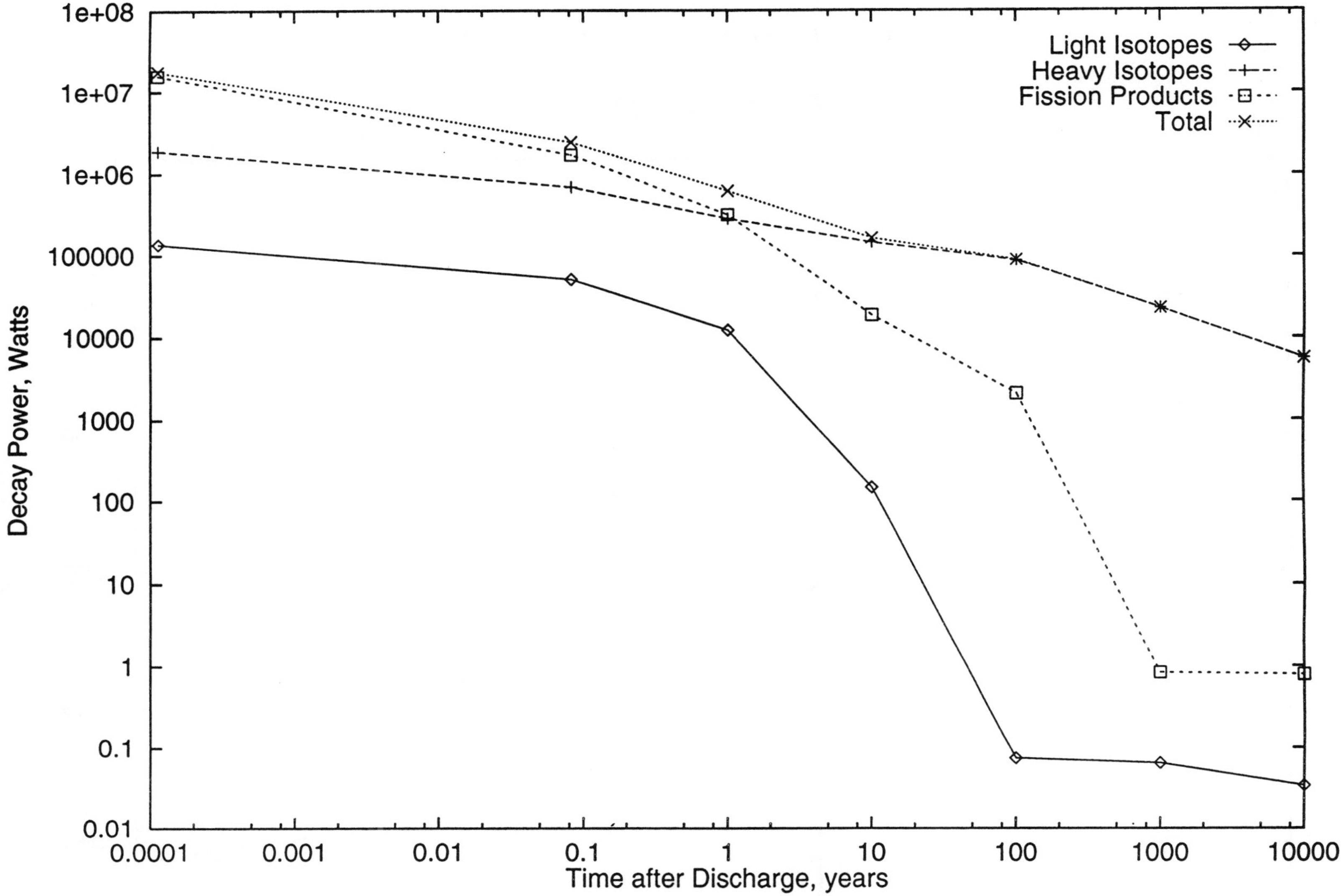

Figure A.7 ***Components of equilibrium cycle EOEC decay power***

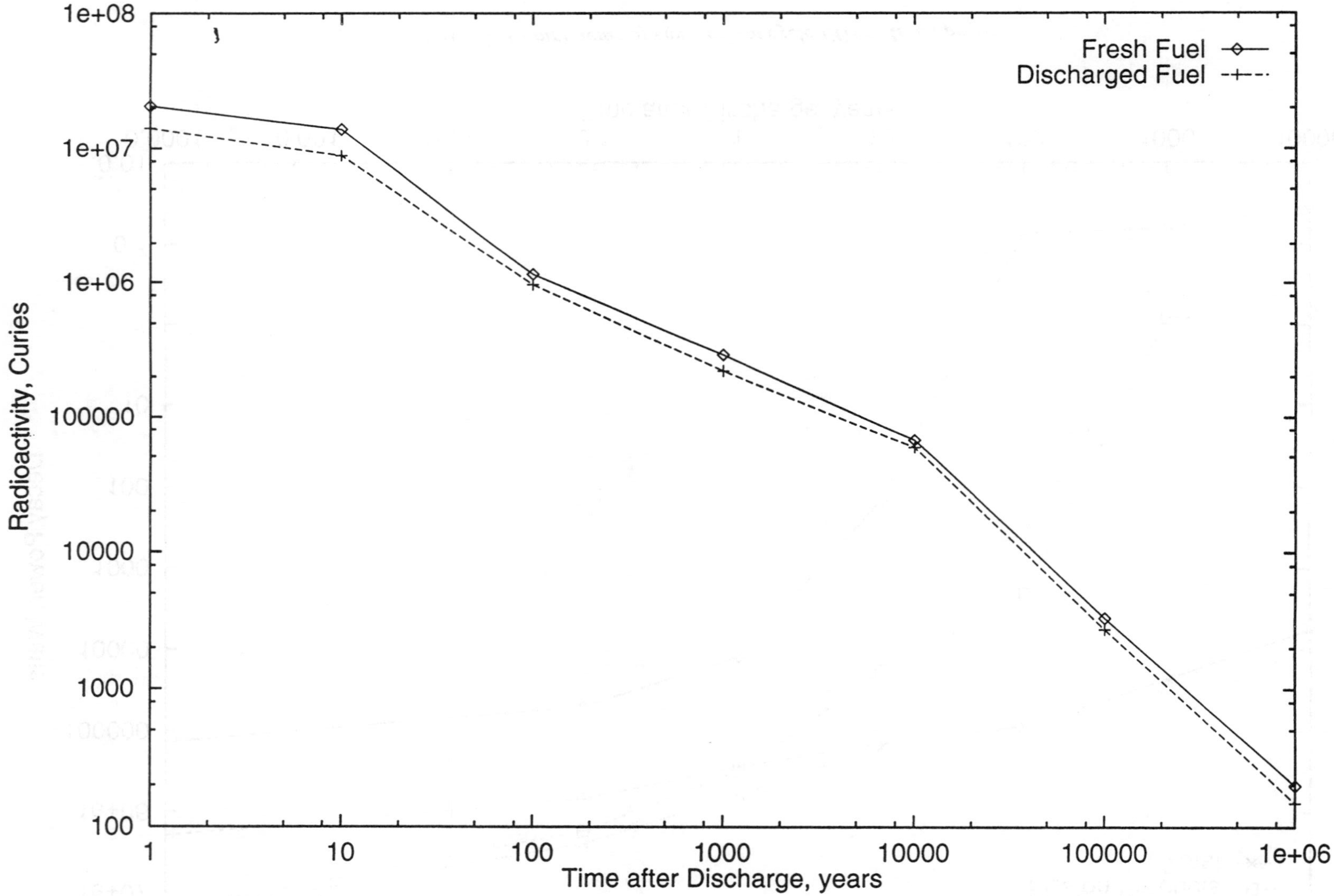

Figure A.8 ***Equilibrium cycle (discharged – fresh) fuel TRU curies***

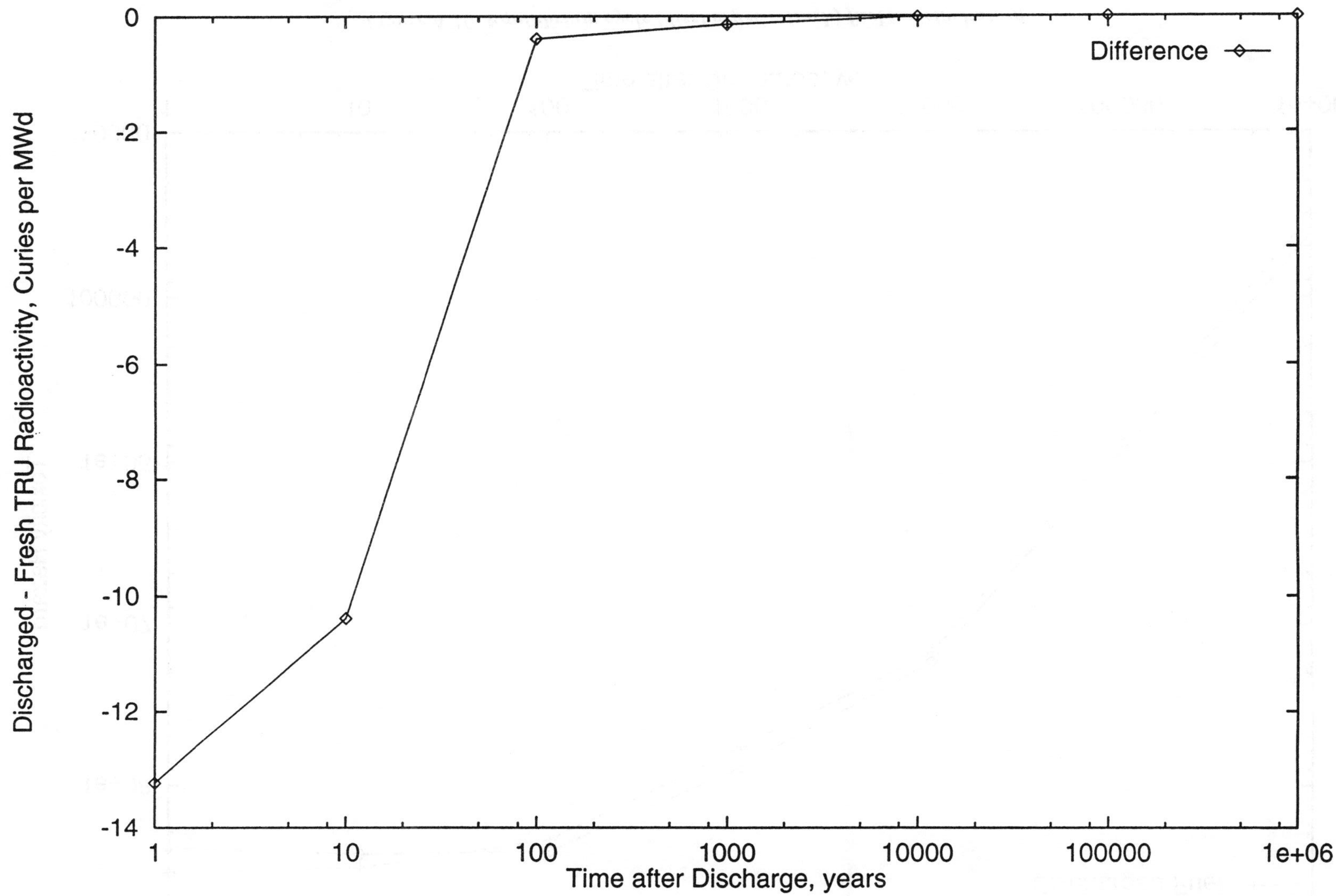

Figure A.9 ***Equilibrium cycle (discharged – fresh) fuel TRU curies /MWd***

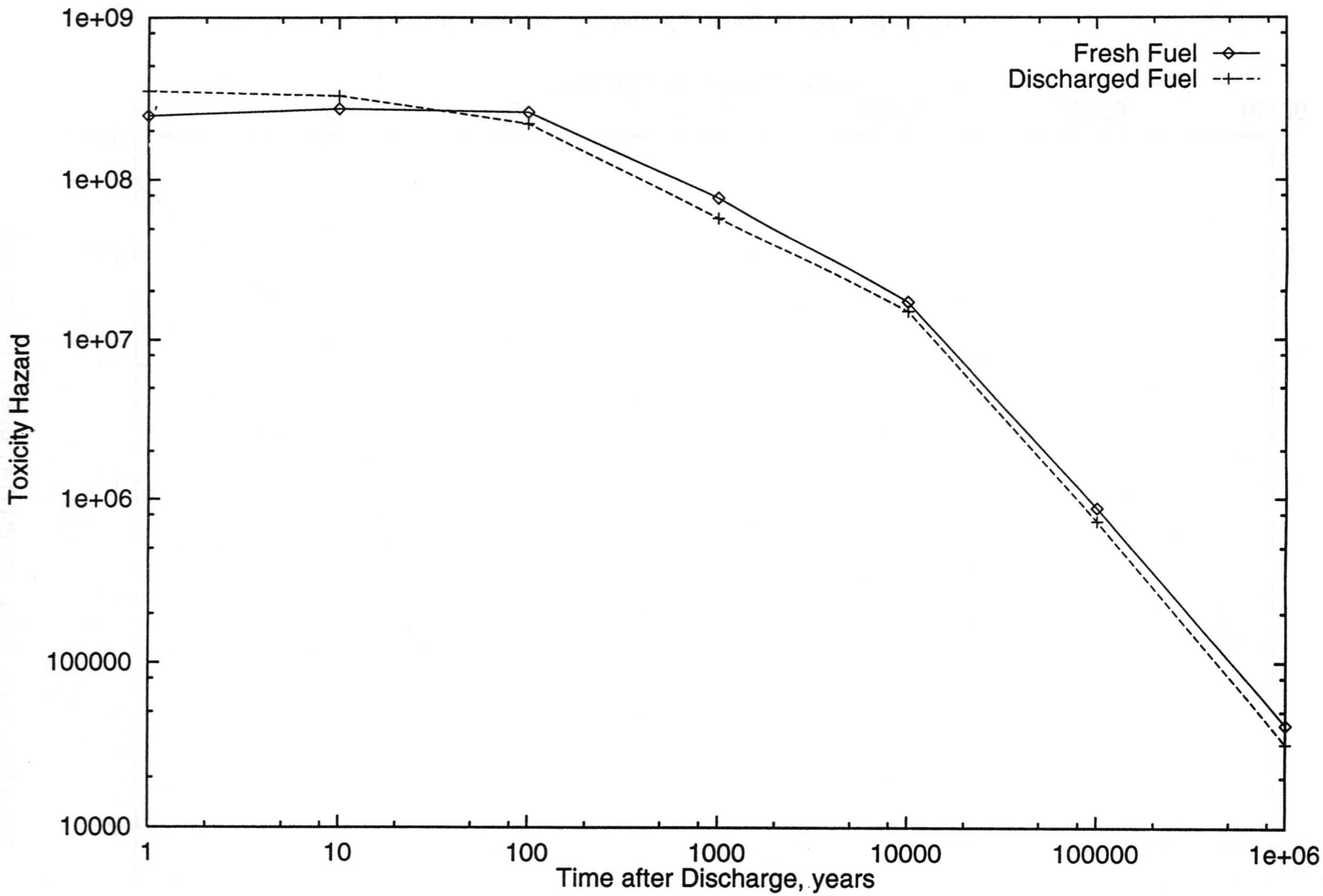

Figure A.10 **Equilibrium cycle (discharged – fresh) fuel toxicity hazard**

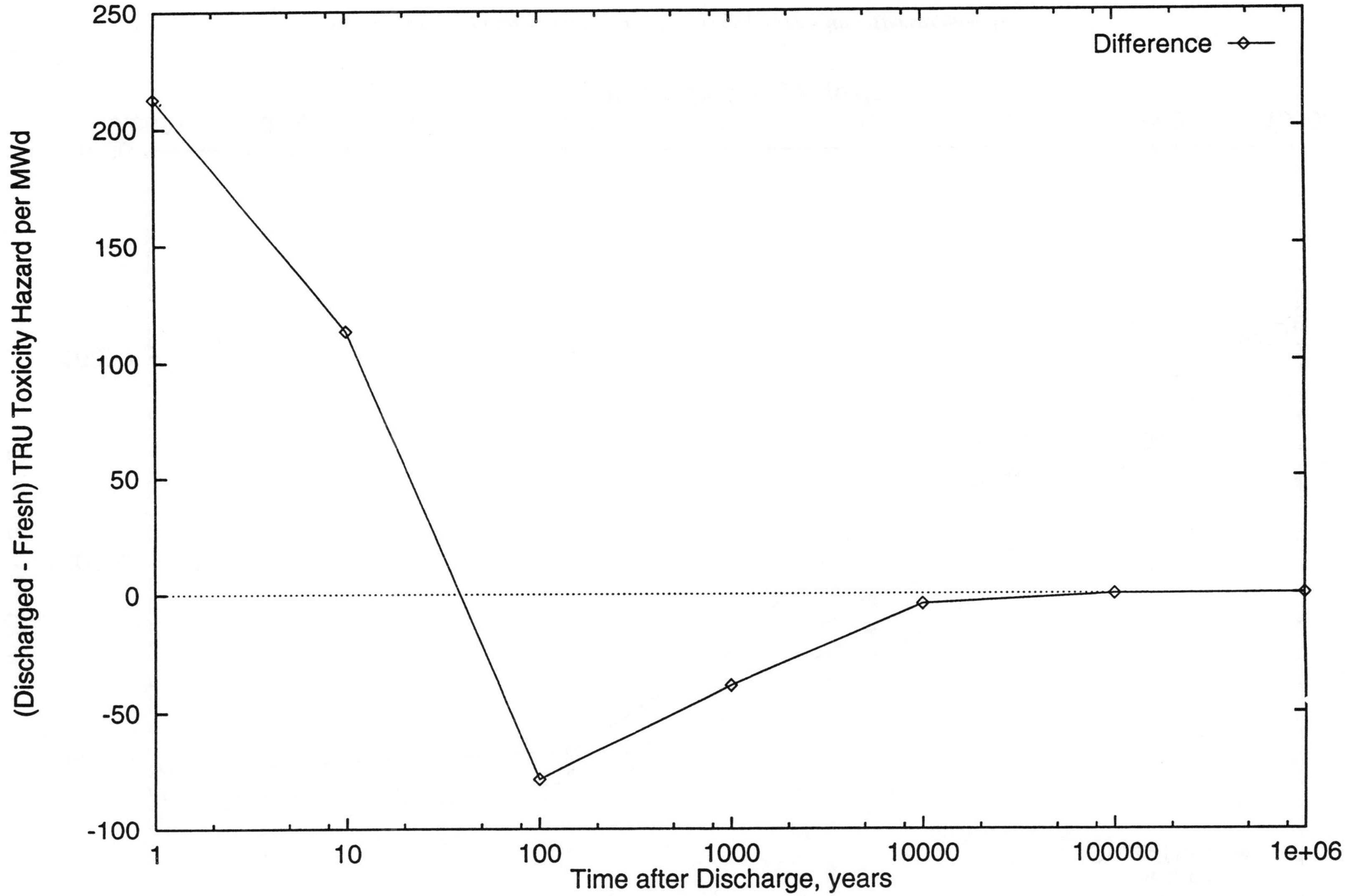

Figure A.11 ***Equilibrium cycle (discharged – fresh) fuel TRU toxicity hazard /MWd***

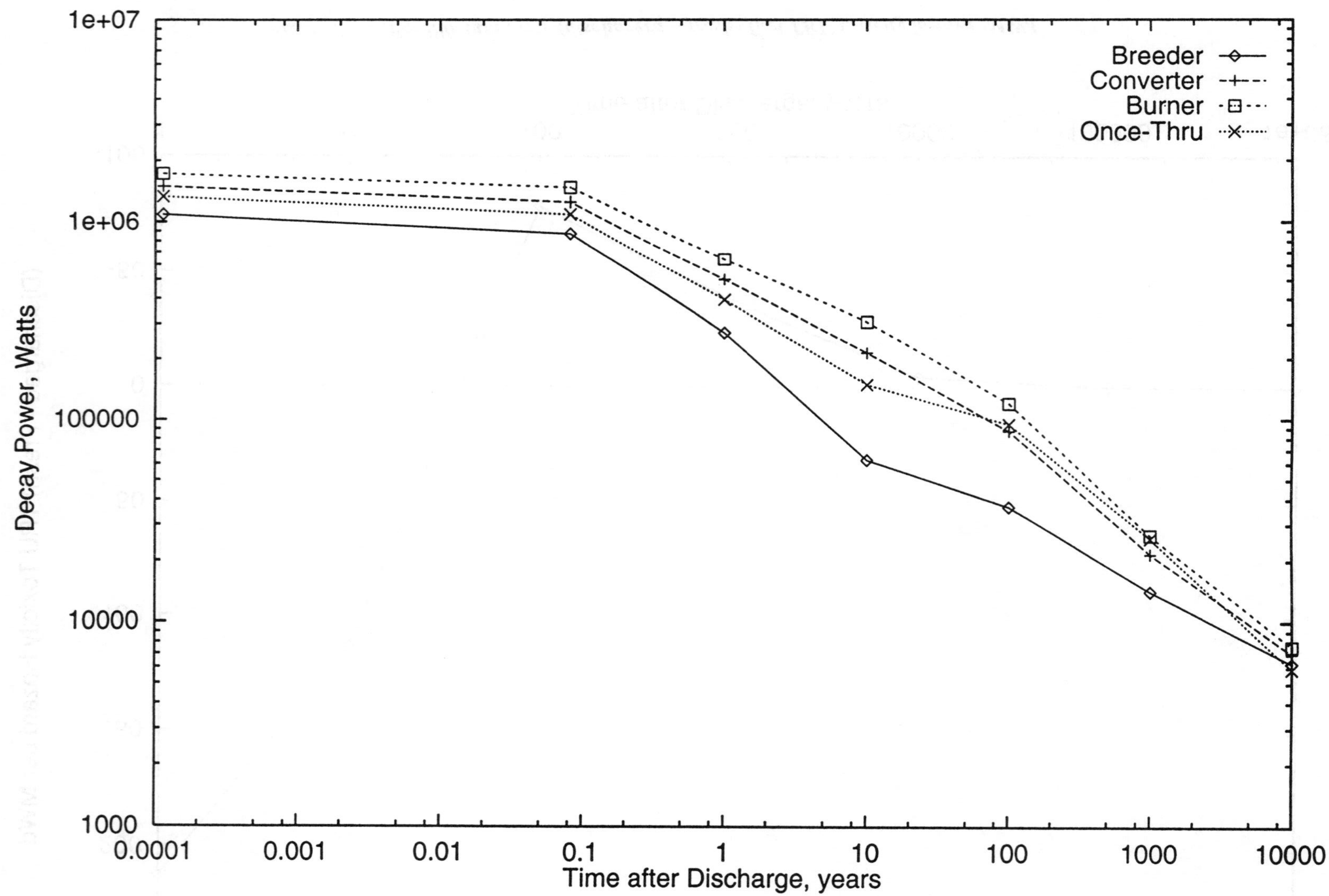

Figure A.12 ***Equilibrium cycle BOEC decay power with different models***

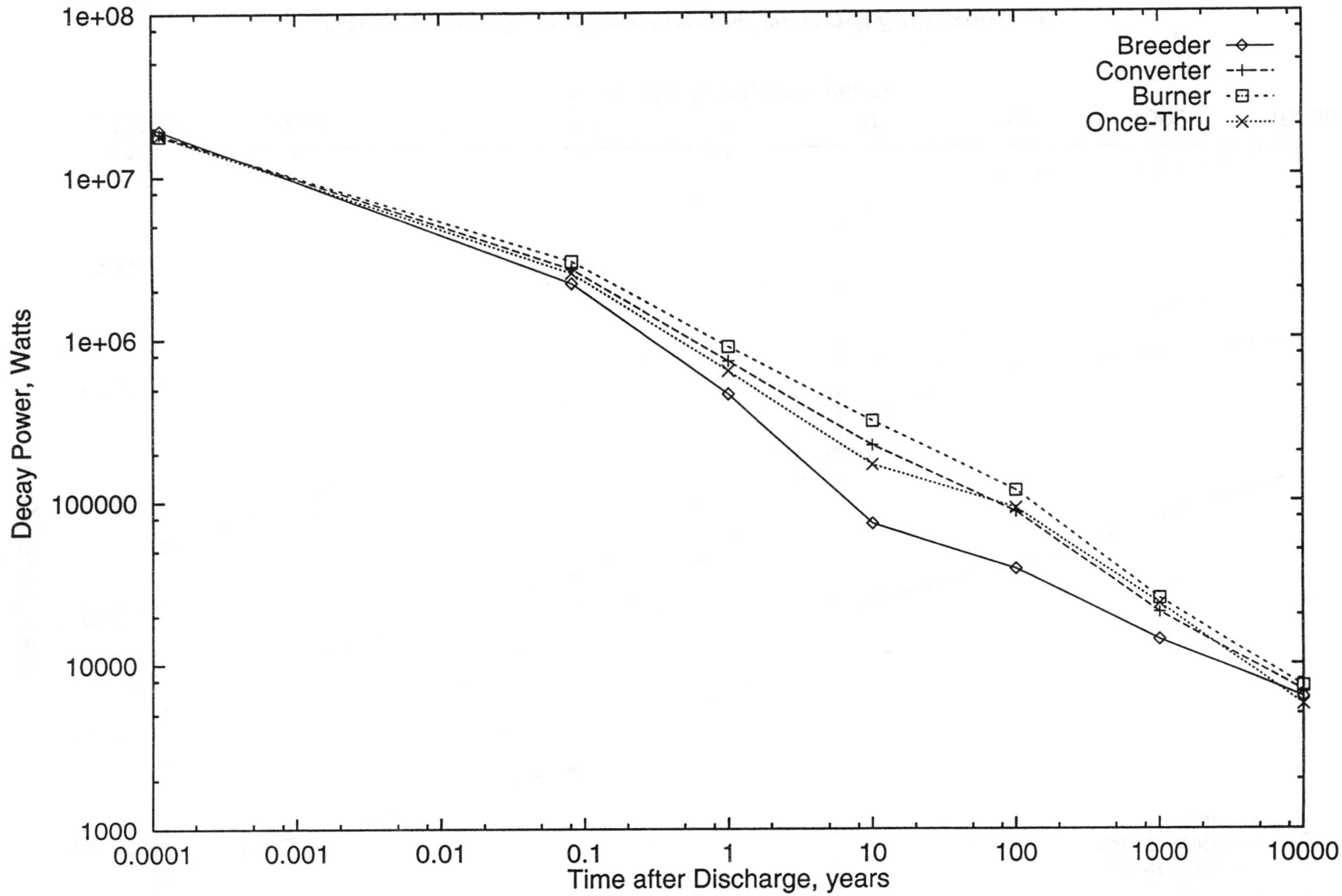

Figure A.13 ***Equilibrium cycle EOEC decay power with different models***

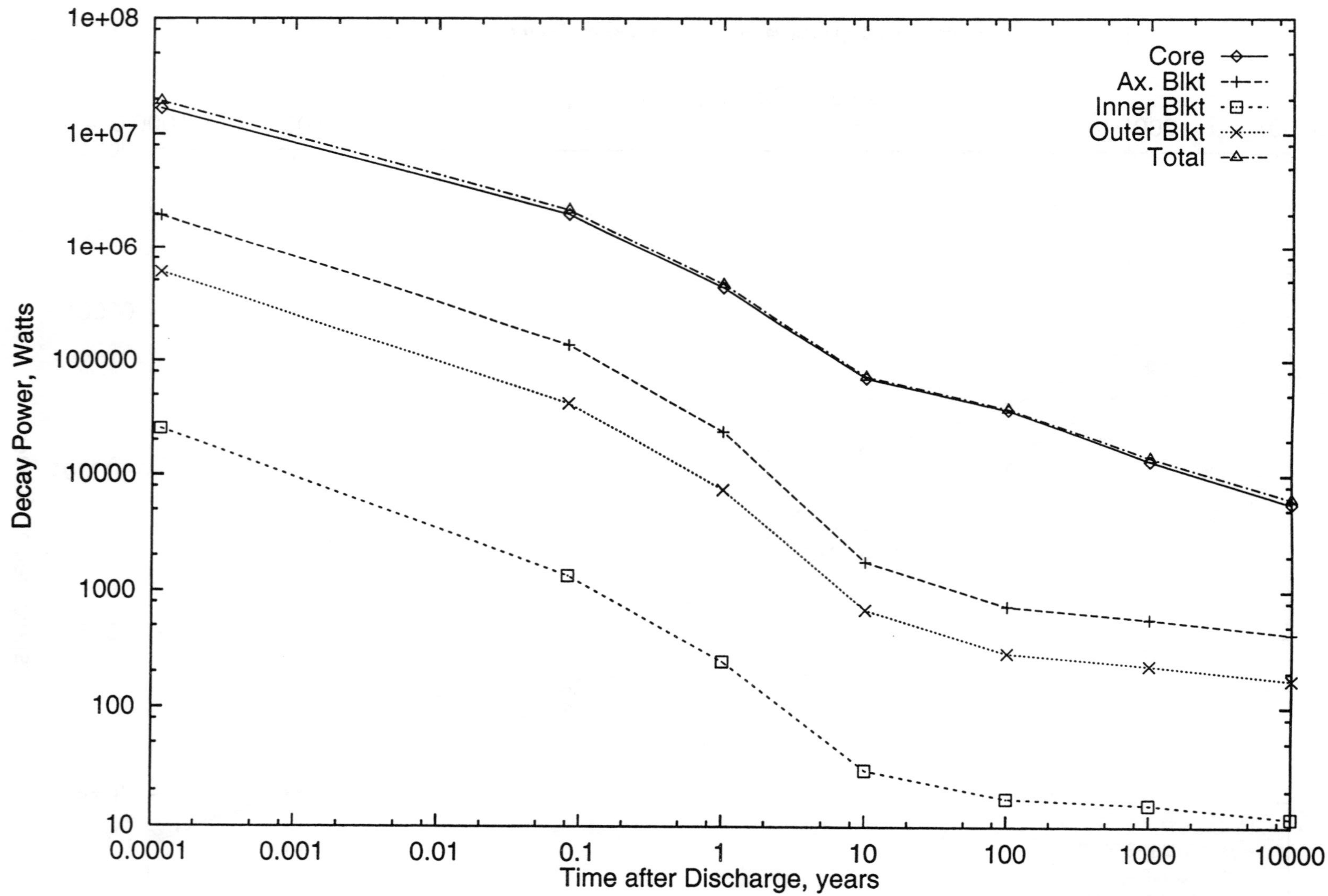

Figure A.14 ***Components of breeder core equilibrium cycle EOEC decay power***

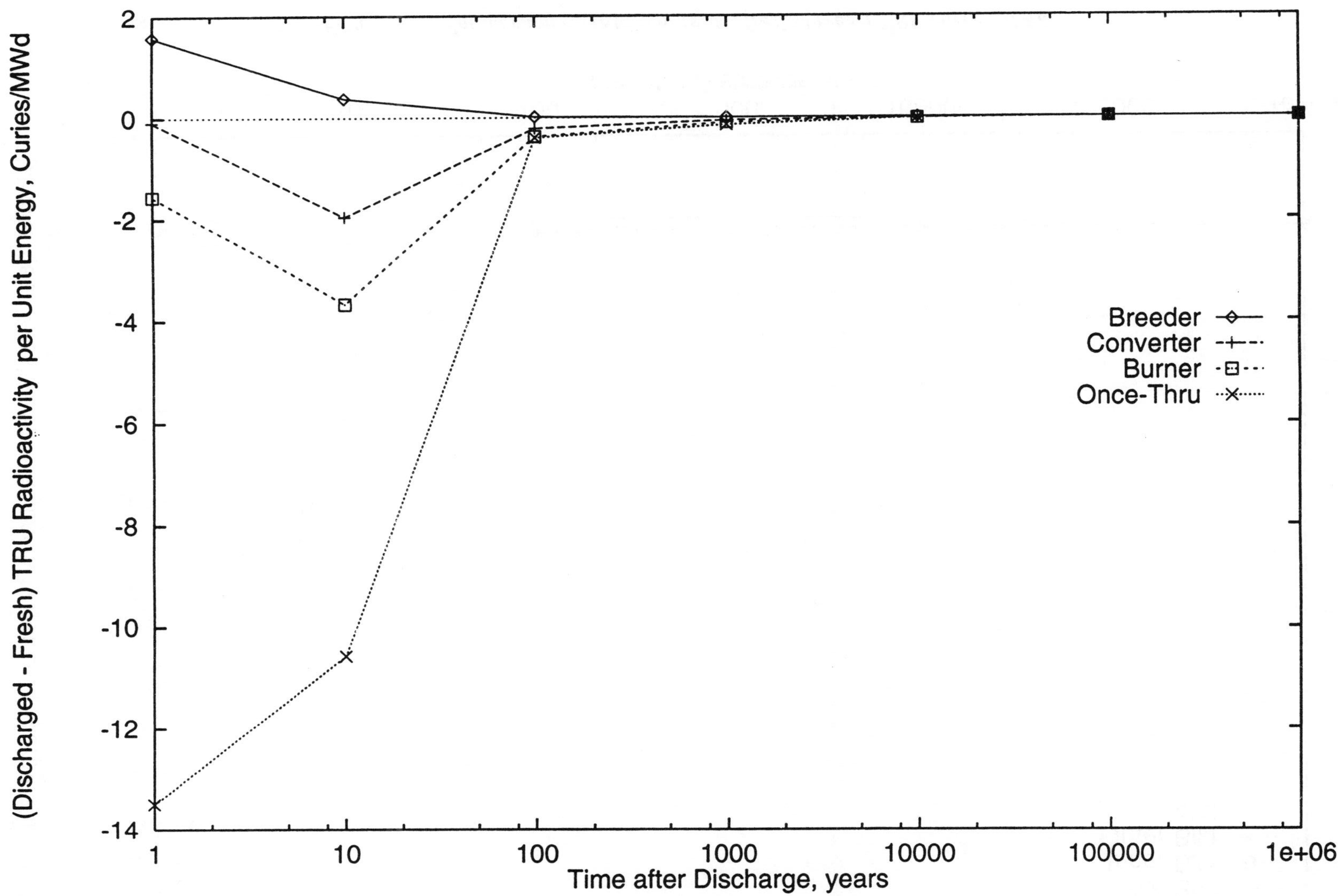

Figure A.15 ***Equilibrium cycle (discharged – fresh) fuel TRU curies /MWd***

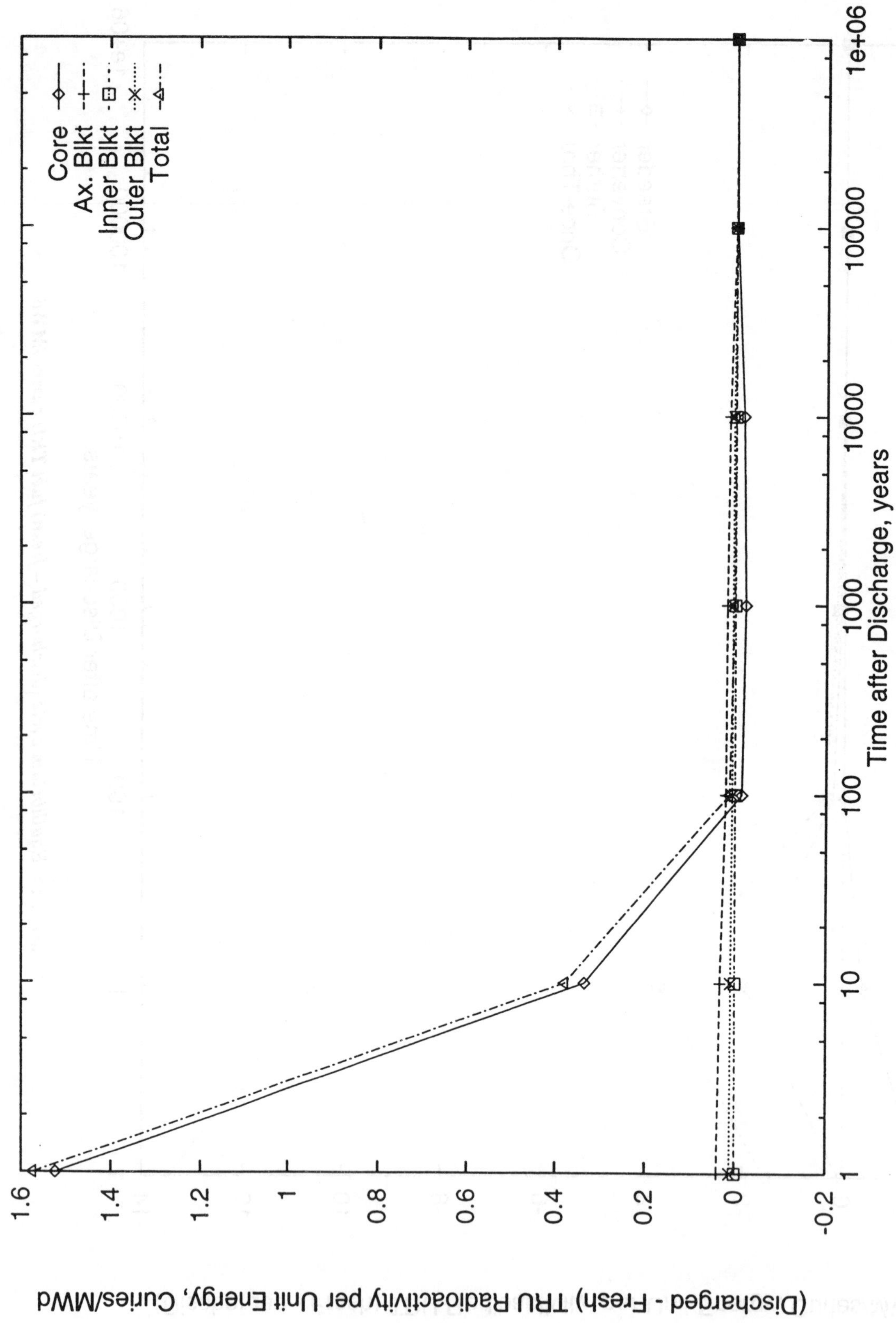

Figure A.16 ***Equilibrium cycle (discharged – fresh) breeder TRU curies /MWd***

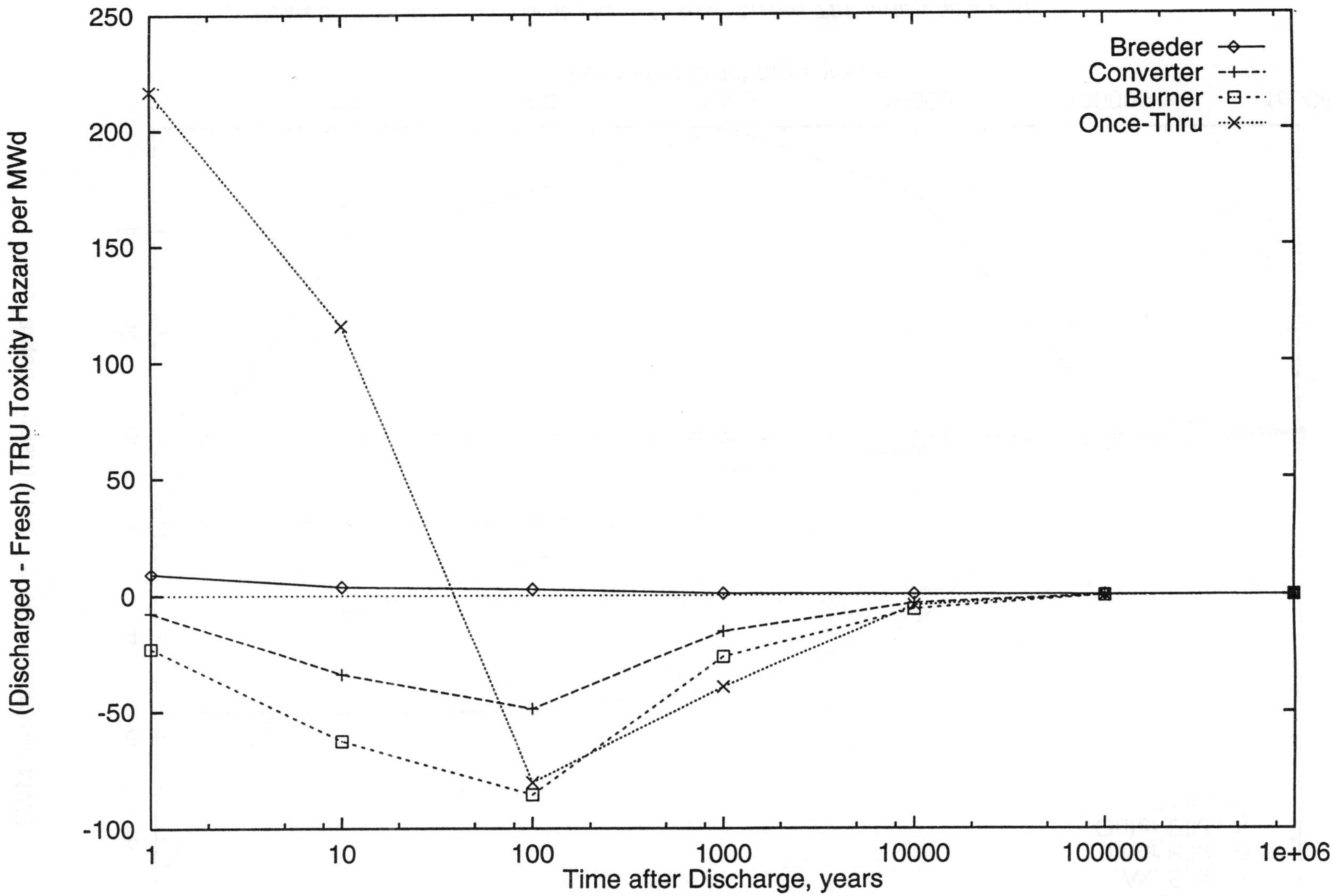

Figure A.17 ***Equilibrium cycle (discharged – fresh) fuel TRU toxicity hazard /MWd***

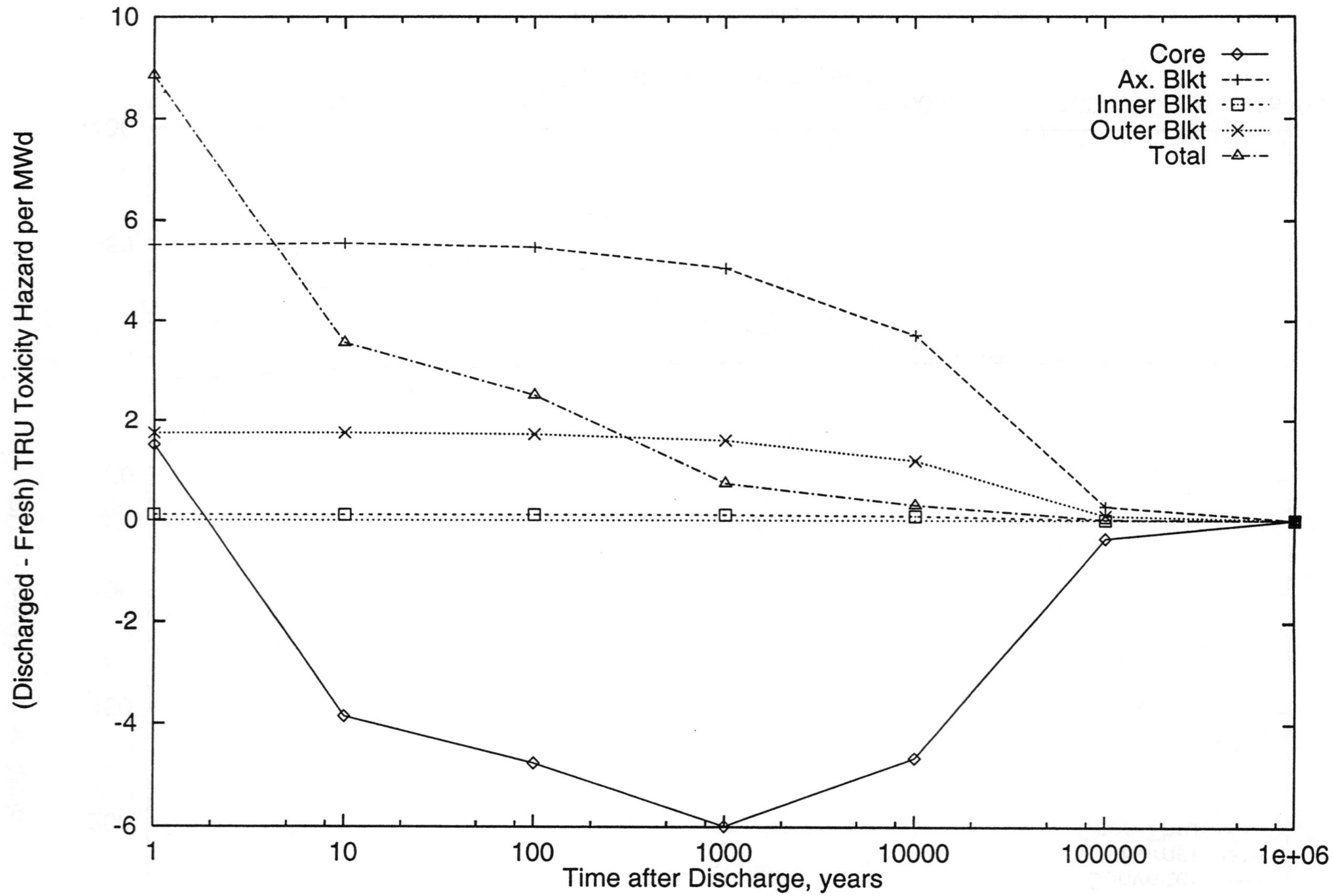

Figure A.18 ***Equilibrium cycke (discharged – fresh) breeder TRU toxicity hazard /MWd***

Table A.1
Cross-section energy group structures

	9 Energy Groups		21 Energy Groups	
Energy Group	**Upper Energy** *eV*	**Lethargy Width**	**Upper Energy** *eV*	**Lethargy Width**
1	1.42E+07	0.85	1.42E+07	0.85
2	6.07E+06	1.50	6.07E+06	0.50
3	1.35E+06	1.00	3.67E+06	0.50
4	4.98E+05	1.00	2.23E+06	0.50
5	1.83E+05	1.00	1.35E+06	0.50
6	6.74E+04	1.00	8.21E+05	0.50
7	2.48E+04	1.00	4.98E+05	0.50
8	9.12E+03	1.00	3.02E+05	0.50
9	3.35E+03	19.63 [a]	1.83E+05	0.50
10			1.11E+05	0.50
11			6.74E+04	0.50
12			4.09E+04	0.50
13			2.48E+04	0.50
14			1.50E+04	0.50
15			9.12E+03	0.50
16			5.53E+03	0.50
17			3.35E+03	0.50
18			2.03E+03	0.50
19			1.23E+03	1.00
20			4.54E+02	2.00
21			6.14E+01	15.63 [a]

[a] Bottom energy of last for each group structure is 0.0 eV.
However, the bottom of the last energy was taken to be 1.0E-05 eV in the lethargy calculation.

Table A.2

Isotopic energy deposition per fission and capture plus assumed isotopic "lumped" fission products

Isotope	Fission Energy *Watt-s/fission*	Capture Energy *Watt-s/capture*	Lumped Fission Product
U-234	3.02098E-11	1.64635E-12	FP35 + RE35 [(a)]
U-235	3.18080E-11	1.10290E-12	FP35 + RE35
U-236	3.05934E-11	8.79120E-13	FP35 + RE35
U-238	3.09939E-11	9.09490E-13	FP38 + RE38
Np-237	3.18082E-11	0.0	FP38 + RE38
Pu-236	3.12258E-11	9.59040E-13	FP35 + RE35
Pu-238	3.14885E-11	1.85412E-12	FP38 + RE38
Pu-239	3.17282E-11	1.04375E-12	FP39 + RE39
Pu-240	3.11828E-11	8.37562E-13	FP40 + RE40
Pu-241	3.20159E-11	1.00859E-12	FP41 + RE41
Pu-242	3.11528E-11	8.37562E-13	FP41 + RE41
Am-241	3.18082E-11	1.11888E-12	FP41 + RE41
Am-242m	3.12258E-11	9.59040E-13	FP41 + RE41
Am-243	3.18082E-11	9.59040E-13	FP41 + RE41
Cm-242	3.12258E-11	9.59040E-13	FP41 + RE41
Cm-243	3.12258E-11	9.59040E-13	FP41 + RE41
Cm-244	3.11688E-11	9.59040E-13	FP41 + RE41
Cm-245	3.12258E-11	9.59040E-13	FP41 + RE41
Cm-246	3.12258E-11	9.59040E-13	FP41 + RE41

[(a)] FPxx + Rexx = Lumped Fission Product (minus rare earths) plus Rare Earth (RE) contribution from isotope xx.
[xx = 35(U-235), 38(U-238), 39(Pu-239), 40(Pu-240), 41(Pu-241)]

Table A.3
Startup core eigenvalue results

ITEM	VALUE
BOL Eigenvalue	1.101176
BOL Eigenvalue Convergence Criterion	1.0000E-6
BOL k-infinity (Central Core Region)	1.61903
EOL Eigenvalue	1.042565
EOL Eigenvalue Convergence Criterion	1.0000E-6
Burnup Swing	-5.1053E-2

Table A.4
BOL and EOL startup core central region flux and spectrum

Energy Group	Upper Energy *eV*	Flux $n/cm^2 s$		Percent in Group	
		BOL	EOL	BOL	EOL
1	1.42E+07	1.24E+13	1.28E+13	0.3	0.3
2	6.07E+06	3.88E+14	4.02E+14	10.4	10.2
3	1.35E+06	7.89E+14	8.24E+14	21.2	21.0
4	4.98E+05	1.09E+15	1.15E+15	29.3	29.2
5	1.83E+05	7.89E+14	8.39E+14	21.2	21.4
6	6.74E+04	3.77E+14	4.04E+14	10.1	10.3
7	2.48E+04	2.22E+14	2.41E+14	6.0	6.1
8	9.12E+03	2.76E+13	3.02E+13	0.7	0.8
9	3.35E+03	2.35E+13	2.61E+13	0.6	0.7
Total	-	3.72E+15	3.93E+15	100.0	100.0

Table A.5
Startup core calculated parameters

ITEM	VALUE
BOL Core leakage/Core absorptions	0.61207
BOL Model leakage/Model absorptions	0.03030
BOL Heavy Metal Captures/Core Absorptions	3.66128E-1
BOL Zirconium Captures/Core Absorptions	7.31266E-3
BOL Structural Captures/Core Absorptions (Structural: Fe, Cr, Ni, Mo, and Mn)	2.61123E-2
BOL Structural Captures/Core Absorptions (Structural: Fe, Cr, Ni, Mo, Mn, and Zr)	3.34249E-2
BOL Coolant Captures/Core Absorptions	1.14561E-3
TRU Breeding Ratio (EOEC/BOEC)	0.9441
TRU Conversion Ratio	0.4482

Table A.6

Startup core 1-group microscopic cross-sections

	CAPTURE		FISSION		ν*FISSION		(n,2n)	
ISOTOPE	BOL	EOL	BOL	EOL	BOL	EOL	BOL	EOL
U-234	- [(a)]	3.55-1	-	4.63-1	-	1.18+0	-	2.22-4
U-235	3.21-1	3.25-1	1.45+0	1.45+0	3.60+0	3.62+0	1.11-3	1.09-3
U-236	-	3.50-1	-	1.39-1	-	3.62-1	-	1.14-3
U-238	1.78-1	1.79-1	5.81-2	5.70-2	1.61-1	1.58-1	2.23-3	2.17-3
Np-237	8.55-1	8.67-1	4.79-1	4.73-1	1.40+0	1.38+0	2.12-4	2.06-4
Pu-236	1.83-1	1.86-1	1.49+0	1.48+0	4.42+0	4.41+0	4.53-4	4.42-4
Pu-238	4.23-1	4.27-1	1.21+0	1.20+0	3.67+0	3.65+0	6.66-4	6.49-4
Pu-239	2.22-1	2.25-1	1.65+0	1.65+0	4.89+0	4.89+0	4.70-4	4.58-4
Pu-240	2.86-1	2.89-1	5.01-1	4.96-1	1.52+0	1.50+0	2.16-4	2.10-4
Pu-241	2.61-1	2.64-1	1.92+0	1.93+0	5.76+0	5.78+0	2.99-3	2.92-3
Pu-242	2.21-1	2.35-1	3.76-1	3.71-1	1.15+0	1.14+0	9.34-4	9.10-4
Am-241	1.01+0	1.02+0	4.08-1	4.02-1	1.38+0	1.36+0	1.77-4	1.73-4
Am-242m	1.48-1	1.51-1	2.80+0	2.81+0	9.37+0	9.40+0	1.70-3	1.65-3
Am-243	6.99-1	7.09-1	3.26-1	3.21-1	1.19+0	1.17+0	1.15-4	1.12-4
Cm-242	1.15-1	1.17-1	2.31-1	2.27-1	8.89-1	8.74-1	2.67-5	2.61-5
Cm-243	1.06-1	1.08-1	2.01+0	2.02+0	7.11+0	7.13+0	1.61-3	1.57-3
Cm-244	4.91-1	4.96-1	5.83-1	5.75-1	2.18+0	2.15+0	5.58-4	5.44-4
Cm-245	1.87-1	1.89-1	2.18+0	2.19+0	8.58+0	8.60+0	2.90-3	2.83-3
Cm-246	1.23-1	1.25-1	3.88-1	3.82-1	1.49+0	1.47+0	7.46-4	7.27-4

[(a)] Isotope not included in BOL description for region.

Table A.7
Depletion decay constants

ISOTOPE BEFORE DECAY	DECAY TYPE	ISOTOPE AFTER DECAY	DECAY CONSTANT (s^{-1})
U-234	alpha	DUMP1	8.978E-14
U-235	alpha	DUMP1	3.120E-17
U-236	alpha	DUMP1	9.379E-16
U-238	alpha	DUMP1	4.915E-18
Np-237	alpha	DUMP1	1.026E-14
Pu-236	alpha	DUMP1	7.703E-09
Pu-238	alpha	U-234	2.503E-10
Pu-239	alpha	U-235	9.109E-13
Pu-240	alpha	U-236	3.353E-12
Pu-241	beta -	Am-241	1.494E-09
Pu-242	alpha	U-238	5.833E-14
Am-241	alpha	Np-237	5.081E-11
Am-242	beta -	Cm-242	1.189E-10
Am-242	beta +	Pu-242	2.487E-11
Am-242	alpha	Pu-238	7.225E-13
Am-243	alpha	Pu-239	2.976E-12
Cm-242	alpha	Pu-238	4.924E-08
Cm-243	beta +	Am-243	2.003E-12
Cm-243	alpha	Pu-239	7.685E-10
Cm-244	alpha	Pu-240	1.213E-09
Cm-245	alpha	Pu-241	2.592E-12
Cm-246	alpha	Pu-242	4.642E-12

DUMP1 = Dump isotope (out of chain)

Table A.8
Branching ratios

Reaction Isotope	Reaction Type	Product Isotope	Branching Ratio
Np-237	(n,2n)	Pu-236	0.346
		U-236	0.374
		DUMP1*	0.280
Am-241	(n,gamma)	Cm-242	0.660
		Am-242	0.200
		Pu-242	0.140
Am-243	(n,2n)	Am-242	0.500
		Pu-242	0.086
		Cm-242	0.414
Cm-242	(n,2n)	Am-241	0.990
		Np-237	0.010

* DUMP1 = Dump isotope (out of chain).

Table A.9
Startup core mass balance, kg

ISOTOPE	BOL	EOL	EOL – BOL
U-234	0.000	0.330	0.330
U-235	25.428	21.936	-3.492
U-236	0.000	0.732	0.732
U-238	12 851.376	12 594.258	-257.118
Np-237	239.004	215.418	-23.586
Pu-236	0.000	0.000	0.000
Pu-238	44.892	58.614	13.182
Pu-239	2 267.676	2 124.582	-143.064
Pu-240	892.080	879.954	-12.126
Pu-241	603.222	501.954	-101.268
Pu-242	175.380	180.564	5.184
Am-241	112.992	120.750	7.758
Am-242	0.504	2.220	1.716
Am-243	112.578	106.650	-5.928
Cm-242	0.042	3.816	3.774
Cm-243	0.354	0.318	-0.036
Cm-244	25.158	28.728	3.570
Cm-245	2.322	2.964	0.642
Cm-246	0.288	0.324	0.036
TRU	4 476.462	4 226.316	-250.146
FISSION PRODUCTS:			
LFP-233	0.000	0.012	0.012
LFP-235	0.000	2.880	2.880
LFP-239	0.000	508.302	508.302

Table A.10

Once-through core enrichment and TRU breeding ratio

ITEM	VALUE
Enrichment	0.263663
TRU Breeding Ratio (EOEC/BOEC)	0.9438
TRU Conversion Ratio	0.4634

Table A.11

Once-through core criticality performance

ITEM	VALUE
BOEC Eigenvalue	1.05572
BOEC Eigenvalue Convergence Criterion	1.0E-6
EOEC Eigenvalue	1.00030
EOEC Eigenvalue Convergence Criterion	1.0E-6
Burnup Swing (rods at constant position)	-5.266E-2

Table A.12
Once-through core mass balance, kg

ISOTOPE	BOL	EOL	EOL – BOL
U-234	0.360	0.756	0.396
U-235	21.774	18.660	-3.114
U-236	0.720	1.350	0.630
U-238	12 485.166	12 221.370	-263.796
Np-237	219.798	197.148	-22.650
Pu-236	0.000	0.000	0.000
Pu-238	59.712	72.246	12.534
Pu-239	2 165.856	2 027.142	-138.714
Pu-240	898.032	883.032	-15.000
Pu-241	516.534	430.050	-86.484
Pu-242	183.972	187.164	3.192
Am-241	119.280	122.136	2.856
Am-242	2.214	3.672	1.458
Am-243	107.400	101.676	-5.724
Cm-242	3.144	4.986	1.842
Cm-243	0.330	0.318	-0.012
Cm-244	28.998	32.082	3.084
Cm-245	3.048	3.744	0.696
Cm-246	0.330	0.378	0.048
TRU	4 250.736	4,006.422	-244.314 [(a)]

(a) Mass increment/energy = -244.314 kg/(1575MW • 310 days) = -5.00E-4 kg/MWd.

Table A.13
Reactor performance and mass flow data

Reactor power, MWe/MWth 600/1575
Fuel residence time, FPD .... 3 x 310
Capacity factor, % 85
BOC k-eff - EOC k-eff 5.542e-02

	U-234	U-235	U-236	U-238	Np-237	Pu-238	Pu-239	Pu-240	Pu-241	Pu-242	Am-241
INITIAL LOADING kg CORE	.0	25.2	.0	12752.9	244.3	45.9	2317.9	911.8	616.5	179.3	115.5
EQUILIBRIUM LOADING kg/y CORE	.0	8.4	.0	4254.4	81.5	15.3	773.2	304.2	205.7	59.8	38.5
EQUILIBRIUM DISCHARGE kg/y CORE	.4	5.3	.6	3990.4	58.8	27.9	634.4	289.2	119.1	63.0	41.7
NET GAIN kg/y CORE	.4	-3.1	.6	-264.0	-22.7	12.5	-138.8	-15.0	-86.6	3.2	3.1

	Am-242	Am-243	Cm-242	Cm-243	Cm-244	Cm-245	Cm-246	TOTAL FISSILE	TOTAL HM
INITIAL LOADING kg CORE	.5	115.1	.0	.4	25.7	2.4	.3	2959.6	17353.6
EQUILIBRIUM LOADING kg/y CORE	.2	38.4	.0	.1	8.6	.8	.1	987.3	5789.2
EQUILIBRIUM DISCHARGE kg/y CORE	1.6	32.7	1.9	.1	11.7	1.5	.1	758.8	5280.4
NET GAIN kg/y CORE	1.5	-5.7	1.9	.0	3.1	.7	.0	-228.5	-508.8

Table A.14
Once-through core safety parameters

ITEM	VALUE
Beta Effective	3.508E-3
Doppler,-Tdk/dT (core fuel HM plus Zr)	1.462E-3
Doppler,-Tdk/dT (core steel)	3.443E-4
Sodium void (all sodium removed), % Δk/kk'	
Total Core	5.813E-1
Plenum	-2.634E+0
Total Core plus Plenum	-2.053E+0
Sodium void (flowing sodium removed), % Δk/kk'	
Total Core	6.059E-1
Plenum	-7.976E-1
Total Core plus Plenum	-1.917E-1

Table A.15
Once-through core BOEC and EOEC decay power, Watts

	BOEC				EOEC			
Time	*Light*	*Heavy*	*FP*	*Total*	Light	Heavy	FP	Total
1 hour	2.85+4	4.67+5	8.26+5	1.33+6	1.40+5	1.95+6	1.60+7	1.80+7
30 days	2.41+4	4.29+5	6.42+5	1.09+6	5.23+4	7.10+5	1.75+6	2.52+6
1 year	6.48+3	2.14+5	1.73+5	3.95+5	1.26+4	2.94+5	3.26+5	6.36+5
10 y	7.71+1	1.39+5	9.79+3	1.49+5	1.49+2	1.49+5	1.92+4	1.68+5
100 y	3.77-2	9.36+4	1.73+3	9.48+4	7.52-2	8.87+4	2.12+3	9.05+4
1000 y	3.26-2	2.55+4	4.38-1	2.57+4	6.55-2	2.33+4	8.50-1	2.33+4
10000 y	1.75-2	5.89+3	4.12-1	5.89+3	3.51-2	5.65+3	8.01-1	5.65+3

Light = ORIGEN Light Isotopes
Heavy = ORIGEN Heavy Isotopes
FP = ORIGEN Fission Products
Total = Sum of Light, Heavy, and FIssion Product Decay Power.

Table A.16

Heavy TRU elements nuclide radioactivity, Curies Basis = EQULIBRIUM FRESH FUEL

	Initial	1 hour	30 days	1 year	10 y	100 y	1000 y	10000 y	100000 y	1000000 y
Np-236	0.000E+00	0.000E+00	0.000E+00	0.000E+00	0.000E+00	0.000E+00	0.000E+00	0.000E+00	0.000E+00	0.000E+00
Np-237	5.632E+01	5.633E+01	5.633E+01	5.638E+01	5.721E+01	7.680E+01	1.879E+02	2.219E+02	2.158E+02	1.612E+02
Np-238	0.000E+00	0.000E+00	0.000E+00	0.000E+00	0.000E+00	0.000E+00	0.000E+00	0.000E+00	0.000E+00	0.000E+00
Np-239	0.000E+00	7.236E+03	7.236E+03	7.236E+03	7.230E+03	7.170E+03	6.606E+03	2.924E+03	8.394E-01	0.000E+00
Np-240m	0.000E+00	0.000E+00	0.000E+00	0.000E+00	0.000E+00	0.000E+00	0.000E+00	0.000E+00	0.000E+00	0.000E+00
Np-240	0.000E+00	0.000E+00	0.000E+00	0.000E+00	0.000E+00	0.000E+00	0.000E+00	0.000E+00	0.000E+00	0.000E+00
Pu-236	8.772E+01	8.454E+01	8.286E+01	6.630E+01	7.428E+00	2.310E-09	0.000E+00	0.000E+00	0.000E+00	0.000E+00
Pu-238	2.533E+05	2.531E+05	2.529E+05	2.512E+05	2.344E+05	1.168E+05	1.194E+02	5.049E-17	0.000E+00	0.000E+00
Pu-239	4.647E+04	4.647E+04	4.647E+04	4.647E+04	4.646E+04	4.637E+04	4.537E+04	3.614E+04	2.905E+03	2.284E-08
Pu-240	6.570E+04	6.570E+04	6.570E+04	6.576E+04	6.624E+04	6.678E+04	6.096E+04	2.422E+04	2.377E+00	0.000E+00
Pu-241	2.048E+07	2.034E+07	2.026E+07	1.940E+07	1.265E+07	1.765E+05	1.264E+02	5.941E+01	3.130E-02	0.000E+00
Pu-242	2.285E+02	2.285E+02	2.285E+02	2.285E+02	2.285E+02	2.285E+02	2.282E+02	2.248E+02	1.907E+02	3.677E+01
Pu-243	0.000E+00	0.000E+00	0.000E+00	0.000E+00	0.000E+00	0.000E+00	0.000E+00	0.000E+00	0.000E+00	0.000E+00
Pu-244	0.000E+00	0.000E+00	0.000E+00	0.000E+00	0.000E+00	0.000E+00	0.000E+00	0.000E+00	0.000E+00	0.000E+00
Pu-245	0.000E+00	0.000E+00	0.000E+00	0.000E+00	0.000E+00	0.000E+00	0.000E+00	0.000E+00	0.000E+00	0.000E+00
Am-241	1.294E+05	1.343E+05	1.369E+05	1.658E+05	3.892E+05	7.128E+05	1.703E+05	5.950E+01	3.302E-02	0.000E+00
Am-242m	1.630E+03	1.629E+03	1.628E+03	1.621E+03	1.556E+03	1.033E+03	1.703E+01	2.545E-17	0.000E+00	0.000E+00
Am-242	0.000E+00	1.629E+03	1.628E+03	1.621E+03	1.556E+03	1.033E+03	1.703E+01	0.000E+00	0.000E+00	0.000E+00
Am-243	7.236E+03	7.236E+03	7.236E+03	7.236E+03	7.230E+03	7.170E+03	6.606E+03	2.924E+03	8.394E-01	0.000E+00
Am-244	0.000E+00	0.000E+00	0.000E+00	0.000E+00	0.000E+00	0.000E+00	0.000E+00	0.000E+00	0.000E+00	0.000E+00
Am-245	0.000E+00	0.000E+00	0.000E+00	0.000E+00	0.000E+00	0.000E+00	0.000E+00	0.000E+00	0.000E+00	0.000E+00
Cm-242	4.867E+04	3.878E+04	3.430E+04	9.252E+03	1.276E+03	8.466E+02	1.396E+01	2.093E-17	0.000E+00	0.000E+00
Cm-243	5.482E+03	5.464E+03	5.455E+03	5.347E+03	4.400E+03	6.264E+02	2.135E-06	0.000E+00	0.000E+00	0.000E+00
Cm-244	6.810E+05	6.768E+05	6.744E+05	6.516E+05	4.615E+05	1.469E+04	1.575E-11	0.000E+00	0.000E+00	0.000E+00
Cm-245	1.372E+02	1.372E+02	1.372E+02	1.372E+02	1.371E+02	1.360E+02	1.262E+02	5.930E+01	3.124E-02	0.000E+00
Cm-246	2.990E+01	2.990E+01	2.990E+01	2.989E+01	2.985E+01	2.946E+01	2.581E+01	6.864E+00	1.216E-05	0.000E+00
Cm-247	0.000E+00	0.000E+00	0.000E+00	0.000E+00	0.000E+00	0.000E+00	0.000E+00	0.000E+00	0.000E+00	0.000E+00
Cm-248	0.000E+00	0.000E+00	0.000E+00	0.000E+00	0.000E+00	0.000E+00	0.000E+00	0.000E+00	0.000E+00	0.000E+00
Cm-249	0.000E+00	0.000E+00	0.000E+00	0.000E+00	0.000E+00	0.000E+00	0.000E+00	0.000E+00	0.000E+00	0.000E+00
Cm-250	0.000E+00	0.000E+00	0.000E+00	0.000E+00	0.000E+00	0.000E+00	0.000E+00	0.000E+00	0.000E+00	0.000E+00
Bk-249	0.000E+00	0.000E+00	0.000E+00	0.000E+00	0.000E+00	0.000E+00	0.000E+00	0.000E+00	0.000E+00	0.000E+00
Bk-250	0.000E+00	0.000E+00	0.000E+00	0.000E+00	0.000E+00	0.000E+00	0.000E+00	0.000E+00	0.000E+00	0.000E+00
Cf-249	0.000E+00	0.000E+00	0.000E+00	0.000E+00	0.000E+00	0.000E+00	0.000E+00	0.000E+00	0.000E+00	0.000E+00
Cf-250	0.000E+00	0.000E+00	0.000E+00	0.000E+00	0.000E+00	0.000E+00	0.000E+00	0.000E+00	0.000E+00	0.000E+00
Cf-251	0.000E+00	0.000E+00	0.000E+00	0.000E+00	0.000E+00	0.000E+00	0.000E+00	0.000E+00	0.000E+00	0.000E+00
Cf-252	0.000E+00	0.000E+00	0.000E+00	0.000E+00	0.000E+00	0.000E+00	0.000E+00	0.000E+00	0.000E+00	0.000E+00
Cf-253	0.000E+00	0.000E+00	0.000E+00	0.000E+00	0.000E+00	0.000E+00	0.000E+00	0.000E+00	0.000E+00	0.000E+00
Cf-254	0.000E+00	0.000E+00	0.000E+00	0.000E+00	0.000E+00	0.000E+00	0.000E+00	0.000E+00	0.000E+00	0.000E+00
Es-253	0.000E+00	0.000E+00	0.000E+00	0.000E+00	0.000E+00	0.000E+00	0.000E+00	0.000E+00	0.000E+00	0.000E+00
Total	2.172E+07	2.158E+07	2.150E+07	2.061E+07	1.388E+07	1.152E+06	2.907E+05	6.684E+04	3.316E+03	1.980E+02

Table A.17

Heavy TRU elements nuclide radioactivity, Curies Basis = EQULIBRIUM DISCHARGED FUEL

	Initial	1 hour	30 days	1 year	10 y	100 y	1000 y	10000 y	100000 y	1000000 y
Np-236	2.470E+03	2.393E+03	3.502E-07	0.000E+00	0.000E+00	0.000E+00	0.000E+00	0.000E+00	0.000E+00	0.000E+00
Np-237	4.066E+01	4.066E+01	4.067E+01	4.072E+01	4.140E+01	5.450E+01	1.274E+02	1.502E+02	1.463E+02	1.093E+02
Np-238	1.185E+07	1.169E+07	5.954E+02	0.000E+00	0.000E+00	0.000E+00	0.000E+00	0.000E+00	0.000E+00	0.000E+00
Np-239	1.637E+08	1.627E+08	2.990E+04	6.156E+03	6.150E+03	6.102E+03	5.623E+03	2.488E+03	7.146E-01	2.564E-04
Np-240m	2.940E+02	2.823E+02	3.244E-06	3.244E-06	3.244E-06	3.245E-06	3.248E-06	3.289E-06	3.653E-06	5.189E-06
Np-240	1.346E+05	6.960E+04	0.000E+00	0.000E+00	0.000E+00	0.000E+00	0.000E+00	0.000E+00	0.000E+00	0.000E+00
Pu-236	6.384E+02	6.384E+02	6.270E+02	5.017E+02	5.620E+01	1.748E-08	0.000E+00	0.000E+00	0.000E+00	0.000E+00
Pu-238	4.649E+05	4.649E+05	4.691E+05	4.865E+05	4.604E+05	2.326E+05	3.229E+02	4.080E-16	0.000E+00	0.000E+00
Pu-239	3.808E+04	3.808E+04	3.813E+04	3.812E+04	3.812E+04	3.804E+04	3.723E+04	2.968E+04	2.389E+03	2.564E-04
Pu-240	6.252E+04	6.252E+04	6.252E+04	6.258E+04	6.324E+04	6.426E+04	5.864E+04	2.330E+04	2.287E+00	5.196E-06
Pu-241	1.170E+07	1.170E+07	1.166E+07	1.116E+07	7.278E+06	1.017E+05	2.363E+02	1.111E+02	5.851E-02	0.000E+00
Pu-242	2.410E+02	2.410E+02	2.411E+02	2.411E+02	2.411E+02	2.414E+02	2.416E+02	2.380E+02	2.020E+02	3.895E+01
Pu-243	3.391E+06	2.950E+06	2.674E-04	2.674E-04	2.674E-04	2.674E-04	2.674E-04	2.673E-04	2.663E-04	2.564E-04
Pu-244	3.248E-06	3.248E-06	3.248E-06	3.248E-06	3.248E-06	3.249E-06	3.253E-06	3.293E-06	3.658E-06	5.196E-06
Pu-245	1.477E+01	1.384E+01	0.000E+00	0.000E+00	0.000E+00	0.000E+00	0.000E+00	0.000E+00	0.000E+00	0.000E+00
Am-241	1.460E+05	1.460E+05	1.475E+05	1.640E+05	2.916E+05	4.688E+05	1.120E+05	1.111E+02	6.174E-02	0.000E+00
Am-242m	1.317E+04	1.317E+04	1.316E+04	1.311E+04	1.258E+04	8.346E+03	1.377E+02	2.056E-16	0.000E+00	0.000E+00
Am-242	8.328E+06	7.974E+06	1.316E+04	1.311E+04	1.258E+04	8.346E+03	1.377E+02	2.056E-16	0.000E+00	0.000E+00
Am-243	6.156E+03	6.156E+03	6.156E+03	6.156E+03	6.150E+03	6.102E+03	5.623E+03	2.488E+03	7.146E-01	2.564E-04
Am-244	5.263E+06	1.064E+06	4.223E-09	4.223E-09	4.223E-09	4.223E-09	4.229E-09	4.281E-09	4.755E-09	6.756E-09
Am-245	3.071E+01	2.604E+01	6.174E-05	2.944E-05	2.076E-08	0.000E+00	0.000E+00	0.000E+00	0.000E+00	0.000E+00
Cm-242	6.162E+06	6.162E+06	5.449E+06	1.318E+06	1.032E+04	6.846E+03	1.129E+02	1.691E-16	0.000E+00	0.000E+00
Cm-243	4.698E+03	4.698E+03	4.690E+03	4.598E+03	3.783E+03	5.384E+02	1.836E-06	0.000E+00	0.000E+00	0.000E+00
Cm-244	9.180E+05	9.180E+05	9.150E+05	8.832E+05	6.258E+05	1.993E+04	4.250E-09	4.281E-09	4.755E-09	6.756E-09
Cm-245	2.565E+02	2.565E+02	2.565E+02	2.564E+02	2.563E+02	2.543E+02	2.359E+02	1.109E+02	5.841E-02	0.000E+00
Cm-246	4.340E+01	4.340E+01	4.340E+01	4.340E+01	4.334E+01	4.277E+01	3.747E+01	9.966E+00	1.766E-05	0.000E+00
Cm-247	2.674E-04	2.674E-04	2.674E-04	2.674E-04	2.674E-04	2.674E-04	2.674E-04	2.673E-04	2.663E-04	2.564E-04
Cm-248	5.921E-04	5.921E-04	5.921E-04	5.921E-04	5.921E-04	5.920E-04	5.910E-04	5.806E-04	4.863E-04	8.262E-05
Cm-249	1.027E+01	5.365E+00	5.633E-09	1.212E-14	0.000E+00	0.000E+00	0.000E+00	0.000E+00	0.000E+00	0.000E+00
Cm-250	2.192E-09	2.192E-09	2.192E-09	2.192E-09	2.191E-09	2.183E-09	2.107E-09	1.472E-09	4.079E-11	0.000E+00
Bk-249	4.394E+00	4.394E+00	4.114E+00	1.963E+00	1.384E-03	0.000E+00	0.000E+00	0.000E+00	0.000E+00	0.000E+00
Bk-250	2.434E-01	1.963E-01	2.192E-09	2.192E-09	2.191E-09	2.183E-09	2.107E-09	1.472E-09	4.079E-11	0.000E+00
Cf-249	5.280E-03	5.281E-03	5.966E-03	1.120E-02	1.573E-02	1.318E-02	2.239E-03	4.492E-11	0.000E+00	0.000E+00
Cf-250	1.525E-02	1.525E-02	1.519E-02	1.447E-02	8.982E-03	7.620E-05	2.107E-09	1.472E-09	4.079E-11	0.000E+00
Cf-251	5.083E-06	5.083E-06	5.083E-06	5.079E-06	5.044E-06	4.706E-06	2.353E-06	2.296E-09	0.000E+00	0.000E+00
Cf-252	3.221E-05	3.221E-05	3.152E-05	2.479E-05	2.345E-06	1.349E-16	0.000E+00	0.000E+00	0.000E+00	0.000E+00
Cf-253	5.823E-06	5.814E-06	1.813E-06	3.900E-12	0.000E+00	0.000E+00	0.000E+00	0.000E+00	0.000E+00	0.000E+00
Cf-254	1.225E-08	1.224E-08	8.688E-09	1.865E-10	0.000E+00	0.000E+00	0.000E+00	0.000E+00	0.000E+00	0.000E+00
Es-253	5.805E-06	5.805E-06	4.100E-06	1.814E-10	0.000E+00	0.000E+00	0.000E+00	0.000E+00	0.000E+00	0.000E+00
Total	2.122E+08	2.059E+08	1.881E+07	1.416E+07	8.809E+06	9.622E+05	2.207E+05	5.869E+04	2.741E+03	1.482E+02

Table A.18

Heavy TRU elements nuclide radioactivity, Curies Basis = EQULIBRIUM DISCHARGED FUEL – EQULIBRIUM FRESH FUEL

	Initial	**1 hour**	**30 days**	**1 year**	**10 y**	**100 y**	**1000 y**	**10000 y**	**100000 y**	**1000000 y**
Np-236	2.470E+03	2.393E+03	3.502E-07	0.000E+00	0.000E+00	0.000E+00	0.000E+00	0.000E+00	0.000E+00	0.000E+00
Np-237	-1.567E+01	-1.567E+01	-1.565E+01	-1.565E+01	-1.581E+01	-2.230E+01	-6.042E+01	-7.170E+01	-6.948E+01	-5.196E+01
Np-238	1.185E+07	1.169E+07	5.954E+02	0.000E+00	0.000E+00	0.000E+00	0.000E+00	0.000E+00	0.000E+00	0.000E+00
Np-239	1.637E+08	1.627E+08	2.267E+04	-1.080E+03	-1.080E+03	-1.068E+03	-9.834E+02	-4.362E+02	-1.248E-01	2.564E-04
Np-240m	2.940E+02	2.823E+02	3.244E-06	3.244E-06	3.244E-06	3.245E-06	3.248E-06	3.289E-06	3.653E-06	5.189E-06
Np-240	1.346E+05	6.960E+04	0.000E+00	0.000E+00	0.000E+00	0.000E+00	0.000E+00	0.000E+00	0.000E+00	0.000E+00
Pu-236	5.507E+02	5.539E+02	5.441E+02	4.354E+02	4.877E+01	1.517E-08	0.000E+00	0.000E+00	0.000E+00	0.000E+00
Pu-238	2.116E+05	2.119E+05	2.162E+05	2.353E+05	2.260E+05	1.158E+05	2.035E+02	3.575E-16	0.000E+00	0.000E+00
Pu-239	-8.388E+03	-8.388E+03	-8.340E+03	-8.346E+03	-8.340E+03	-8.328E+03	-8.142E+03	-6.456E+03	-5.160E+02	2.564E-04
Pu-240	-3.180E+03	-3.180E+03	-3.180E+03	-3.180E+03	-3.000E+03	-2.520E+03	-2.316E+03	-9.180E+02	-9.000E-02	5.196E-06
Pu-241	-8.784E+06	-8.640E+06	-8.604E+06	-8.238E+06	-5.376E+06	-7.476E+04	1.099E+02	5.165E+01	2.722E-02	0.000E+00
Pu-242	1.254E+01	1.254E+01	1.260E+01	1.260E+01	1.260E+01	1.284E+01	1.338E+01	1.326E+01	1.128E+01	2.178E+00
Pu-243	3.391E+06	2.950E+06	2.674E-04	2.674E-04	2.674E-04	2.674E-04	2.674E-04	2.673E-04	2.663E-04	2.564E-04
Pu-244	3.248E-06	3.248E-06	3.248E-06	3.248E-06	3.248E-06	3.249E-06	3.253E-06	3.293E-06	3.658E-06	5.196E-06
Pu-245	1.477E+01	1.384E+01	0.000E+00	0.000E+00	0.000E+00	0.000E+00	0.000E+00	0.000E+00	0.000E+00	0.000E+00
Am-241	1.668E+04	1.176E+04	1.062E+04	-1.800E+03	-9.756E+04	-2.440E+05	-5.832E+04	5.162E+01	2.872E-02	0.000E+00
Am-242m	1.154E+04	1.154E+04	1.154E+04	1.149E+04	1.103E+04	7.313E+03	1.207E+02	1.802E-16	0.000E+00	0.000E+00
Am-242	8.328E+06	7.972E+06	1.154E+04	1.149E+04	1.103E+04	7.313E+03	1.207E+02	2.056E-16	0.000E+00	0.000E+00
Am-243	-1.080E+03	-1.080E+03	-1.080E+03	-1.080E+03	-1.080E+03	-1.068E+03	-9.834E+02	-4.362E+02	-1.248E-01	2.564E-04
Am-244	5.263E+06	1.064E+06	4.223E-09	4.223E-09	4.223E-09	4.223E-09	4.229E-09	4.281E-09	4.755E-09	6.756E-09
Am-245	3.071E+01	2.604E+01	6.174E-05	2.944E-05	2.076E-08	0.000E+00	0.000E+00	0.000E+00	0.000E+00	0.000E+00
Cm-242	6.113E+06	6.123E+06	5.414E+06	1.308E+06	9.044E+03	5.999E+03	9.896E+01	1.482E-16	0.000E+00	0.000E+00
Cm-243	-7.842E+02	-7.662E+02	-7.644E+02	-7.494E+02	-6.174E+02	-8.796E+01	-2.994E-07	0.000E+00	0.000E+00	0.000E+00
Cm-244	2.370E+05	2.412E+05	2.406E+05	2.316E+05	1.643E+05	5.238E+03	4.234E-09	4.281E-09	4.755E-09	6.756E-09
Cm-245	1.193E+02	1.193E+02	1.193E+02	1.193E+02	1.192E+02	1.183E+02	1.097E+02	5.158E+01	2.717E-02	0.000E+00
Cm-246	1.351E+01	1.351E+01	1.351E+01	1.351E+01	1.349E+01	1.331E+01	1.166E+01	3.102E+00	5.496E-06	0.000E+00
Cm-247	2.674E-04	2.674E-04	2.674E-04	2.674E-04	2.674E-04	2.674E-04	2.674E-04	2.673E-04	2.663E-04	2.564E-04
Cm-248	5.921E-04	5.921E-04	5.921E-04	5.921E-04	5.921E-04	5.920E-04	5.910E-04	5.806E-04	4.863E-04	8.262E-05
Cm-249	1.027E+01	5.365E+00	5.633E-09	1.212E-14	0.000E+00	0.000E+00	0.000E+00	0.000E+00	0.000E+00	0.000E+00
Cm-250	2.192E-09	2.192E-09	2.192E-09	2.192E-09	2.191E-09	2.183E-09	2.107E-09	1.472E-09	4.079E-11	0.000E+00
Bk-249	4.394E+00	4.394E+00	4.114E+00	1.963E+00	1.384E-03	0.000E+00	0.000E+00	0.000E+00	0.000E+00	0.000E+00
Bk-250	2.434E-01	1.963E-01	2.192E-09	2.192E-09	2.191E-09	2.183E-09	2.107E-09	1.472E-09	4.079E-11	0.000E+00
Cf-249	5.280E-03	5.281E-03	5.966E-03	1.120E-02	1.573E-02	1.318E-02	2.239E-03	4.492E-11	0.000E+00	0.000E+00
Cf-250	1.525E-02	1.525E-02	1.519E-02	1.447E-02	8.982E-03	7.620E-05	2.107E-09	1.472E-09	4.079E-11	0.000E+00
Cf-251	5.083E-06	5.083E-06	5.083E-06	5.079E-06	5.044E-06	4.706E-06	2.353E-06	2.296E-09	0.000E+00	0.000E+00
Cf-252	3.221E-05	3.221E-05	3.152E-05	2.479E-05	2.345E-06	1.349E-16	0.000E+00	0.000E+00	0.000E+00	0.000E+00
Cf-253	5.823E-06	5.814E-06	1.813E-06	3.900E-12	0.000E+00	0.000E+00	0.000E+00	0.000E+00	0.000E+00	0.000E+00
Cf-254	1.225E-08	1.224E-08	8.688E-09	1.865E-10	0.000E+00	0.000E+00	0.000E+00	0.000E+00	0.000E+00	0.000E+00
Es-253	5.805E-06	5.805E-06	4.100E-06	1.814E-10	0.000E+00	0.000E+00	0.000E+00	0.000E+00	0.000E+00	0.000E+00
Total	1.905E+08	1.843E+08	-2.689E+06	-6.455E+06	-5.066E+06	-1.900E+05	-7.002E+04	-8.147E+03	-5.745E+02	-4.978E+01
Total/MWd	3.902E+02	3.776E+02	-5.507E+00	-1.322E+01	-1.038E+01	-3.892E-01	-1.434E-01	-1.669E-02	-1.177E-03	-1.020E-04

Table A.19

Heavy TRU elements nuclide toxicity hazard Basis = EQULIBRIUM FRESH FUEL

	Initial	1 hour	30 days	1 year	10 y	100 y	1000 y	10000 y	100000 y	1000000 y	
Np-236	0.000E+00	0.000E+00	0.000E+00	0.000E+00	0.000E+00	0.000E+00	0.000E+00	0.000E+00	0.000E+00	0.000E+00	0.000E+00
Np-237	1.111E+04	1.111E+04	1.111E+04	1.112E+04	1.128E+04	1.514E+04	3.705E+04	4.375E+04	4.255E+04	3.179E+04	1.972E+02
Np-238	0.000E+00	0.000E+00	0.000E+00	0.000E+00	0.000E+00	0.000E+00	0.000E+00	0.000E+00	0.000E+00	0.000E+00	0.000E+00
Np-239	0.000E+00	0.000E+00	0.000E+00	0.000E+00	0.000E+00	0.000E+00	0.000E+00	0.000E+00	0.000E+00	0.000E+00	0.000E+00
Np-240m	0.000E+00	0.000E+00	0.000E+00	0.000E+00	0.000E+00	0.000E+00	0.000E+00	0.000E+00	0.000E+00	0.000E+00	0.000E+00
Np-240	0.000E+00	0.000E+00	0.000E+00	0.000E+00	0.000E+00	0.000E+00	0.000E+00	0.000E+00	0.000E+00	0.000E+00	0.000E+00
Pu-236	0.000E+00	0.000E+00	0.000E+00	0.000E+00	0.000E+00	0.000E+00	0.000E+00	0.000E+00	0.000E+00	0.000E+00	0.000E+00
Pu-238	6.234E+07	6.228E+07	6.224E+07	6.183E+07	5.768E+07	2.873E+07	2.938E+04	1.243E-14	0.000E+00	0.000E+00	2.461E+02
Pu-239	1.243E+07	1.243E+07	1.243E+07	1.243E+07	1.243E+07	1.240E+07	1.214E+07	9.667E+06	7.771E+05	6.109E-06	2.675E+02
Pu-240	1.757E+07	1.757E+07	1.757E+07	1.759E+07	1.772E+07	1.786E+07	1.631E+07	6.479E+06	6.357E+02	0.000E+00	2.675E+02
Pu-241	0.000E+00	0.000E+00	0.000E+00	0.000E+00	0.000E+00	0.000E+00	0.000E+00	0.000E+00	0.000E+00	0.000E+00	0.000E+00
Pu-242	6.112E+04	6.112E+04	6.112E+04	6.112E+04	6.112E+04	6.113E+04	6.105E+04	6.012E+04	5.102E+04	9.835E+03	2.675E+02
Pu-243	0.000E+00	0.000E+00	0.000E+00	0.000E+00	0.000E+00	0.000E+00	0.000E+00	0.000E+00	0.000E+00	0.000E+00	0.000E+00
Pu-244	0.000E+00	0.000E+00	0.000E+00	0.000E+00	0.000E+00	0.000E+00	0.000E+00	0.000E+00	0.000E+00	0.000E+00	0.000E+00
Pu-245	0.000E+00	0.000E+00	0.000E+00	0.000E+00	0.000E+00	0.000E+00	0.000E+00	0.000E+00	0.000E+00	0.000E+00	0.000E+00
Am-241	3.530E+07	3.665E+07	3.737E+07	4.526E+07	1.062E+08	1.945E+08	4.649E+07	1.624E+04	9.012E+00	0.000E+00	2.729E+02
Am-242m	4.361E+05	4.358E+05	4.356E+05	4.337E+05	4.163E+05	2.762E+05	4.555E+03	6.807E-15	0.000E+00	0.000E+00	2.675E+02
Am-242	0.000E+00	0.000E+00	0.000E+00	0.000E+00	0.000E+00	0.000E+00	0.000E+00	0.000E+00	0.000E+00	0.000E+00	0.000E+00
Am-243	1.975E+06	1.975E+06	1.975E+06	1.975E+06	1.973E+06	1.957E+06	1.803E+06	7.979E+05	2.291E+02	0.000E+00	2.729E+02
Am-244	0.000E+00	0.000E+00	0.000E+00	0.000E+00	0.000E+00	0.000E+00	0.000E+00	0.000E+00	0.000E+00	0.000E+00	0.000E+00
Am-245	0.000E+00	0.000E+00	0.000E+00	0.000E+00	0.000E+00	0.000E+00	0.000E+00	0.000E+00	0.000E+00	0.000E+00	0.000E+00
Cm-242	3.358E+05	2.676E+05	2.367E+05	6.384E+04	8.806E+03	5.842E+03	9.634E+01	1.444E-16	0.000E+00	0.000E+00	6.900E+00
Cm-243	1.079E+06	1.076E+06	1.074E+06	1.053E+06	8.664E+05	1.233E+05	4.205E-04	0.000E+00	0.000E+00	0.000E+00	1.969E+02
Cm-244	1.110E+08	1.103E+08	1.099E+08	1.062E+08	7.522E+07	2.395E+06	2.567E-09	0.000E+00	0.000E+00	0.000E+00	1.630E+02
Cm-245	3.897E+04	3.895E+04	3.895E+04	3.895E+04	3.894E+04	3.863E+04	3.584E+04	1.684E+04	8.873E+00	0.000E+00	2.840E+02
Cm-246	8.506E+03	8.506E+03	8.506E+03	8.504E+03	8.492E+03	8.381E+03	7.342E+03	1.953E+03	3.460E-03	0.000E+00	2.845E+02
Cm-247	0.000E+00	0.000E+00	0.000E+00	0.000E+00	0.000E+00	0.000E+00	0.000E+00	0.000E+00	0.000E+00	0.000E+00	0.000E+00
Cm-248	0.000E+00	0.000E+00	0.000E+00	0.000E+00	0.000E+00	0.000E+00	0.000E+00	0.000E+00	0.000E+00	0.000E+00	0.000E+00
Cm-249	0.000E+00	0.000E+00	0.000E+00	0.000E+00	0.000E+00	0.000E+00	0.000E+00	0.000E+00	0.000E+00	0.000E+00	0.000E+00
Cm-250	0.000E+00	0.000E+00	0.000E+00	0.000E+00	0.000E+00	0.000E+00	0.000E+00	0.000E+00	0.000E+00	0.000E+00	0.000E+00
Bk-249	0.000E+00	0.000E+00	0.000E+00	0.000E+00	0.000E+00	0.000E+00	0.000E+00	0.000E+00	0.000E+00	0.000E+00	0.000E+00
Bk-250	0.000E+00	0.000E+00	0.000E+00	0.000E+00	0.000E+00	0.000E+00	0.000E+00	0.000E+00	0.000E+00	0.000E+00	0.000E+00
Cf-249	0.000E+00	0.000E+00	0.000E+00	0.000E+00	0.000E+00	0.000E+00	0.000E+00	0.000E+00	0.000E+00	0.000E+00	0.000E+00
Cf-250	0.000E+00	0.000E+00	0.000E+00	0.000E+00	0.000E+00	0.000E+00	0.000E+00	0.000E+00	0.000E+00	0.000E+00	0.000E+00
Cf-251	0.000E+00	0.000E+00	0.000E+00	0.000E+00	0.000E+00	0.000E+00	0.000E+00	0.000E+00	0.000E+00	0.000E+00	0.000E+00
Cf-252	0.000E+00	0.000E+00	0.000E+00	0.000E+00	0.000E+00	0.000E+00	0.000E+00	0.000E+00	0.000E+00	0.000E+00	0.000E+00
Cf-253	0.000E+00	0.000E+00	0.000E+00	0.000E+00	0.000E+00	0.000E+00	0.000E+00	0.000E+00	0.000E+00	0.000E+00	0.000E+00
Cf-254	0.000E+00	0.000E+00	0.000E+00	0.000E+00	0.000E+00	0.000E+00	0.000E+00	0.000E+00	0.000E+00	0.000E+00	0.000E+00
Es-253	0.000E+00	0.000E+00	0.000E+00	0.000E+00	0.000E+00	0.000E+00	0.000E+00	0.000E+00	0.000E+00	0.000E+00	0.000E+00
Total	2.426E+08	2.431E+08	2.434E+08	2.470E+08	2.726E+08	2.584E+08	7.691E+07	1.708E+07	8.716E+05	4.163E+04	0.000E+00

Table A.20

Heavy TRU elements nuclide toxicity hazard Basis = EQULIBRIUM DISCHARGED FUEL

	Initial	1 hour	30 days	1 year	10 y	100 y	1000 y	10000 y	100000 y	1000000 y	
Np-236	0.000E+00	0.000E+00	0.000E+00	0.000E+00	0.000E+00	0.000E+00	0.000E+00	0.000E+00	0.000E+00	0.000E+00	0.000E+00
Np-237	8.017E+03	8.017E+03	8.021E+03	8.030E+03	8.164E+03	1.075E+04	2.513E+04	2.962E+04	2.885E+04	2.155E+04	1.972E+02
Np-238	0.000E+00	0.000E+00	0.000E+00	0.000E+00	0.000E+00	0.000E+00	0.000E+00	0.000E+00	0.000E+00	0.000E+00	0.000E+00
Np-239	0.000E+00	0.000E+00	0.000E+00	0.000E+00	0.000E+00	0.000E+00	0.000E+00	0.000E+00	0.000E+00	0.000E+00	0.000E+00
Np-240m	0.000E+00	0.000E+00	0.000E+00	0.000E+00	0.000E+00	0.000E+00	0.000E+00	0.000E+00	0.000E+00	0.000E+00	0.000E+00
Np-240	0.000E+00	0.000E+00	0.000E+00	0.000E+00	0.000E+00	0.000E+00	0.000E+00	0.000E+00	0.000E+00	0.000E+00	0.000E+00
Pu-236	0.000E+00	0.000E+00	0.000E+00	0.000E+00	0.000E+00	0.000E+00	0.000E+00	0.000E+00	0.000E+00	0.000E+00	0.000E+00
Pu-238	1.144E+08	1.144E+08	1.155E+08	1.197E+08	1.133E+08	5.723E+07	7.947E+04	1.004E-13	0.000E+00	0.000E+00	2.461E+02
Pu-239	1.019E+07	1.019E+07	1.020E+07	1.020E+07	1.020E+07	1.018E+07	9.959E+06	7.940E+06	6.391E+05	6.858E-02	2.675E+02
Pu-240	1.672E+07	1.672E+07	1.672E+07	1.674E+07	1.692E+07	1.719E+07	1.569E+07	6.234E+06	6.117E+02	1.390E-03	2.675E+02
Pu-241	0.000E+00	0.000E+00	0.000E+00	0.000E+00	0.000E+00	0.000E+00	0.000E+00	0.000E+00	0.000E+00	0.000E+00	0.000E+00
Pu-242	6.447E+04	6.447E+04	6.449E+04	6.449E+04	6.449E+04	6.457E+04	6.463E+04	6.367E+04	5.404E+04	1.042E+04	2.675E+02
Pu-243	0.000E+00	0.000E+00	0.000E+00	0.000E+00	0.000E+00	0.000E+00	0.000E+00	0.000E+00	0.000E+00	0.000E+00	0.000E+00
Pu-244	0.000E+00	0.000E+00	0.000E+00	0.000E+00	0.000E+00	0.000E+00	0.000E+00	0.000E+00	0.000E+00	0.000E+00	0.000E+00
Pu-245	0.000E+00	0.000E+00	0.000E+00	0.000E+00	0.000E+00	0.000E+00	0.000E+00	0.000E+00	0.000E+00	0.000E+00	0.000E+00
Am-241	3.985E+07	3.985E+07	4.026E+07	4.477E+07	7.958E+07	1.279E+08	3.057E+07	3.032E+04	1.685E+01	0.000E+00	2.729E+02
Am-242m	3.523E+06	3.523E+06	3.521E+06	3.507E+06	3.366E+06	2.233E+06	3.683E+04	5.500E-14	0.000E+00	0.000E+00	2.675E+02
Am-242	0.000E+00	0.000E+00	0.000E+00	0.000E+00	0.000E+00	0.000E+00	0.000E+00	0.000E+00	0.000E+00	0.000E+00	0.000E+00
Am-243	1.680E+06	1.680E+06	1.680E+06	1.680E+06	1.678E+06	1.665E+06	1.534E+06	6.789E+05	1.950E+02	6.997E-02	2.729E+02
Am-244	0.000E+00	0.000E+00	0.000E+00	0.000E+00	0.000E+00	0.000E+00	0.000E+00	0.000E+00	0.000E+00	0.000E+00	0.000E+00
Am-245	0.000E+00	0.000E+00	0.000E+00	0.000E+00	0.000E+00	0.000E+00	0.000E+00	0.000E+00	0.000E+00	0.000E+00	0.000E+00
Cm-242	4.252E+07	4.252E+07	3.760E+07	9.091E+06	7.121E+04	4.724E+04	7.791E+02	1.167E-15	0.000E+00	0.000E+00	6.900E+00
Cm-243	9.250E+05	9.250E+05	9.235E+05	9.053E+05	7.449E+05	1.060E+05	3.615E-04	0.000E+00	0.000E+00	0.000E+00	1.969E+02
Cm-244	1.496E+08	1.496E+08	1.491E+08	1.440E+08	1.020E+08	3.249E+06	6.927E-07	6.978E-07	7.751E-07	1.101E-06	1.630E+02
Cm-245	7.285E+04	7.285E+04	7.285E+04	7.283E+04	7.278E+04	7.223E+04	6.698E+04	3.149E+04	1.659E+01	0.000E+00	2.840E+02
Cm-246	1.235E+04	1.235E+04	1.235E+04	1.235E+04	1.233E+04	1.217E+04	1.066E+04	2.835E+03	5.024E-03	0.000E+00	2.845E+02
Cm-247	0.000E+00	0.000E+00	0.000E+00	0.000E+00	0.000E+00	0.000E+00	0.000E+00	0.000E+00	0.000E+00	0.000E+00	0.000E+00
Cm-248	0.000E+00	0.000E+00	0.000E+00	0.000E+00	0.000E+00	0.000E+00	0.000E+00	0.000E+00	0.000E+00	0.000E+00	0.000E+00
Cm-249	0.000E+00	0.000E+00	0.000E+00	0.000E+00	0.000E+00	0.000E+00	0.000E+00	0.000E+00	0.000E+00	0.000E+00	0.000E+00
Cm-250	0.000E+00	0.000E+00	0.000E+00	0.000E+00	0.000E+00	0.000E+00	0.000E+00	0.000E+00	0.000E+00	0.000E+00	0.000E+00
Bk-249	0.000E+00	0.000E+00	0.000E+00	0.000E+00	0.000E+00	0.000E+00	0.000E+00	0.000E+00	0.000E+00	0.000E+00	0.000E+00
Bk-250	0.000E+00	0.000E+00	0.000E+00	0.000E+00	0.000E+00	0.000E+00	0.000E+00	0.000E+00	0.000E+00	0.000E+00	0.000E+00
Cf-249	0.000E+00	0.000E+00	0.000E+00	0.000E+00	0.000E+00	0.000E+00	0.000E+00	0.000E+00	0.000E+00	0.000E+00	0.000E+00
Cf-250	0.000E+00	0.000E+00	0.000E+00	0.000E+00	0.000E+00	0.000E+00	0.000E+00	0.000E+00	0.000E+00	0.000E+00	0.000E+00
Cf-251	0.000E+00	0.000E+00	0.000E+00	0.000E+00	0.000E+00	0.000E+00	0.000E+00	0.000E+00	0.000E+00	0.000E+00	0.000E+00
Cf-252	0.000E+00	0.000E+00	0.000E+00	0.000E+00	0.000E+00	0.000E+00	0.000E+00	0.000E+00	0.000E+00	0.000E+00	0.000E+00
Cf-253	0.000E+00	0.000E+00	0.000E+00	0.000E+00	0.000E+00	0.000E+00	0.000E+00	0.000E+00	0.000E+00	0.000E+00	0.000E+00
Cf-254	0.000E+00	0.000E+00	0.000E+00	0.000E+00	0.000E+00	0.000E+00	0.000E+00	0.000E+00	0.000E+00	0.000E+00	0.000E+00
Es-253	0.000E+00	0.000E+00	0.000E+00	0.000E+00	0.000E+00	0.000E+00	0.000E+00	0.000E+00	0.000E+00	0.000E+00	0.000E+00
Total	3.796E+08	3.796E+08	3.757E+08	3.507E+08	3.280E+08	2.200E+08	5.804E+07	1.501E+07	7.228E+05	3.196E+04	0.000E+00

Table A.21

Heavy TRU Elements Nuclide Toxicity Hazard Basis = EQULIBRIUM DISCHARGED FUEL - EQULIBRIUM FRESH FUEL

	Initial	1 hour	30 days	1 year	10 y	100 y	1000 y	10000 y	100000 y	1000000 y	
Np-236	0.000E+00	0.000E+00	0.000E+00	0.000E+00	0.000E+00	0.000E+00	0.000E+00	0.000E+00	0.000E+00	0.000E+00	0.000E+00
Np-237	-3.089E+03	-3.091E+03	-3.087E+03	-3.087E+03	-3.118E+03	-4.397E+03	-1.191E+04	-1.414E+04	-1.370E+04	-1.025E+04	1.972E+02
Np-238	0.000E+00	0.000E+00	0.000E+00	0.000E+00	0.000E+00	0.000E+00	0.000E+00	0.000E+00	0.000E+00	0.000E+00	0.000E+00
Np-239	0.000E+00	0.000E+00	0.000E+00	0.000E+00	0.000E+00	0.000E+00	0.000E+00	0.000E+00	0.000E+00	0.000E+00	0.000E+00
Np-240m	0.000E+00	0.000E+00	0.000E+00	0.000E+00	0.000E+00	0.000E+00	0.000E+00	0.000E+00	0.000E+00	0.000E+00	0.000E+00
Np-240	0.000E+00	0.000E+00	0.000E+00	0.000E+00	0.000E+00	0.000E+00	0.000E+00	0.000E+00	0.000E+00	0.000E+00	0.000E+00
Pu-236	0.000E+00	0.000E+00	0.000E+00	0.000E+00	0.000E+00	0.000E+00	0.000E+00	0.000E+00	0.000E+00	0.000E+00	0.000E+00
Pu-238	5.208E+07	5.214E+07	5.322E+07	5.790E+07	5.562E+07	2.850E+07	5.009E+04	8.798E-14	0.000E+00	0.000E+00	2.461E+02
Pu-239	-2.244E+06	-2.244E+06	-2.231E+06	-2.233E+06	-2.231E+06	-2.228E+06	-2.178E+06	-1.727E+06	-1.380E+05	6.858E-02	2.675E+02
Pu-240	-8.506E+05	-8.506E+05	-8.506E+05	-8.506E+05	-8.025E+05	-6.741E+05	-6.195E+05	-2.456E+05	-2.407E+01	1.390E-03	2.675E+02
Pu-241	0.000E+00	0.000E+00	0.000E+00	0.000E+00	0.000E+00	0.000E+00	0.000E+00	0.000E+00	0.000E+00	0.000E+00	0.000E+00
Pu-242	3.354E+03	3.354E+03	3.370E+03	3.370E+03	3.370E+03	3.435E+03	3.579E+03	3.547E+03	3.017E+03	5.826E+02	2.675E+02
Pu-243	0.000E+00	0.000E+00	0.000E+00	0.000E+00	0.000E+00	0.000E+00	0.000E+00	0.000E+00	0.000E+00	0.000E+00	0.000E+00
Pu-244	0.000E+00	0.000E+00	0.000E+00	0.000E+00	0.000E+00	0.000E+00	0.000E+00	0.000E+00	0.000E+00	0.000E+00	0.000E+00
Pu-245	0.000E+00	0.000E+00	0.000E+00	0.000E+00	0.000E+00	0.000E+00	0.000E+00	0.000E+00	0.000E+00	0.000E+00	0.000E+00
Am241	4.552E+06	3.209E+06	2.898E+06	-4.912E+05	-2.662E+07	-6.658E+07	-1.592E+07	1.409E+04	7.837E+00	0.000E+00	2.729E+02
Am-242m	3.087E+06	3.087E+06	3.086E+06	3.073E+06	2.949E+06	1.956E+06	3.228E+04	4.820E-14	0.000E+00	0.000E+00	2.675E+02
Am-242	0.000E+00	0.000E+00	0.000E+00	0.000E+00	0.000E+00	0.000E+00	0.000E+00	0.000E+00	0.000E+00	0.000E+00	0.000E+00
Am-243	-2.947E+05	-2.947E+05	-2.947E+05	-2.947E+05	-2.947E+05	-2.915E+05	-2.684E+05	-1.190E+05	-3.406E+01	6.997E-02	2.729E+02
Am-244	0.000E+00	0.000E+00	0.000E+00	0.000E+00	0.000E+00	0.000E+00	0.000E+00	0.000E+00	0.000E+00	0.000E+00	0.000E+00
Am-245	0.000E+00	0.000E+00	0.000E+00	0.000E+00	0.000E+00	0.000E+00	0.000E+00	0.000E+00	0.000E+00	0.000E+00	0.000E+00
Cm-242	4.218E+07	4.225E+07	3.736E+07	9.028E+06	6.240E+04	4.140E+04	6.828E+02	1.023E-15	0.000E+00	0.000E+00	0.000E+00
Cm-243	-1.544E+05	-1.509E+05	-1.505E+05	-1.476E+05	-1.216E+05	-1.732E+04	-5.895E-05	0.000E+00	0.000E+00	0.000E+00	6.900E+00
Cm-244	3.863E+07	3.932E+07	3.922E+07	3.775E+07	2.679E+07	8.538E+05	6.902E-07	6.978E-07	7.751E-07	1.101E-06	1.969E+02
Cm-245	3.388E+04	3.389E+04	3.389E+04	3.388E+04	3.384E+04	3.360E+04	3.115E+04	1.465E+04	7.716E+00	0.000E+00	1.630E+02
Cm-246	3.842E+03	3.842E+03	3.842E+03	3.842E+03	3.839E+03	3.788E+03	3.318E+03	8.825E+02	1.564E-03	0.000E+00	2.840E+02
Cm-247	0.000E+00	0.000E+00	0.000E+00	0.000E+00	0.000E+00	0.000E+00	0.000E+00	0.000E+00	0.000E+00	0.000E+00	2.845E+02
Cm-248	0.000E+00	0.000E+00	0.000E+00	0.000E+00	0.000E+00	0.000E+00	0.000E+00	0.000E+00	0.000E+00	0.000E+00	0.000E+00
Cm-249	0.000E+00	0.000E+00	0.000E+00	0.000E+00	0.000E+00	0.000E+00	0.000E+00	0.000E+00	0.000E+00	0.000E+00	0.000E+00
Cm-250	0.000E+00	0.000E+00	0.000E+00	0.000E+00	0.000E+00	0.000E+00	0.000E+00	0.000E+00	0.000E+00	0.000E+00	0.000E+00
Bk-249	0.000E+00	0.000E+00	0.000E+00	0.000E+00	0.000E+00	0.000E+00	0.000E+00	0.000E+00	0.000E+00	0.000E+00	0.000E+00
Bk-250	0.000E+00	0.000E+00	0.000E+00	0.000E+00	0.000E+00	0.000E+00	0.000E+00	0.000E+00	0.000E+00	0.000E+00	0.000E+00
Cf-249	0.000E+00	0.000E+00	0.000E+00	0.000E+00	0.000E+00	0.000E+00	0.000E+00	0.000E+00	0.000E+00	0.000E+00	0.000E+00
Cf-250	0.000E+00	0.000E+00	0.000E+00	0.000E+00	0.000E+00	0.000E+00	0.000E+00	0.000E+00	0.000E+00	0.000E+00	0.000E+00
Cf-251	0.000E+00	0.000E+00	0.000E+00	0.000E+00	0.000E+00	0.000E+00	0.000E+00	0.000E+00	0.000E+00	0.000E+00	0.000E+00
Cf-252	0.000E+00	0.000E+00	0.000E+00	0.000E+00	0.000E+00	0.000E+00	0.000E+00	0.000E+00	0.000E+00	0.000E+00	0.000E+00
Cf-253	0.000E+00	0.000E+00	0.000E+00	0.000E+00	0.000E+00	0.000E+00	0.000E+00	0.000E+00	0.000E+00	0.000E+00	0.000E+00
Cf-254	0.000E+00	0.000E+00	0.000E+00	0.000E+00	0.000E+00	0.000E+00	0.000E+00	0.000E+00	0.000E+00	0.000E+00	0.000E+00
Es-253	0.000E+00	0.000E+00	0.000E+00	0.000E+00	0.000E+00	0.000E+00	0.000E+00	0.000E+00	0.000E+00	0.000E+00	0.000E+00
Total	1.370E+08	1.365E+08	1.323E+08	1.038E+08	5.539E+07	-3.840E+07	-1.887E+07	-2.073E+06	-1.488E+05	-9.664E+03	0.000E+00
Total/MWd	2.806E+02	2.796E+02	2.709E+02	2.125E+02	1.134E+02	-7.865E+01	-3.865E+01	-4.245E+00	-3.047E-01	-1.979E-02	0.000E+00

Table A.22
Enrichment and blending ratio for once-through and recycle configurations

Item	Once-Through	Recycle: Burner	Recycle: Converter	Recycle: Breeder
Enrichment	0.2637	0.3021	0.2691	0.2248
Enrichment (no Rare Earth)	-	0.3005	0.2679	0.2236
Blending Ratio (Recycle mass/Total mass)	-	0.86	0.91	1.00
Feed mass/Recycle mass	-	0.16	0.099	0.0
TRU Feed Mass, kg/y	1506.5	255.2	148.0	0.0

Table A.23
Recycled TRU masses for burner, converter, and breeder cycles

	BURNER		*CONVERTER*		*BREEDER*	
Isotope	**Mass, kg**	**%**	**Mass, kg**	**%**	**Mass, kg**	**%**
Np-237	38.055	2.53	25.409	1.79	7.753	0.59
Pu-236	0.000	0.00	0.000	0.00	0.000	0.00
Pu-238	55.724	3.71	39.161	2.76	11.650	0.89
Pu-239	609.378	40.57	716.712	50.59	927.414	70.48
Pu-240	454.636	30.26	396.169	27.96	294.956	22.42
Pu-241	76.108	5.07	58.610	4.14	30.352	2.31
Pu-242	116.744	7.77	79.009	5.58	18.562	1.41
Am-241	63.273	4.21	43.748	3.09	16.380	1.24
Am-242	4.543	0.30	3.263	0.23	1.151	0.09
Am-243	49.139	3.27	31.698	2.24	4.597	0.35
Cm-242	0.614	0.04	0.443	0.03	0.158	0.01
Cm-243	0.162	0.01	0.118	0.01	0.037	0.00
Cm-244	25.407	1.69	16.724	1.18	2.091	0.16
Cm-245	5.916	0.39	3.972	0.28	0.473	0.04
Cm-246	2.527	0.17	1.782	0.13	0.196	0.01
Total	1502.226	100.00	1416.818	100.00	1315.770	100.00

Table A.24
Feed TRU masses for burner, converter, and breeder cycles

	BURNER		*CONVERTER*		*BREEDER*	
Isotope	**Mass, kg**	**%**	**Mass, kg**	**%**	**Mass, kg**	**%**
Np-237	13.628	5.34	7.905	5.34	0.000	0.00
Pu-236	0.000	0.00	0.000	0.00	0.000	0.00
Pu-238	2.560	1.00	1.485	1.00	0.000	0.00
Pu-239	129.286	50.66	74.995	50.66	0.000	0.00
Pu-240	50.857	19.93	29.501	19.93	0.000	0.00
Pu-241	34.389	13.47	19.948	13.47	0.000	0.00
Pu-242	9.999	3.92	5.800	3.92	0.000	0.00
Am-241	6.441	2.52	3.737	2.52	0.000	0.00
Am-242	0.029	0.01	0.017	0.01	0.000	0.00
Am-243	6.417	2.51	3.723	2.51	0.000	0.00
Cm-242	0.003	0.00	0.001	0.00	0.000	0.00
Cm-243	0.020	0.01	0.012	0.01	0.000	0.00
Cm-244	1.434	0.56	0.832	0.56	0.000	0.00
Cm-245	0.133	0.05	0.077	0.05	0.000	0.00
Cm-246	0.017	0.01	0.010	0.01	0.000	0.00
Total	255.213	100.00	148.043	100.00	0.000	0.00

Table A.25
Fabricated fresh TRU masses for burner, converter, and breeder cycles

	BURNER		*CONVERTER*		*BREEDER*	
Isotope	**Mass, kg**	**%**	**Mass, kg**	**%**	**Mass, kg**	**%**
Np-237	51.738	2.94	33.352	2.13	7.711	0.59
Pu-236	0.000	0.00	0.000	0.00	0.000	0.00
Pu-238	58.383	3.32	40.723	2.60	11.605	0.89
Pu-239	738.660	42.04	791.700	50.60	920.790	70.49
Pu-240	505.968	28.79	425.976	27.22	292.877	22.42
Pu-241	107.918	6.14	76.724	4.90	29.433	2.25
Pu-242	126.745	7.21	84.811	5.42	18.430	1.41
Am-241	72.237	4.11	49.280	3.15	16.954	1.30
Am-242	4.561	0.26	3.272	0.21	1.141	0.09
Am-243	55.554	3.16	35.419	2.26	4.564	0.35
Cm-242	0.289	0.02	0.208	0.01	0.074	0.01
Cm-243	0.180	0.01	0.129	0.01	0.036	0.00
Cm-244	26.331	1.50	17.222	1.10	2.036	0.16
Cm-245	6.049	0.34	4.049	0.26	0.470	0.04
Cm-246	2.543	0.14	1.792	0.11	0.194	0.01
Total	1757.156	100.00	1564.657	100.00	1306.315	100.00

Table A.26

Criticality performance for once-through and recycle configurations

Item	Once-Through	*RECYCLE* Burner	*RECYCLE* Converter	*RECYCLE* Breeder
BOEC Eigenvalue	1.05572	1.04538	1.03747	1.03599
EOEC Eigenvalue	1.00030	0.99997	0.99992	0.99942
Burnup Swing Δk/kk', %	-5.248	-4.344	-3.620	-3.532
TRU C.R.	0.4634	0.4514	0.6740	1.0375

Table A.27

BOEC, EOEC, and (EOEC-BOEC) masses for burner, converter, and breeder cycles

	BURNER			*CONVERTER*			*BREEDER*		
	BOEC	**EOEC**	**DIFF.**	**BOEC**	**EOEC**	**DIFF.**	**BOEC**	**EOEC**	**DIFF.**
Isotope	Mass, kg	Mass, kg	Mass, kg	Mass, kg	Mass, kg	Mass, kg	Mass, kg	Mass, kg	Mass, kg
Np-237	140.454	126.648	-13.806	91.488	83.466	-8.022	23.250	23.268	0.018
Pu-236	0.000	0.000	0.000	0.000	0.000	0.000	0.000	0.000	0.000
Pu-238	169.296	164.964	-4.332	118.242	115.458	-2.784	33.918	33.486	-0.432
Pu-239	2075.394	1946.742	-128.652	2294.040	2219.790	-74.250	2789.856	2797.446	7.590
Pu-240	1466.400	1414.128	-52.272	1248.066	1217.760	-30.306	881.994	884.304	2.310
Pu-241	287.574	262.032	-25.542	209.820	196.386	-13.434	89.832	92.934	3.102
Pu-242	370.548	360.660	-9.888	248.826	243.102	-5.724	55.416	55.566	0.150
Am-241	208.404	193.470	-14.934	142.464	132.468	-9.996	49.860	47.190	-2.670
Am-242	13.866	13.884	0.018	9.954	9.966	0.012	3.474	3.492	0.018
Am-243	160.032	153.678	-6.354	102.414	98.730	-3.684	13.728	13.770	0.042
Cm-242	5.592	8.328	2.736	4.032	6.006	1.974	1.404	2.112	0.708
Cm-243	0.504	0.492	-0.012	0.366	0.360	-0.006	0.102	0.108	0.006
Cm-244	78.444	79.038	0.594	51.408	51.906	0.498	6.186	6.360	0.174
Cm-245	17.982	17.856	-0.126	12.048	11.976	-0.072	1.410	1.410	0.000
Cm-246	7.614	7.602	-0.012	5.370	5.358	-0.012	0.588	0.588	0.000
Total	5002.104	4749.522	-252.582	4538.538	4392.732	-145.806	3951.018	3962.034	11.016

BURNER	mass increment/energy	-0.517 g/MWd
CONVERTER	mass increment/energy	-0.299 g/MWd
BREEDER	mass increment/energy	0.023 g/MWd

Table A.28
Safety parameters for once-through and recycle configurations

		Recycle		
Item	**Once-Through**	**Burner**	**Converter**	**Breeder**
Beta, 10^{-3}	3.51	3.15	3.25	3.33
Core Doppler (Drivers only) -Tdk/dT • 10^3 (HM plus ZR)	1.462	1.354	1.511	1.799
Na Void (Core flowing),% Δk/kk'				
Below Driver Fuel	-0.4395	-0.3921	-0.1555	-0.0051
Driver Fuel	0.6059	1.0952	0.9980	0.7987
Above Driver Fuel (Plenum)	-0.7976	-0.7266	-0.7322	-0.7713
Total Driver	-0.6312	-0.0236	0.1103	0.0223
Driver Fuel and Plenum	-0.1917	0.3686	0.2658	0.0274

Table A.29
Recycle burner-core BOEC and EOEC decay power (all regions), Watts

	BOEC				*EOEC*			
Time	**Light**	**Heavy**	**FP**	**Total**	**Light**	**Heavy**	**FP**	**Total**
1 hour	2.88E+04	9.10E+05	7.82E+05	1.72E+06	1.41E+05	2.50E+06	1.53E+07	1.79E+07
30 days	2.42E+04	8.44E+05	6.04E+05	1.47E+06	5.26E+04	1.26E+06	1.67E+06	2.98E+06
1 year	6.54E+03	4.72E+05	1.64E+05	6.42E+05	1.27E+04	5.71E+05	3.10E+05	8.94E+05
10 y	7.78E+01	2.98E+05	9.22E+03	3.07E+05	1.50E+02	2.95E+05	1.81E+04	3.14E+05
100 y	3.79E-02	1.19E+05	1.02E+03	1.20E+05	7.57E-02	1.14E+05	2.00E+03	1.16E+05
1000 y	3.28E-02	2.66E+04	4.11E-01	2.66E+04	6.55E-02	2.51E+04	8.04E-01	2.51E+04
10000 y	1.77E-02	7.62E+03	3.88E-01	7.62E+03	3.53E-02	7.28E+03	7.58E-01	7.28E+03

Light = ORIGEN Light Isotopes
Heavy = ORIGEN Heavy Isotopes
FP = ORIGEN Fission Products
Total = Sum of Light, Heavy, and Fission Product Decay Power

Table A.30
Recycle converter-core BOEC and EOEC decay power (all regions), Watts

	BOEC				EOEC			
Time	**Light**	**Heavy**	**FP**	**Total**	**Light**	**Heavy**	**FP**	**Total**
1 hour	3.46E+04	6.45E+05	8.06E+05	1.49E+06	1.73E+05	2.40E+06	1.58E+07	1.84E+07
30 days	2.91E+04	5.97E+05	6.23E+05	1.25E+06	6.26E+04	8.93E+05	1.72E+06	2.67E+06
1 year	7.81E+03	3.27E+05	1.68E+05	5.03E+05	1.50E+04	3.98E+05	3.16E+05	7.29E+05
10 y	9.29E+01	2.05E+05	9.50E+03	2.15E+05	1.76E+02	2.04E+05	1.85E+04	2.22E+05
100 y	4.53E-02	8.64E+04	1.06E+03	8.75E+04	8.93E-02	8.36E+04	2.05E+03	8.57E+04
1000 y	3.93E-02	2.16E+04	4.19E-01	2.16E+04	7.73E-02	2.07E+04	8.11E-01	2.07E+04
10000 y	2.12E-02	7.03E+03	3.95E-01	7.03E+03	4.17E-02	6.84E+03	7.64E-01	6.84E+03

Table A.31
Recycle breeder-core BOEC and EOEC decay power (all regions), Watts

	BOEC				EOEC			
Time	**Light**	**Heavy**	**FP**	**Total**	**Light**	**Heavy**	**FP**	**Total**
1 hour	4.31E+04	1.95E+05	8.51E+05	1.09E+06	2.15E+05	2.25E+06	1.68E+07	1.93E+07
30 days	3.62E+04	1.78E+05	6.58E+05	8.72E+05	7.73E+04	2.84E+05	1.82E+06	2.18E+06
1 year	9.56E+03	8.50E+04	1.77E+05	2.71E+05	1.82E+04	1.11E+05	3.33E+05	4.62E+05
10 y	1.14E+02	5.26E+04	1.01E+04	6.28E+04	2.14E+02	5.30E+04	1.96E+04	7.28E+04
100 y	5.61E-02	3.58E+04	1.12E+03	3.69E+04	1.09E-01	3.60E+04	2.17E+03	3.81E+04
1000 y	4.86E-02	1.41E+04	4.39E-01	1.41E+04	9.46E-02	1.42E+04	8.51E-01	1.42E+04
10000 y	2.63E-02	6.27E+03	4.14E-01	6.27E+03	5.11E-02	6.29E+03	8.02E-01	6.29E+03

Light = ORIGEN Light Isotopes
Heavy = ORIGEN Heavy Isotopes
FP = ORIGEN Fission Products
Total = Sum of Light, Heavy, and Fission Product Decay Power

Table A.32
Recycle converter-core BOEC and EOEC decay power (core), Watts

	BOEC				EOEC			
Time	**Light**	**Heavy**	**FP**	**Total**	**Light**	**Heavy**	**FP**	**Total**
1 hour	2.85E+04	6.45E+05	7.75E+05	1.45E+06	1.38E+05	2.05E+06	1.50E+07	1.72E+07
30 days	2.39E+04	5.97E+05	5.99E+05	1.22E+06	5.14E+04	8.92E+05	1.63E+06	2.58E+06
1 year	6.47E+03	3.27E+05	1.62E+05	4.95E+05	1.24E+04	3.97E+05	3.03E+05	7.13E+05
10 y	7.71E+01	2.05E+05	9.16E+03	2.14E+05	1.46E+02	2.03E+05	1.77E+04	2.21E+05
100 y	3.76E-02	8.63E+04	1.02E+03	8.73E+04	7.39E-02	8.33E+04	1.96E+03	8.52E+04
1000 y	3.26E-02	2.14E+04	4.08E-01	2.14E+04	6.40E-02	2.04E+04	7.85E-01	2.04E+04
10000 y	1.76E-02	6.92E+03	3.84E-01	6.92E+03	3.45E-02	6.62E+03	7.40E-01	6.62E+03

Table A.33
Recycle converter-core BOEC and EOEC decay power (axial blanket), Watts

	BOEC				EOEC			
Time	**Light**	**Heavy**	**FP**	**Total**	**Light**	**Heavy**	**FP**	**Total**
1 hour	6.17E+03	1.65E+02	3.19E+04	3.83E+04	3.46E+04	3.58E+05	8.36E+05	1.23E+06
30 days	5.17E+03	1.64E+02	2.41E+04	2.95E+04	1.13E+04	4.03E+02	8.22E+04	9.39E+04
1 year	1.33E+03	1.63E+02	5.73E+03	7.23E+03	2.59E+03	3.31E+02	1.31E+04	1.60E+04
10 y	1.58E+01	1.63E+02	3.49E+02	5.28E+02	3.03E+01	3.31E+02	7.78E+02	1.14E+03
100 y	7.70E-03	1.59E+02	3.90E+01	1.98E+02	1.54E-02	3.23E+02	8.69E+01	4.10E+02
1000 y	6.67E-03	1.47E+02	1.11E-02	1.47E+02	1.33E-02	2.90E+02	2.63E-02	2.90E+02
10000 y	3.60E-03	1.10E+02	1.05E-02	1.10E+02	7.20E-03	2.13E+02	2.48E-02	2.13E+02

Light = ORIGEN Light Isotopes
Heavy = ORIGEN Heavy Isotopes
FP = ORIGEN Fission Products
Total = Sum of Light, Heavy, and Fission Product Decay Power

Table A.34

Recycle breeder-core BOEC and EOEC decay power (core), Watts

	BOEC				EOEC			
Time	**Light**	**Heavy**	**FP**	**Total**	**Light**	**Heavy**	**FP**	**Total**
1 hour	2.80E+04	1.94E+05	7.89E+05	1.01E+06	1.37E+05	1.31E+06	1.52E+07	1.67E+07
30 day	2.36E+04	1.78E+05	6.10E+05	8.12E+05	5.06E+04	2.83E+05	1.66E+06	2.00E+06
1E0 y	6.38E+03	8.45E+04	1.65E+05	2.56E+05	1.22E+04	1.10E+05	3.08E+05	4.31E+05
1E1 y	7.60E+01	5.21E+04	9.32E+03	6.15E+04	1.44E+02	5.21E+04	1.80E+04	7.03E+04
1E2 y	3.71E-02	3.53E+04	1.04E+03	3.64E+04	7.28E-02	3.51E+04	2.00E+03	3.71E+04
1E3 y	3.21E-02	1.37E+04	4.16E-01	1.37E+04	6.31E-02	1.33E+04	7.99E-01	1.33E+04
1E4 y	1.73E-02	5.94E+03	3.92E-01	5.94E+03	3.40E-02	5.68E+03	7.53E-01	5.68E+03

Table A.35

Recycle breeder-core BOEC and EOEC decay power (axial blanket), Watts

	BOEC				EOEC			
Time	**Light**	**Heavy**	**FP**	**Total**	**Light**	**Heavy**	**FP**	**Total**
1 hour	1.13E+04	3.16E+02	4.52E+04	5.68E+04	6.31E+04	7.06E+05	1.19E+06	1.96E+06
30 day	9.48E+03	3.14E+02	3.41E+04	4.39E+04	2.07E+04	7.48E+02	1.17E+05	1.38E+05
1E0 y	2.45E+03	3.13E+02	8.06E+03	1.08E+04	4.76E+03	6.30E+02	1.84E+04	2.38E+04
1E1 y	2.89E+01	3.13E+02	4.95E+02	8.37E+02	5.56E+01	6.31E+02	1.11E+03	1.79E+03
1E2 y	1.41E-02	3.08E+02	5.55E+01	3.64E+02	2.82E-02	6.20E+02	1.24E+02	7.44E+02
1E3 y	1.22E-02	2.92E+02	1.58E-02	2.92E+02	2.45E-02	5.77E+02	3.75E-02	5.77E+02
1E4 y	6.61E-03	2.20E+02	1.49E-02	2.20E+02	1.32E-02	4.30E+02	3.54E-02	4.30E+02

Light = ORIGEN Light Isotopes
Heavy = ORIGEN Heavy Isotopes
FP = ORIGEN Fission Products
Total = Sum of Light, Heavy, and Fission Product Decay Power

Table A.36

Recycle breeder-core BOEC and EOEC decay power (internal blanket), Watts

	BOEC				EOEC			
Time	**Light**	**Heavy**	**FP**	**Total**	**Light**	**Heavy**	**FP**	**Total**
1 hour	2.00E+02	9.59E+00	4.95E+02	7.04E+02	8.58E+02	1.48E+04	9.78E+03	2.54E+04
30 days	1.66E+02	9.56E+00	3.79E+02	5.55E+02	3.19E+02	1.80E+01	1.01E+03	1.35E+03
1 year	3.89E+01	9.56E+00	9.38E+01	1.42E+02	6.49E+01	1.59E+01	1.66E+02	2.47E+02
10 y	4.93E-01	9.56E+00	7.48E+00	1.75E+01	7.93E-01	1.59E+01	1.29E+01	2.95E+01
100 y	2.63E-04	9.50E+00	8.47E-01	1.04E+01	4.37E-04	1.58E+01	1.46E+00	1.72E+01
1000 y	2.28E-04	9.23E+00	2.08E-04	9.23E+00	3.80E-04	1.53E+01	3.70E-04	1.53E+01
10000y	1.24E-04	7.11E+00	1.96E-04	7.11E+00	2.07E-04	1.18E+01	3.50E-04	1.18E+01

Table A.37

Recycle breeder-core BOEC and EOEC decay power (radial blanket), Watts

	BOEC				EOEC			
Time	**Light**	**Heavy**	**FP**	**Total**	**Light**	**Heavy**	**FP**	**Total**
1 hour	3.56E+03	1.50E+02	1.72E+04	2.09E+04	1.47E+04	2.25E+05	3.57E+05	5.97E+05
30 days	2.96E+03	1.50E+02	1.32E+04	1.63E+04	5.68E+03	2.89E+02	3.64E+04	4.24E+04
1 year	6.94E+02	1.49E+02	3.33E+03	4.17E+03	1.16E+03	2.50E+02	6.11E+03	7.52E+03
10 y	8.79E+00	1.49E+02	2.41E+02	3.99E+02	1.41E+01	2.49E+02	4.28E+02	6.92E+02
100 y	4.69E-03	1.47E+02	2.72E+01	1.74E+02	7.79E-03	2.45E+02	4.82E+01	2.93E+02
1000 y	4.07E-03	1.39E+02	7.48E-03	1.39E+02	6.76E-03	2.29E+02	1.39E-02	2.29E+02
10000 y	2.21E-03	1.05E+02	7.06E-03	1.05E+02	3.68E-03	1.72E+02	1.31E-02	1.72E+02

Light = ORIGEN Light Isotopes
Heavy = ORIGEN Heavy Isotopes
FP = ORIGEN Fission Products
Total = Sum of Light, Heavy, and Fission Product Decay Power

Table A.38

Burner [illegible] TRU [illegible]nts nuclide radioactivity (core), Curies Basis = EQULIBRIUM FRESH FUEL BURNER CORE

	Initial	1 hour	30 days	1 year	10 y	100 y	1000 y	10000 y	100000 y	1000000 y
Np-236	0.000E+00	0.000E+00	0.000E+00	0.000E+00	0.000E+00	0.000E+00	0.000E+00	0.000E+00	0.000E+00	0.000E+00
Np-237	3.652E+01	3.652E+01	3.653E+01	3.661E+01	3.758E+01	5.284E+01	1.358E+02	1.631E+02	1.603E+02	1.198E+02
Np-238	0.000E+00	0.000E+00	0.000E+00	0.000E+00	0.000E+00	0.000E+00	0.000E+00	0.000E+00	0.000E+00	0.000E+00
Np-239	0.000E+00	1.304E+02	1.069E+04	1.069E+04	1.067E+04	1.059E+04	9.762E+03	4.318E+03	1.240E+00	0.000E+00
Np-240m	0.000E+00	0.000E+00	0.000E+00	0.000E+00	0.000E+00	0.000E+00	0.000E+00	0.000E+00	0.000E+00	0.000E+00
Np-240	0.000E+00	0.000E+00	0.000E+00	0.000E+00	0.000E+00	0.000E+00	0.000E+00	0.000E+00	0.000E+00	0.000E+00
Pu-236	2.044E+02	2.044E+02	2.004E+02	1.603E+02	1.796E+01	5.585E-09	0.000E+00	0.000E+00	0.000E+00	0.000E+00
Pu-238	9.846E+05	9.846E+05	9.840E+05	9.786E+05	9.150E+05	4.680E+05	8.016E+02	1.370E-15	0.000E+00	0.000E+00
Pu-239	4.529E+04	4.529E+04	4.529E+04	4.529E+04	4.528E+04	4.520E+04	4.432E+04	3.580E+04	2.929E+03	2.302E-08
Pu-240	1.116E+05	1.116E+05	1.117E+05	1.118E+05	1.133E+05	1.159E+05	1.058E+05	4.204E+04	4.126E+00	0.000E+00
Pu-241	1.071E+07	1.071E+07	1.067E+07	1.022E+07	6.666E+06	9.390E+04	9.840E+02	4.625E+02	2.437E-01	0.000E+00
Pu-242	4.941E+02	4.941E+02	4.941E+02	4.942E+02	4.943E+02	4.954E+02	4.977E+02	4.957E+02	4.224E+02	8.142E+01
Pu-243	0.000E+00	0.000E+00	0.000E+00	0.000E+00	0.000E+00	0.000E+00	0.000E+00	0.000E+00	0.000E+00	0.000E+00
Pu-244	0.000E+00	0.000E+00	0.000E+00	0.000E+00	0.000E+00	0.000E+00	0.000E+00	0.000E+00	0.000E+00	0.000E+00
Pu-245	0.000E+00	0.000E+00	0.000E+00	0.000E+00	0.000E+00	0.000E+00	0.000E+00	0.000E+00	0.000E+00	0.000E+00
Am-241	2.559E+05	2.559E+05	2.573E+05	2.722E+05	3.872E+05	5.334E+05	1.278E+05	4.625E+02	2.571E-01	0.000E+00
Am-242m	4.424E+04	4.424E+04	4.422E+04	4.404E+04	4.226E+04	2.804E+04	4.625E+02	6.906E-16	0.000E+00	0.000E+00
Am-242	0.000E+00	1.874E+03	4.422E+04	4.404E+04	4.227E+04	2.804E+04	4.625E+02	6.906E-16	0.000E+00	0.000E+00
Am-243	1.069E+04	1.069E+04	1.069E+04	1.069E+04	1.067E+04	1.059E+04	9.762E+03	4.318E+03	1.240E+00	0.000E+00
Am-244	0.000E+00	0.000E+00	0.000E+00	0.000E+00	0.000E+00	0.000E+00	0.000E+00	0.000E+00	0.000E+00	0.000E+00
Am-245	0.000E+00	0.000E+00	0.000E+00	0.000E+00	0.000E+00	0.000E+00	0.000E+00	0.000E+00	0.000E+00	0.000E+00
Cm-242	4.606E+05	4.604E+05	4.096E+05	1.259E+05	3.466E+04	2.299E+04	3.793E+02	5.680E-16	0.000E+00	0.000E+00
Cm-243	8.172E+03	8.172E+03	8.160E+03	7.998E+03	6.582E+03	9.366E+02	3.193E-06	0.000E+00	0.000E+00	0.000E+00
Cm-244	2.092E+06	2.092E+06	2.085E+06	2.013E+06	1.426E+06	4.541E+04	4.868E-11	0.000E+00	0.000E+00	0.000E+00
Cm-245	1.068E+03	1.068E+03	1.068E+03	1.068E+03	1.067E+03	1.059E+03	9.822E+02	4.616E+02	2.432E-01	0.000E+00
Cm-246	7.842E+02	7.842E+02	7.842E+02	7.842E+02	7.830E+02	7.728E+02	6.768E+02	1.801E+02	3.191E-04	0.000E+00
Cm-247	0.000E+00	0.000E+00	0.000E+00	0.000E+00	0.000E+00	0.000E+00	0.000E+00	0.000E+00	0.000E+00	0.000E+00
Cm-248	0.000E+00	0.000E+00	0.000E+00	0.000E+00	0.000E+00	0.000E+00	0.000E+00	0.000E+00	0.000E+00	0.000E+00
Cm-249	0.000E+00	0.000E+00	0.000E+00	0.000E+00	0.000E+00	0.000E+00	0.000E+00	0.000E+00	0.000E+00	0.000E+00
Cm-250	0.000E+00	0.000E+00	0.000E+00	0.000E+00	0.000E+00	0.000E+00	0.000E+00	0.000E+00	0.000E+00	0.000E+00
Bk-249	0.000E+00	0.000E+00	0.000E+00	0.000E+00	0.000E+00	0.000E+00	0.000E+00	0.000E+00	0.000E+00	0.000E+00
Bk-250	0.000E+00	0.000E+00	0.000E+00	0.000E+00	0.000E+00	0.000E+00	0.000E+00	0.000E+00	0.000E+00	0.000E+00
Cf-249	0.000E+00	0.000E+00	0.000E+00	0.000E+00	0.000E+00	0.000E+00	0.000E+00	0.000E+00	0.000E+00	0.000E+00
Cf-250	0.000E+00	0.000E+00	0.000E+00	0.000E+00	0.000E+00	0.000E+00	0.000E+00	0.000E+00	0.000E+00	0.000E+00
Cf-251	0.000E+00	0.000E+00	0.000E+00	0.000E+00	0.000E+00	0.000E+00	0.000E+00	0.000E+00	0.000E+00	0.000E+00
Cf-252	0.000E+00	0.000E+00	0.000E+00	0.000E+00	0.000E+00	0.000E+00	0.000E+00	0.000E+00	0.000E+00	0.000E+00
Cf-253	0.000E+00	0.000E+00	0.000E+00	0.000E+00	0.000E+00	0.000E+00	0.000E+00	0.000E+00	0.000E+00	0.000E+00
Cf-254	0.000E+00	0.000E+00	0.000E+00	0.000E+00	0.000E+00	0.000E+00	0.000E+00	0.000E+00	0.000E+00	0.000E+00
Es-253	0.000E+00	0.000E+00	0.000E+00	0.000E+00	0.000E+00	0.000E+00	0.000E+00	0.000E+00	0.000E+00	0.000E+00
Total	1.473E+07	1.473E+07	1.468E+07	1.388E+07	9.702E+06	1.405E+06	3.028E+05	8.870E+04	3.519E+03	2.012E+02

Table A.39

Burner core heavy TRU elements nuclide radioactivity (core), Curies Basis = EQULIBRIUM DISCHARGE FUEL BURNER CORE

	Initial	1 hour	30 days	1 year	10 y	100 y	1000 y	10000 y	100000 y	1000000 y
Np-236	1.613E+03	1.564E+03	2.288E-07	0.000E+00	0.000E+00	0.000E+00	0.000E+00	0.000E+00	0.000E+00	0.000E+00
Np-237	2.678E+01	2.678E+01	2.680E+01	2.687E+01	2.765E+01	3.952E+01	1.039E+02	1.255E+02	1.237E+02	9.240E+01
Np-238	7.968E+06	7.854E+06	4.002E+02	0.000E+00	0.000E+00	0.000E+00	0.000E+00	0.000E+00	0.000E+00	0.000E+00
Np-239	1.622E+08	1.612E+08	3.299E+04	9.462E+03	9.450E+03	9.378E+03	8.640E+03	3.823E+03	1.103E+00	5.510E-03
Np-240m	2.939E+02	2.823E+02	6.642E-06	6.642E-06	6.642E-06	6.648E-06	6.744E-06	7.692E-06	1.625E-05	5.266E-05
Np-240	1.346E+05	6.960E+04	0.000E+00	0.000E+00	0.000E+00	0.000E+00	0.000E+00	0.000E+00	0.000E+00	0.000E+00
Pu236	4.846E+02	4.846E+02	4.759E+02	3.806E+02	4.264E+01	1.326E-08	0.000E+00	0.000E+00	0.000E+00	0.000E+00
Pu-238	9.168E+05	9.168E+05	9.228E+05	9.474E+05	8.952E+05	4.566E+05	7.542E+02	1.235E-15	0.000E+00	0.000E+00
Pu-239	3.736E+04	3.736E+04	3.740E+04	3.740E+04	3.740E+04	3.733E+04	3.662E+04	2.966E+04	2.435E+03	5.510E-03
Pu-240	1.002E+05	1.002E+05	1.002E+05	1.004E+05	1.019E+05	1.046E+05	9.552E+04	3.797E+04	3.725E+00	5.272E-05
Pu-241	8.022E+06	8.022E+06	7.986E+06	7.650E+06	4.989E+06	7.056E+04	9.606E+02	4.516E+02	2.380E-01	0.000E+00
Pu-242	4.562E+02	4.562E+02	4.562E+02	4.562E+02	4.564E+02	4.574E+02	4.596E+02	4.582E+02	3.905E+02	7.530E+01
Pu-243	6.510E+06	5.665E+06	5.747E-03	5.747E-03	5.747E-03	5.747E-03	5.747E-03	5.745E-03	5.723E-03	5.510E-03
Pu-244	6.648E-06	6.648E-06	6.648E-06	6.648E-06	6.648E-06	6.660E-06	6.756E-06	7.698E-06	1.628E-05	5.272E-05
Pu-245	3.052E+01	2.859E+01	0.000E+00	0.000E+00	0.000E+00	0.000E+00	0.000E+00	0.000E+00	0.000E+00	0.000E+00
Am-241	2.083E+05	2.083E+05	2.093E+05	2.204E+05	3.063E+05	4.136E+05	9.924E+04	4.517E+02	2.511E-01	0.000E+00
Am-242m	3.989E+04	3.989E+04	3.988E+04	3.971E+04	3.811E+04	2.528E+04	4.171E+02	6.228E-16	0.000E+00	0.000E+00
Am-242	1.210E+07	1.159E+07	3.988E+04	3.971E+04	3.811E+04	2.528E+04	4.171E+02	6.228E-16	0.000E+00	0.000E+00
Am243	9.462E+03	9.462E+03	9.462E+03	9.462E+03	9.450E+03	9.378E+03	8.640E+03	3.823E+03	1.103E+00	5.510E-03
Am-244	8.256E+06	1.669E+06	8.646E-09	8.646E-09	8.646E-09	8.658E-09	8.784E-09	1.001E-08	2.116E-08	6.852E-08
Am-245	5.579E+01	4.831E+01	1.495E-03	7.134E-04	5.029E-07	0.000E+00	0.000E+00	0.000E+00	0.000E+00	0.000E+00
Cm-242	9.390E+06	9.390E+06	8.304E+06	2.021E+06	3.125E+04	2.073E+04	3.420E+02	5.122E-16	0.000E+00	0.000E+00
Cm-243	7.458E+03	7.458E+03	7.446E+03	7.302E+03	6.006E+03	8.550E+02	2.915E-06	0.000E+00	0.000E+00	0.000E+00
Cm-244	2.119E+06	2.119E+06	2.113E+06	2.039E+06	1.445E+06	4.601E+04	8.832E-09	1.001E-08	2.116E-08	6.852E-08
Cm245	1.043E+03	1.043E+03	1.043E+03	1.043E+03	1.042E+03	1.034E+03	9.588E+02	4.508E+02	2.375E-01	0.000E+00
Cm-246	7.800E+02	7.800E+02	7.800E+02	7.800E+02	7.794E+02	7.686E+02	6.738E+02	1.792E+02	3.175E-04	0.000E+00
Cm-247	5.747E-03	5.747E-03	5.747E-03	5.747E-03	5.747E-03	5.747E-03	5.747E-03	5.745E-03	5.723E-03	5.510E-03
Cm-248	1.383E-02	1.383E-02	1.383E-02	1.383E-02	1.383E-02	1.383E-02	1.381E-02	1.356E-02	1.136E-02	1.930E-03
Cm-249	2.413E+02	1.261E+02	1.523E-07	3.277E-13	0.000E+00	0.000E+00	0.000E+00	0.000E+00	0.000E+00	0.000E+00
Cm-250	5.434E-08	5.434E-08	5.434E-08	5.433E-08	5.431E-08	5.412E-08	5.221E-08	3.648E-08	1.011E-09	0.000E+00
Bk-249	1.064E+02	1.064E+02	9.966E+01	4.753E+01	3.352E-02	0.000E+00	0.000E+00	0.000E+00	0.000E+00	0.000E+00
Bk-250	5.954E+00	4.802E+00	5.434E-08	5.433E-08	5.431E-08	5.412E-08	5.221E-08	3.648E-08	1.011E-09	0.000E+00
Cf-249	1.312E-01	1.312E-01	1.478E-01	2.747E-01	3.841E-01	3.218E-01	5.468E-02	1.097E-09	0.000E+00	0.000E+00
Cf-250	3.864E-01	3.864E-01	3.849E-01	3.666E-01	2.275E-01	1.930E-03	5.221E-08	3.648E-08	1.011E-09	0.000E+00
Cf-251	1.322E-04	1.322E-04	1.322E-04	1.321E-04	1.312E-04	1.224E-04	6.120E-05	5.972E-08	0.000E+00	0.000E+00
Cf-252	8.580E-04	8.580E-04	8.400E-04	6.606E-04	6.246E-05	3.595E-15	0.000E+00	0.000E+00	0.000E+00	0.000E+00
Cf-253	1.574E-04	1.572E-04	4.900E-05	1.054E-10	0.000E+00	0.000E+00	0.000E+00	0.000E+00	0.000E+00	0.000E+00
Cf-254	3.379E-07	3.377E-07	2.396E-07	5.143E-09	0.000E+00	0.000E+00	0.000E+00	0.000E+00	0.000E+00	0.000E+00
Es-253	1.570E-04	1.570E-04	1.108E-04	4.905E-09	0.000E+00	0.000E+00	0.000E+00	0.000E+00	0.000E+00	0.000E+00
Total	2.181E+08	2.089E+08	1.981E+07	1.312E+07	7.909E+06	1.222E+06	2.537E+05	7.739E+04	2.956E+03	1.677E+02

Table A.40

Burner core heavy TRU elements nuclide radioactivity (core), Curies Basis = (EQULIBRIUM DISCHARGE FUEL - EQULIBRIUM FRESH FUEL) BURNER CORE

	Initial	1 hour	30 days	1 year	10 y	100 y	1000 y	10000 y	100000 y	1000000 y
Np-236	1.613E+03	1.564E+03	2.288E-07	0.000E+00	0.000E+00	0.000E+00	0.000E+00	0.000E+00	0.000E+00	0.000E+00
Np-237	-9.738E+00	-9.738E+00	-9.726E+00	-9.738E+00	-9.930E+00	-1.333E+01	-3.198E+01	-3.768E+01	-3.660E+01	-2.736E+01
Np-238	7.968E+06	7.854E+06	4.002E+02	0.000E+00	0.000E+00	0.000E+00	0.000E+00	0.000E+00	0.000E+00	0.000E+00
Np-239	1.622E+08	1.612E+08	2.230E+04	-1.224E+03	-1.224E+03	-1.212E+03	-1.122E+03	-4.950E+02	-1.362E-01	5.510E-03
Np-240m	2.939E+02	2.823E+02	6.642E-06	6.642E-06	6.642E-06	6.648E-06	6.744E-06	7.692E-06	1.625E-05	5.266E-05
Np-240	1.346E+05	6.960E+04	0.000E+00	0.000E+00	0.000E+00	0.000E+00	0.000E+00	0.000E+00	0.000E+00	0.000E+00
Pu-236	2.802E+02	2.802E+02	2.755E+02	2.203E+02	2.468E+01	7.675E-09	0.000E+00	0.000E+00	0.000E+00	0.000E+00
Pu-238	-6.780E+04	-6.780E+04	-6.120E+04	-3.120E+04	-1.980E+04	-1.140E+04	-4.740E+01	-1.350E-16	0.000E+00	0.000E+00
Pu-239	-7.938E+03	-7.938E+03	-7.890E+03	-7.896E+03	-7.884E+03	-7.872E+03	-7.704E+03	-6.138E+03	-4.932E+02	5.510E-03
Pu-240	-1.140E+04	-1.140E+04	-1.146E+04	-1.146E+04	-1.140E+04	-1.128E+04	-1.026E+04	-4.074E+03	-4.002E-01	5.272E-05
Pu-241	-2.688E+06	-2.688E+06	-2.682E+06	-2.568E+06	-1.677E+06	-2.334E+04	-2.340E+01	-1.086E+01	-5.700E-03	0.000E+00
Pu-242	-3.786E+01	-3.786E+01	-3.786E+01	-3.792E+01	-3.792E+01	-3.798E+01	-3.810E+01	-3.756E+01	-3.186E+01	-6.120E+00
Pu-243	6.510E+06	5.665E+06	5.747E-03	5.747E-03	5.747E-03	5.747E-03	5.747E-03	5.745E-03	5.723E-03	5.510E-03
Pu-244	6.648E-06	6.648E-06	6.648E-06	6.648E-06	6.648E-06	6.660E-06	6.756E-06	7.698E-06	1.628E-05	5.272E-05
Pu-245	3.052E+01	2.859E+01	0.000E+00	0.000E+00	0.000E+00	0.000E+00	0.000E+00	0.000E+00	0.000E+00	0.000E+00
Am-241	-4.764E+04	-4.764E+04	-4.800E+04	-5.178E+04	-8.088E+04	-1.198E+05	-2.856E+04	-1.086E+01	-6.000E-03	0.000E+00
Am-242m	-4.350E+03	-4.350E+03	-4.344E+03	-4.332E+03	-4.152E+03	-2.760E+03	-4.548E+01	-6.780E-17	0.000E+00	0.000E+00
Am-242	1.210E+07	1.158E+07	-4.344E+03	-4.332E+03	-4.158E+03	-2.754E+03	-4.548E+01	-6.780E-17	0.000E+00	0.000E+00
Am-243	-1.224E+03	-1.224E+03	-1.224E+03	-1.224E+03	-1.224E+03	-1.212E+03	-1.122E+03	-4.950E+02	-1.362E-01	5.510E-03
Am-244	8.256E+06	1.669E+06	8.646E-09	8.646E-09	8.646E-09	8.658E-09	8.784E-09	1.001E-08	2.116E-08	6.852E-08
Am-245	5.579E+01	4.831E+01	1.495E-03	7.134E-04	5.029E-07	0.000E+00	0.000E+00	0.000E+00	0.000E+00	0.000E+00
Cm-242	8.929E+06	8.930E+06	7.894E+06	1.895E+06	-3.408E+03	-2.262E+03	-3.726E+01	-5.580E-17	0.000E+00	0.000E+00
Cm-243	-7.140E+02	-7.140E+02	-7.140E+02	-6.960E+02	-5.760E+02	-8.160E+01	-2.784E-07	0.000E+00	0.000E+00	0.000E+00
Cm-244	2.760E+04	2.760E+04	2.760E+04	2.640E+04	1.860E+04	6.060E+02	8.783E-09	1.001E-08	2.116E-08	6.852E-08
Cm-245	-2.520E+01	-2.520E+01	-2.520E+01	-2.520E+01	-2.460E+01	-2.460E+01	-2.340E+01	-1.080E+01	-5.640E-03	0.000E+00
Cm-246	-4.200E+00	-4.200E+00	-4.200E+00	-4.200E+00	-3.600E+00	-4.200E+00	-3.000E+00	-9.600E-01	-1.680E-06	0.000E+00
Cm-247	5.747E-03	5.747E-03	5.747E-03	5.747E-03	5.747E-03	5.747E-03	5.747E-03	5.745E-03	5.723E-03	5.510E-03
Cm-248	1.383E-02	1.383E-02	1.383E-02	1.383E-02	1.383E-02	1.383E-02	1.381E-02	1.356E-02	1.136E-02	1.930E-03
Cm-249	2.413E+02	1.261E+02	1.523E-07	3.277E-13	0.000E+00	0.000E+00	0.000E+00	0.000E+00	0.000E+00	0.000E+00
Cm-250	5.434E-08	5.434E-08	5.434E-08	5.433E-08	5.431E-08	5.412E-08	5.221E-08	3.648E-08	1.011E-09	0.000E+00
Bk-249	1.064E+02	1.064E+02	9.966E+01	4.753E+01	3.352E-02	0.000E+00	0.000E+00	0.000E+00	0.000E+00	0.000E+00
Bk-250	5.954E+00	4.802E+00	5.434E-08	5.433E-08	5.431E-08	5.412E-08	5.221E-08	3.648E-08	1.011E-09	0.000E+00
Cf-249	1.312E-01	1.312E-01	1.478E-01	2.747E-01	3.841E-01	3.218E-01	5.468E-02	1.097E-09	0.000E+00	0.000E+00
Cf-250	3.864E-01	3.864E-01	3.849E-01	3.666E-01	2.275E-01	1.930E-03	5.221E-08	3.648E-08	1.011E-09	0.000E+00
Cf-251	1.322E-04	1.322E-04	1.322E-04	1.321E-04	1.312E-04	1.224E-04	6.120E-05	5.972E-08	0.000E+00	0.000E+00
Cf-252	8.580E-04	8.580E-04	8.400E-04	6.606E-04	6.246E-05	3.595E-15	0.000E+00	0.000E+00	0.000E+00	0.000E+00
Cf-253	1.574E-04	1.572E-04	4.900E-05	1.054E-10	0.000E+00	0.000E+00	0.000E+00	0.000E+00	0.000E+00	0.000E+00
Cf-254	3.379E-07	3.377E-07	2.396E-07	5.143E-09	0.000E+00	0.000E+00	0.000E+00	0.000E+00	0.000E+00	0.000E+00
Es-253	1.570E-04	1.570E-04	1.108E-04	4.905E-09	0.000E+00	0.000E+00	0.000E+00	0.000E+00	0.000E+00	0.000E+00
Total	2.033E+08	1.941E+08	5.124E+06	-7.606E+05	-1.793E+06	-1.835E+05	-4.906E+04	-1.131E+04	-5.623E+02	-3.345E+01
Total/MWd	4.165E+02	3.976E+02	1.049E+01	-1.558E+00	-3.673E+00	-3.758E-01	-1.005E-01	-2.317E-02	-1.152E-03	-6.851E-05

Table A.41

Converter core heavy TRU elements nuclide radioactivity (all regions), Curies Basis = EQULIBRIUM FRESH FUEL CONVERTER CORE

	Initial	**1 hour**	**30 days**	**1 year**	**10 y**	**100 y**	**1000 y**	**10000 y**	**100000 y**	**1000000 y**
Np-236	0.000E+00	0.000E+00	0.000E+00	0.000E+00	0.000E+00	0.000E+00	0.000E+00	0.000E+00	0.000E+00	0.000E+00
Np-237	2.354E+01	2.354E+01	2.355E+01	2.360E+01	2.428E+01	3.494E+01	9.300E+01	1.120E+02	1.100E+02	8.220E+01
Np-238	0.000E+00	0.000E+00	0.000E+00	0.000E+00	0.000E+00	0.000E+00	0.000E+00	0.000E+00	0.000E+00	0.000E+00
Np-239	0.000E+00	8.316E+01	6.810E+03	6.810E+03	6.804E+03	6.750E+03	6.222E+03	2.753E+03	7.902E-01	0.000E+00
Np-240m	0.000E+00	0.000E+00	0.000E+00	0.000E+00	0.000E+00	0.000E+00	0.000E+00	0.000E+00	0.000E+00	0.000E+00
Np-240	0.000E+00	0.000E+00	0.000E+00	0.000E+00	0.000E+00	0.000E+00	0.000E+00	0.000E+00	0.000E+00	0.000E+00
Pu-236	1.227E+02	1.227E+02	1.202E+02	9.618E+01	1.078E+01	3.352E-09	0.000E+00	0.000E+00	0.000E+00	0.000E+00
Pu-238	6.864E+05	6.864E+05	6.864E+05	6.828E+05	6.384E+05	3.268E+05	5.672E+02	9.828E-16	0.000E+00	0.000E+00
Pu-239	4.855E+04	4.855E+04	4.855E+04	4.855E+04	4.853E+04	4.843E+04	4.738E+04	3.763E+04	3.015E+03	2.369E-08
Pu-240	9.396E+04	9.396E+04	9.396E+04	9.408E+04	9.504E+04	9.654E+04	8.808E+04	3.501E+04	3.436E+00	0.000E+00
Pu-241	7.614E+06	7.614E+06	7.584E+06	7.260E+06	4.738E+06	6.672E+04	6.588E+02	3.096E+02	1.631E-01	0.000E+00
Pu-242	3.307E+02	3.307E+02	3.307E+02	3.307E+02	3.308E+02	3.315E+02	3.332E+02	3.322E+02	2.831E+02	5.458E+01
Pu-243	0.000E+00	0.000E+00	0.000E+00	0.000E+00	0.000E+00	0.000E+00	0.000E+00	0.000E+00	0.000E+00	0.000E+00
Pu-244	0.000E+00	0.000E+00	0.000E+00	0.000E+00	0.000E+00	0.000E+00	0.000E+00	0.000E+00	0.000E+00	0.000E+00
Pu-245	0.000E+00	0.000E+00	0.000E+00	0.000E+00	0.000E+00	0.000E+00	0.000E+00	0.000E+00	0.000E+00	0.000E+00
Am-241	1.748E+05	1.748E+05	1.758E+05	1.864E+05	2.683E+05	3.732E+05	8.940E+04	3.096E+02	1.721E-01	0.000E+00
Am-242m	3.173E+04	3.173E+04	3.172E+04	3.159E+04	3.032E+04	2.011E+04	3.318E+02	4.954E-16	0.000E+00	0.000E+00
Am-242	0.000E+00	1.345E+03	3.172E+04	3.159E+04	3.032E+04	2.011E+04	3.318E+02	4.954E-16	0.000E+00	0.000E+00
Am-243	6.810E+03	6.810E+03	6.810E+03	6.810E+03	6.804E+03	6.750E+03	6.222E+03	2.753E+03	7.902E-01	0.000E+00
Am-244	0.000E+00	0.000E+00	0.000E+00	0.000E+00	0.000E+00	0.000E+00	0.000E+00	0.000E+00	0.000E+00	0.000E+00
Am-245	0.000E+00	0.000E+00	0.000E+00	0.000E+00	0.000E+00	0.000E+00	0.000E+00	0.000E+00	0.000E+00	0.000E+00
Cm-242	3.323E+05	3.322E+05	2.955E+05	9.072E+04	2.486E+04	1.649E+04	2.721E+02	4.075E-16	0.000E+00	0.000E+00
Cm-243	5.846E+03	5.846E+03	5.836E+03	5.720E+03	4.707E+03	6.702E+02	2.284E-06	0.000E+00	0.000E+00	0.000E+00
Cm-244	1.368E+06	1.368E+06	1.364E+06	1.316E+06	9.330E+05	2.970E+04	3.184E-11	0.000E+00	0.000E+00	0.000E+00
Cm-245	7.152E+02	7.152E+02	7.152E+02	7.146E+02	7.146E+02	7.092E+02	6.576E+02	3.091E+02	1.628E-01	0.000E+00
Cm-246	5.527E+02	5.527E+02	5.527E+02	5.526E+02	5.519E+02	5.446E+02	4.771E+02	1.269E+02	2.249E-04	0.000E+00
Cm-247	0.000E+00	0.000E+00	0.000E+00	0.000E+00	0.000E+00	0.000E+00	0.000E+00	0.000E+00	0.000E+00	0.000E+00
Cm-248	0.000E+00	0.000E+00	0.000E+00	0.000E+00	0.000E+00	0.000E+00	0.000E+00	0.000E+00	0.000E+00	0.000E+00
Cm-249	0.000E+00	0.000E+00	0.000E+00	0.000E+00	0.000E+00	0.000E+00	0.000E+00	0.000E+00	0.000E+00	0.000E+00
Cm-250	0.000E+00	0.000E+00	0.000E+00	0.000E+00	0.000E+00	0.000E+00	0.000E+00	0.000E+00	0.000E+00	0.000E+00
Bk-249	0.000E+00	0.000E+00	0.000E+00	0.000E+00	0.000E+00	0.000E+00	0.000E+00	0.000E+00	0.000E+00	0.000E+00
Bk-250	0.000E+00	0.000E+00	0.000E+00	0.000E+00	0.000E+00	0.000E+00	0.000E+00	0.000E+00	0.000E+00	0.000E+00
Cf-249	0.000E+00	0.000E+00	0.000E+00	0.000E+00	0.000E+00	0.000E+00	0.000E+00	0.000E+00	0.000E+00	0.000E+00
Cf-250	0.000E+00	0.000E+00	0.000E+00	0.000E+00	0.000E+00	0.000E+00	0.000E+00	0.000E+00	0.000E+00	0.000E+00
Cf-251	0.000E+00	0.000E+00	0.000E+00	0.000E+00	0.000E+00	0.000E+00	0.000E+00	0.000E+00	0.000E+00	0.000E+00
Cf-252	0.000E+00	0.000E+00	0.000E+00	0.000E+00	0.000E+00	0.000E+00	0.000E+00	0.000E+00	0.000E+00	0.000E+00
Cf-253	0.000E+00	0.000E+00	0.000E+00	0.000E+00	0.000E+00	0.000E+00	0.000E+00	0.000E+00	0.000E+00	0.000E+00
Cf-254	0.000E+00	0.000E+00	0.000E+00	0.000E+00	0.000E+00	0.000E+00	0.000E+00	0.000E+00	0.000E+00	0.000E+00
Es-253	0.000E+00	0.000E+00	0.000E+00	0.000E+00	0.000E+00	0.000E+00	0.000E+00	0.000E+00	0.000E+00	0.000E+00
Total	1.036E+07	1.037E+07	1.033E+07	9.763E+06	6.826E+06	1.014E+06	2.410E+05	7.965E+04	3.414E+03	1.368E+02

Table A.42

Converter core heavy TRU elements nuclide radioactivity (all regions), Curies Basis = EQULIBRIUM DISCHARGE FUEL CONVERTER CORE

	Initial	1 hour	30 days	1 year	10 y	100 y	1000 y	10000 y	100000 y	1000000 y
Np-236	9.762E+02	9.462E+02	1.384E-07	0.000E+00	0.000E+00	0.000E+00	0.000E+00	0.000E+00	0.000E+00	0.000E+00
Np237	1.788E+01	1.788E+01	1.790E+01	1.795E+01	1.849E+01	2.718E+01	7.446E+01	9.024E+01	8.886E+01	6.636E+01
Np-238	5.485E+06	5.410E+06	2.756E+02	0.000E+00	0.000E+00	0.000E+00	0.000E+00	0.000E+00	0.000E+00	0.000E+00
Np-239	2.345E+08	2.330E+08	4.012E+04	6.102E+03	6.096E+03	6.048E+03	5.574E+03	2.466E+03	7.122E-01	3.983E-03
Np-240m	3.718E+02	3.570E+02	1.795E-06	1.795E-06	1.796E-06	1.802E-06	1.870E-06	2.537E-06	8.592E-06	3.433E-05
Np-240	1.703E+05	8.802E+04	0.000E+00	0.000E+00	0.000E+00	0.000E+00	0.000E+00	0.000E+00	0.000E+00	0.000E+00
Pu-236	2.926E+02	2.926E+02	2.873E+02	2.298E+02	2.575E+01	8.010E-09	0.000E+00	0.000E+00	0.000E+00	0.000E+00
Pu-238	6.444E+05	6.444E+05	6.486E+05	6.666E+05	6.300E+05	3.215E+05	5.360E+02	8.880E-16	0.000E+00	0.000E+00
Pu-239	4.393E+04	4.393E+04	4.399E+04	4.399E+04	4.399E+04	4.390E+04	4.293E+04	3.409E+04	2.731E+03	3.983E-03
Pu-240	8.736E+04	8.736E+04	8.736E+04	8.748E+04	8.844E+04	9.006E+04	8.220E+04	3.266E+04	3.206E+00	3.437E-05
Pu-241	6.180E+06	6.180E+06	6.156E+06	5.892E+06	3.844E+06	5.425E+04	6.450E+02	3.032E+02	1.598E-01	0.000E+00
Pu-242	3.088E+02	3.088E+02	3.088E+02	3.088E+02	3.089E+02	3.096E+02	3.112E+02	3.104E+02	2.647E+02	5.103E+01
Pu-243	4.508E+06	3.923E+06	4.155E-03	4.155E-03	4.155E-03	4.155E-03	4.155E-03	4.154E-03	4.138E-03	3.983E-03
Pu-244	1.798E-06	1.798E-06	1.798E-06	1.798E-06	1.798E-06	1.805E-06	1.873E-06	2.540E-06	8.598E-06	3.437E-05
Pu-245	2.149E+00	2.013E+00	0.000E+00	0.000E+00	0.000E+00	0.000E+00	0.000E+00	0.000E+00	0.000E+00	0.000E+00
Am-241	1.432E+05	1.432E+05	1.440E+05	1.526E+05	2.190E+05	3.040E+05	7.290E+04	3.033E+02	1.686E-01	0.000E+00
Am-242m	2.866E+04	2.866E+04	2.865E+04	2.853E+04	2.738E+04	1.816E+04	2.996E+02	4.474E-16	0.000E+00	0.000E+00
Am-242	8.730E+06	8.364E+06	2.865E+04	2.853E+04	2.738E+04	1.816E+04	2.996E+02	4.474E-16	0.000E+00	0.000E+00
Am-243	6.102E+03	6.102E+03	6.102E+03	6.102E+03	6.096E+03	6.048E+03	5.574E+03	2.466E+03	7.122E-01	3.983E-03
Am-244	5.459E+06	1.104E+06	2.337E-09	2.337E-09	2.338E-09	2.347E-09	2.434E-09	3.302E-09	1.118E-08	4.468E-08
Am-245	1.852E+01	1.384E+01	1.035E-03	4.937E-04	3.481E-07	0.000E+00	0.000E+00	0.000E+00	0.000E+00	0.000E+00
Cm-242	6.768E+06	6.768E+06	5.985E+06	1.456E+06	2.246E+04	1.490E+04	2.457E+02	3.680E-16	0.000E+00	0.000E+00
Cm-243	5.459E+03	5.459E+03	5.449E+03	5.342E+03	4.396E+03	6.258E+02	2.133E-06	0.000E+00	0.000E+00	0.000E+00
Cm-244	1.395E+06	1.395E+06	1.390E+06	1.342E+06	9.510E+05	3.028E+04	2.467E-09	3.302E-09	1.118E-08	4.468E-08
Cm-245	7.002E+02	7.002E+02	7.002E+02	7.002E+02	6.996E+02	6.942E+02	6.438E+02	3.027E+02	1.595E-01	0.000E+00
Cm-246	5.504E+02	5.504E+02	5.504E+02	5.503E+02	5.496E+02	5.424E+02	4.751E+02	1.264E+02	2.239E-04	0.000E+00
Cm-247	4.155E-03	4.155E-03	4.155E-03	4.155E-03	4.155E-03	4.155E-03	4.155E-03	4.154E-03	4.138E-03	3.983E-03
Cm-248	9.774E-03	9.774E-03	9.774E-03	9.774E-03	9.774E-03	9.768E-03	9.756E-03	9.582E-03	8.028E-03	1.364E-03
Cm-249	1.672E+02	8.730E+01	9.810E-08	2.111E-13	0.000E+00	0.000E+00	0.000E+00	0.000E+00	0.000E+00	0.000E+00
Cm-250	3.687E-08	3.687E-08	3.687E-08	3.687E-08	3.686E-08	3.673E-08	3.543E-08	2.476E-08	6.864E-10	0.000E+00
Bk-249	7.368E+01	7.368E+01	6.900E+01	3.290E+01	2.320E-02	0.000E+00	0.000E+00	0.000E+00	0.000E+00	0.000E+00
Bk-250	4.045E+00	3.262E+00	3.687E-08	3.687E-08	3.686E-08	3.673E-08	3.543E-08	2.476E-08	6.864E-10	0.000E+00
Cf-249	9.108E-02	9.108E-02	1.026E-01	1.904E-01	2.662E-01	2.230E-01	3.788E-02	7.602E-10	0.000E+00	0.000E+00
Cf-250	2.630E-01	2.630E-01	2.620E-01	2.495E-01	1.549E-01	1.313E-03	3.543E-08	2.476E-08	6.864E-10	0.000E+00
Cf-251	8.838E-05	8.838E-05	8.838E-05	8.832E-05	8.772E-05	8.184E-05	4.093E-05	3.994E-08	0.000E+00	0.000E+00
Cf-252	5.632E-04	5.632E-04	5.512E-04	4.334E-04	4.101E-05	2.359E-15	0.000E+00	0.000E+00	0.000E+00	0.000E+00
Cf-253	1.014E-04	1.012E-04	3.156E-05	6.792E-11	0.000E+00	0.000E+00	0.000E+00	0.000E+00	0.000E+00	0.000E+00
Cf-254	2.132E-07	2.131E-07	1.512E-07	3.246E-09	0.000E+00	0.000E+00	0.000E+00	0.000E+00	0.000E+00	0.000E+00
Es-253	1.011E-04	1.011E-04	7.140E-05	3.160E-09	0.000E+00	0.000E+00	0.000E+00	0.000E+00	0.000E+00	0.000E+00
Total	2.742E+08	2.672E+08	1.457E+07	9.718E+06	5.871E+06	9.094E+05	2.127E+05	7.312E+04	3.089E+03	1.174E+02

Table A.43

Converter Core Heavy TRU Elements Nuclide RadioActivity (All Regions), Curies

Basis = (EQULIBRIUM DISCHARGE FUEL - EQULIBRIUM FRESH FUEL) CONVERTER CORE

	Initial	1 hour	30 days	1 year	10 y	100 y	1000 y	10000 y	100000 y	1000000 y
Np-236	9.762E+02	9.462E+02	1.384E-07	0.000E+00	0.000E+00	0.000E+00	0.000E+00	0.000E+00	0.000E+00	0.000E+00
Np-237	-5.664E+00	-5.664E+00	-5.652E+00	-5.658E+00	-5.784E+00	-7.758E+00	-1.854E+01	-2.178E+01	-2.118E+01	-1.584E+01
Np-238	5.485E+06	5.410E+06	2.756E+02	0.000E+00	0.000E+00	0.000E+00	0.000E+00	0.000E+00	0.000E+00	0.000E+00
Np-239	2.345E+08	2.330E+08	3.331E+04	-7.080E+02	-7.080E+02	-7.020E+02	-6.480E+02	-2.868E+02	-7.800E-02	3.983E-03
Np-240m	3.718E+02	3.570E+02	1.795E-06	1.795E-06	1.796E-06	1.802E-06	1.870E-06	2.537E-06	8.592E-06	3.433E-05
Np-240	1.703E+05	8.802E+04	0.000E+00	0.000E+00	0.000E+00	0.000E+00	0.000E+00	0.000E+00	0.000E+00	0.000E+00
Pu-236	1.699E+02	1.699E+02	1.671E+02	1.336E+02	1.497E+01	4.658E-09	0.000E+00	0.000E+00	0.000E+00	0.000E+00
Pu-238	-4.200E+04	-4.200E+04	-3.780E+04	-1.620E+04	-8.400E+03	-5.340E+03	-3.114E+01	-9.480E-17	0.000E+00	0.000E+00
Pu-239	-4.614E+03	-4.614E+03	-4.554E+03	-4.554E+03	-4.548E+03	-4.536E+03	-4.446E+03	-3.540E+03	-2.844E+02	3.983E-03
Pu-240	-6.600E+03	-6.600E+03	-6.600E+03	-6.600E+03	-6.600E+03	-6.480E+03	-5.880E+03	-2.346E+03	-2.298E-01	3.437E-05
Pu-241	-1.434E+06	-1.434E+06	-1.428E+06	-1.368E+06	-8.940E+05	-1.247E+04	-1.380E+0	1-6.360E+00	-3.300E-03	0.000E+00
Pu-242	-2.184E+01	-2.184E+01	-2.184E+01	-2.184E+01	-2.184E+01	-2.190E+01	-2.202E+01	-2.172E+01	-1.836E+01	-3.546E+00
Pu-243	4.508E+06	3.923E+06	4.155E-03	4.155E-03	4.155E-03	4.155E-03	4.155E-03	4.154E-03	4.138E-03	3.983E-03
Pu-244	1.798E-06	1.798E-06	1.798E-06	1.798E-06	1.798E-06	1.805E-06	1.873E-06	2.540E-06	8.598E-06	3.437E-05
Pu-245	2.149E+00	2.013E+00	0.000E+00	0.000E+00	0.000E+00	0.000E+00	0.000E+00	0.000E+00	0.000E+00	0.000E+00
Am-241	-3.162E+04	-3.162E+04	-3.180E+04	-3.378E+04	-4.926E+04	-6.924E+04	-1.650E+04	-6.300E+00	-3.540E-03	0.000E+00
Am-242m	-3.072E+03	-3.072E+03	-3.072E+03	-3.060E+03	-2.934E+03	-1.950E+03	-3.216E+01	-4.800E-17	0.000E+00	0.000E+00
Am-242	8.730E+06	8.363E+06	-3.072E+03	-3.060E+03	-2.934E+03	-1.950E+03	-3.216E+01	-4.800E-17	0.000E+00	0.000E+00
Am-243	-7.080E+02	-7.080E+02	-7.080E+02	-7.080E+02	-7.080E+02	-7.020E+02	-6.480E+02	-2.868E+02	-7.800E-02	3.983E-03
Am-244	5.459E+06	1.104E+06	2.337E-09	2.337E-09	2.338E-09	2.347E-09	2.434E-09	3.302E-09	1.118E-08	4.468E-08
Am-245	1.852E+01	1.384E+01	1.035E-03	4.937E-04	3.481E-07	0.000E+00	0.000E+00	0.000E+00	0.000E+00	0.000E+00
Cm-242	6.436E+06	6.436E+06	5.690E+06	1.365E+06	-2.406E+03	-1.596E+03	-2.640E+01	-3.948E-17	0.000E+00	0.000E+00
Cm-243	-3.864E+02	-3.864E+02	-3.864E+02	-3.780E+02	-3.114E+02	-4.440E+01	-1.512E-07	0.000E+00	0.000E+00	0.000E+00
Cm-244	2.700E+04	2.700E+04	2.640E+04	2.580E+04	1.800E+04	5.820E+02	2.435E-09	3.302E-09	1.118E-08	4.468E-08
Cm-245	-1.500E+01	-1.500E+01	-1.500E+01	-1.440E+01	-1.500E+01	-1.500E+01	-1.380E+01	-6.360E+00	-3.360E-03	0.000E+00
Cm-246	-2.280E+00	-2.280E+00	-2.340E+00	-2.280E+00	-2.280E+00	-2.220E+00	-1.980E+00	-5.400E-01	-9.600E-07	0.000E+00
Cm-247	4.155E-03	4.155E-03	4.155E-03	4.155E-03	4.155E-03	4.155E-03	4.155E-03	4.154E-03	4.138E-03	3.983E-03
Cm-248	9.774E-03	9.774E-03	9.774E-03	9.774E-03	9.774E-03	9.768E-03	9.756E-03	9.582E-03	8.028E-03	1.364E-03
Cm-249	1.672E+02	8.730E+01	9.810E-08	2.111E-13	0.000E+00	0.000E+00	0.000E+00	0.000E+00	0.000E+00	0.000E+00
Cm-250	3.687E-08	3.687E-08	3.687E-08	3.687E-08	3.686E-08	3.673E-08	3.543E-08	2.476E-08	6.864E-10	0.000E+00
Bk-249	7.368E+01	7.368E+01	6.900E+01	3.290E+01	2.320E-02	0.000E+00	0.000E+00	0.000E+00	0.000E+00	0.000E+00
Bk-250	4.045E+00	3.262E+00	3.687E-08	3.687E-08	3.686E-08	3.673E-08	3.543E-08	2.476E-08	6.864E-10	0.000E+00
Cf-249	9.108E-02	9.108E-02	1.026E-01	1.904E-01	2.662E-01	2.230E-01	3.788E-02	7.602E-10	0.000E+00	0.000E+00
Cf-250	2.630E-01	2.630E-01	2.620E-01	2.495E-01	1.549E-01	1.313E-03	3.543E-08	2.476E-08	6.864E-10	0.000E+00
Cf-251	8.838E-05	8.838E-05	8.838E-05	8.832E-05	8.772E-05	8.184E-05	4.093E-05	3.994E-08	0.000E+00	0.000E+00
Cf-252	5.632E-04	5.632E-04	5.512E-04	4.334E-04	4.101E-05	2.359E-15	0.000E+00	0.000E+00	0.000E+00	0.000E+00
Cf-253	1.014E-04	1.012E-04	3.156E-05	6.792E-11	0.000E+00	0.000E+00	0.000E+00	0.000E+00	0.000E+00	0.000E+00
Cf-254	2.132E-07	2.131E-07	1.512E-07	3.246E-09	0.000E+00	0.000E+00	0.000E+00	0.000E+00	0.000E+00	0.000E+00
Es-253	1.011E-04	1.011E-04	7.140E-05	3.160E-09	0.000E+00	0.000E+00	0.000E+00	0.000E+00	0.000E+00	0.000E+00
Total	2.638E+08	2.569E+08	4.234E+06	-4.564E+04	-9.548E+05	-1.045E+05	-2.831E+04	-6.523E+03	-3.243E+02	-1.936E+01
Total/MWd	5.404E+02	5.261E+02	8.671E+00	-9.349E-02	-1.956E+00	-2.140E-01	-5.799E-02	-1.336E-02	-6.642E-04	-3.966E-05

Table A.44

Breeder core heavy TRU elements nuclide radioactivity (all regions), Curies Basis = EQULIBRIUM FRESH FUEL BREEDER CORE

	Initial	1 hour	30 days	1 year	10 y	100 y	1000 y	10000 y	100000 y	1000000 y
Np-236	0.000E+00	0.000E+00	0.000E+00	0.000E+00	0.000E+00	0.000E+00	0.000E+00	0.000E+00	0.000E+00	0.000E+00
Np-237	5.446E+00	5.446E+00	5.447E+00	5.467E+00	5.705E+00	9.618E+00	3.098E+01	3.766E+01	3.672E+01	2.744E+01
Np-238	0.000E+00	0.000E+00	0.000E+00	0.000E+00	0.000E+00	0.000E+00	0.000E+00	0.000E+00	0.000E+00	0.000E+00
Np-239	0.000E+00	1.072E+01	8.778E+02	8.778E+02	8.772E+02	8.700E+02	8.016E+02	3.547E+02	1.019E-01	0.000E+00
Np-240m	0.000E+00	0.000E+00	0.000E+00	0.000E+00	0.000E+00	0.000E+00	0.000E+00	0.000E+00	0.000E+00	0.000E+00
Np-240	0.000E+00	0.000E+00	0.000E+00	0.000E+00	0.000E+00	0.000E+00	0.000E+00	0.000E+00	0.000E+00	0.000E+00
Pu-236	2.896E+01	2.896E+01	2.839E+01	2.271E+01	2.543E+00	7.914E-10	0.000E+00	0.000E+00	0.000E+00	0.000E+00
Pu-238	1.958E+05	1.958E+05	1.958E+05	1.948E+05	1.823E+05	9.396E+04	1.796E+02	3.427E-16	0.000E+00	0.000E+00
Pu-239	5.646E+04	5.646E+04	5.646E+04	5.646E+04	5.645E+04	5.630E+04	5.491E+04	4.264E+04	3.319E+03	2.608E-08
Pu-240	6.456E+04	6.456E+04	6.456E+04	6.456E+04	6.462E+04	6.432E+04	5.866E+04	2.330E+04	2.287E+00	0.000E+00
Pu-241	2.921E+06	2.921E+06	2.910E+06	2.786E+06	1.817E+06	2.541E+04	7.638E+01	3.590E+01	1.892E-02	0.000E+00
Pu-242	7.188E+01	7.188E+01	7.188E+01	7.188E+01	7.188E+01	7.212E+01	7.260E+01	7.188E+01	6.114E+01	1.178E+01
Pu-243	0.000E+00	0.000E+00	0.000E+00	0.000E+00	0.000E+00	0.000E+00	0.000E+00	0.000E+00	0.000E+00	0.000E+00
Pu-244	0.000E+00	0.000E+00	0.000E+00	0.000E+00	0.000E+00	0.000E+00	0.000E+00	0.000E+00	0.000E+00	0.000E+00
Pu-245	0.000E+00	0.000E+00	0.000E+00	0.000E+00	0.000E+00	0.000E+00	0.000E+00	0.000E+00	0.000E+00	0.000E+00
Am-241	6.036E+04	6.036E+04	6.078E+04	6.486E+04	9.636E+04	1.375E+05	3.280E+04	3.592E+01	1.996E-02	0.000E+00
Am-242m	1.106E+04	1.106E+04	1.106E+04	1.101E+04	1.057E+04	7.008E+03	1.157E+02	1.727E-16	0.000E+00	0.000E+00
Am-242	0.000E+00	4.687E+02	1.106E+04	1.101E+04	1.057E+04	7.008E+03	1.157E+02	1.727E-16	0.000E+00	0.000E+00
Am-243	8.778E+02	8.778E+02	8.778E+02	8.778E+02	8.772E+02	8.700E+02	8.016E+02	3.547E+02	1.019E-01	0.000E+00
Am-244	0.000E+00	0.000E+00	0.000E+00	0.000E+00	0.000E+00	0.000E+00	0.000E+00	0.000E+00	0.000E+00	0.000E+00
Am-245	0.000E+00	0.000E+00	0.000E+00	0.000E+00	0.000E+00	0.000E+00	0.000E+00	0.000E+00	0.000E+00	0.000E+00
Cm-242	1.174E+05	1.174E+05	1.044E+05	3.194E+04	8.664E+03	5.749E+03	9.486E+01	1.420E-16	0.000E+00	0.000E+00
Cm-243	1.630E+03	1.630E+03	1.628E+03	1.595E+03	1.313E+03	1.868E+02	6.372E-07	0.000E+00	0.000E+00	0.000E+00
Cm-244	1.618E+05	1.618E+05	1.613E+05	1.557E+05	1.103E+05	3.512E+03	3.765E-12	0.000E+00	0.000E+00	0.000E+00
Cm-245	8.292E+01	8.292E+01	8.292E+01	8.292E+01	8.286E+01	8.220E+01	7.626E+01	3.584E+01	1.888E-02	0.000E+00
Cm-246	6.000E+01	6.000E+01	6.000E+01	6.000E+01	5.994E+01	5.915E+01	5.182E+01	1.378E+01	2.442E-05	0.000E+00
Cm-247	0.000E+00	0.000E+00	0.000E+00	0.000E+00	0.000E+00	0.000E+00	0.000E+00	0.000E+00	0.000E+00	0.000E+00
Cm-248	0.000E+00	0.000E+00	0.000E+00	0.000E+00	0.000E+00	0.000E+00	0.000E+00	0.000E+00	0.000E+00	0.000E+00
Cm-249	0.000E+00	0.000E+00	0.000E+00	0.000E+00	0.000E+00	0.000E+00	0.000E+00	0.000E+00	0.000E+00	0.000E+00
Cm-250	0.000E+00	0.000E+00	0.000E+00	0.000E+00	0.000E+00	0.000E+00	0.000E+00	0.000E+00	0.000E+00	0.000E+00
Bk-249	0.000E+00	0.000E+00	0.000E+00	0.000E+00	0.000E+00	0.000E+00	0.000E+00	0.000E+00	0.000E+00	0.000E+00
Bk-250	0.000E+00	0.000E+00	0.000E+00	0.000E+00	0.000E+00	0.000E+00	0.000E+00	0.000E+00	0.000E+00	0.000E+00
Cf-249	0.000E+00	0.000E+00	0.000E+00	0.000E+00	0.000E+00	0.000E+00	0.000E+00	0.000E+00	0.000E+00	0.000E+00
Cf-250	0.000E+00	0.000E+00	0.000E+00	0.000E+00	0.000E+00	0.000E+00	0.000E+00	0.000E+00	0.000E+00	0.000E+00
Cf-251	0.000E+00	0.000E+00	0.000E+00	0.000E+00	0.000E+00	0.000E+00	0.000E+00	0.000E+00	0.000E+00	0.000E+00
Cf-252	0.000E+00	0.000E+00	0.000E+00	0.000E+00	0.000E+00	0.000E+00	0.000E+00	0.000E+00	0.000E+00	0.000E+00
Cf-253	0.000E+00	0.000E+00	0.000E+00	0.000E+00	0.000E+00	0.000E+00	0.000E+00	0.000E+00	0.000E+00	0.000E+00
Cf-254	0.000E+00	0.000E+00	0.000E+00	0.000E+00	0.000E+00	0.000E+00	0.000E+00	0.000E+00	0.000E+00	0.000E+00
Es-253	0.000E+00	0.000E+00	0.000E+00	0.000E+00	0.000E+00	0.000E+00	0.000E+00	0.000E+00	0.000E+00	0.000E+00
Total	3.592E+06	3.592E+06	3.579E+06	3.380E+06	2.360E+06	4.029E+05	1.488E+05	6.688E+04	3.419E+03	3.922E+01

Table A.45

Breeder core heavy TRU elements nuclide radioactivity (all regions), Curies Basis = EQULIBRIUM DISCHARGE FUEL BREEDER CORE

	Initial	1 hour	30 days	1 year	10 y	100 y	1000 y	10000 y	100000 y	1000000 y
Np-236	2.751E+02	2.666E+02	3.901E-08	0.000E+00	0.000E+00	0.000E+00	0.000E+00	0.000E+00	0.000E+00	0.000E+00
Np-237	5.440E+00	5.440E+00	5.460E+00	5.477E+00	5.699E+00	9.618E+00	3.124E+01	3.800E+01	3.705E+01	2.768E+01
Np-238	1.650E+06	1.627E+06	8.292E+01	0.000E+00	0.000E+00	0.000E+00	0.000E+00	0.000E+00	0.000E+00	0.000E+00
Np-239	3.436E+08	3.414E+08	5.072E+04	8.850E+02	8.844E+02	8.772E+02	8.082E+02	3.576E+02	1.031E-01	4.352E-04
Np-240m	4.293E+02	4.123E+02	4.045E-07	4.045E-07	4.046E-07	4.053E-07	4.125E-07	4.840E-07	1.132E-06	3.889E-06
Np-240	1.967E+05	1.016E+05	0.000E+00	0.000E+00	0.000E+00	0.000E+00	0.000E+00	0.000E+00	0.000E+00	0.000E+00
Pu-236	7.464E+01	7.464E+01	7.332E+01	5.864E+01	6.570E+00	2.043E-09	0.000E+00	0.000E+00	0.000E+00	0.000E+00
Pu-238	1.904E+05	1.904E+05	1.919E+05	1.987E+05	1.882E+05	9.660E+04	1.741E+02	3.137E-16	0.000E+00	0.000E+00
Pu-239	5.681E+04	5.681E+04	5.690E+04	5.690E+04	5.689E+04	5.675E+04	5.533E+04	4.297E+04	3.345E+03	4.353E-04
Pu-240	6.510E+04	6.510E+04	6.510E+04	6.510E+04	6.516E+04	6.486E+04	5.918E+04	2.351E+04	2.308E+00	3.893E-06
Pu-241	3.204E+06	3.204E+06	3.191E+06	3.055E+06	1.993E+06	2.785E+04	7.680E+01	3.610E+01	1.902E-02	0.000E+00
Pu-242	7.260E+01	7.260E+01	7.260E+01	7.260E+01	7.266E+01	7.284E+01	7.332E+01	7.260E+01	6.168E+01	1.190E+01
Pu-243	1.060E+06	9.228E+05	4.540E-04	4.540E-04	4.540E-04	4.540E-04	4.540E-04	4.538E-04	4.521E-04	4.352E-04
Pu-244	4.051E-07	4.051E-07	4.051E-07	4.051E-07	4.051E-07	4.058E-07	4.130E-07	4.846E-07	1.134E-06	3.893E-06
Pu-245	4.849E-01	4.543E-01	0.000E+00	0.000E+00	0.000E+00	0.000E+00	0.000E+00	0.000E+00	0.000E+00	0.000E+00
Am-241	5.245E+04	5.245E+04	5.287E+04	5.738E+04	9.210E+04	1.390E+05	3.319E+04	3.612E+01	2.007E-02	0.000E+00
Am-242m	1.013E+04	1.013E+04	1.012E+04	1.008E+04	9.672E+03	6.420E+03	1.058E+02	1.581E-16	0.000E+00	0.000E+00
Am-242	3.191E+06	3.056E+06	1.012E+04	1.008E+04	9.672E+03	6.420E+03	1.058E+02	1.581E-16	0.000E+00	0.000E+00
Am-243	8.850E+02	8.850E+02	8.850E+02	8.850E+02	8.844E+02	8.772E+02	8.082E+02	3.576E+02	1.031E-01	4.352E-04
Am-244	7.932E+05	1.604E+05	5.266E-10	5.266E-10	5.266E-10	5.276E-10	5.370E-10	6.300E-10	1.474E-09	5.062E-09
Am-245	2.832E+00	2.160E+00	1.083E-04	5.167E-05	3.643E-08	0.000E+00	0.000E+00	0.000E+00	0.000E+00	0.000E+00
Cm-242	2.434E+06	2.434E+06	2.153E+06	5.237E+05	7.932E+03	5.262E+03	8.682E+01	1.300E-16	0.000E+00	0.000E+00
Cm-243	1.683E+03	1.683E+03	1.680E+03	1.647E+03	1.355E+03	1.928E+02	6.576E-07	0.000E+00	0.000E+00	0.000E+00
Cm-244	1.744E+05	1.744E+05	1.738E+05	1.678E+05	1.189E+05	3.787E+03	5.410E-10	6.300E-10	1.474E-09	5.062E-09
Cm-245	8.340E+01	8.340E+01	8.340E+01	8.334E+01	8.328E+01	8.268E+01	7.668E+01	3.604E+01	1.899E-02	0.000E+00
Cm-246	6.048E+01	6.048E+01	6.048E+01	6.048E+01	6.042E+01	5.962E+01	5.222E+01	1.389E+01	2.461E-05	0.000E+00
Cm-247	4.540E-04	4.540E-04	4.540E-04	4.540E-04	4.540E-04	4.540E-04	4.540E-04	4.538E-04	4.521E-04	4.352E-04
Cm-248	1.047E-03	1.047E-03	1.047E-03	1.047E-03	1.047E-03	1.047E-03	1.045E-03	1.027E-03	8.598E-04	1.461E-04
Cm-249	1.752E+01	9.150E+00	9.162E-09	1.971E-14	0.000E+00	0.000E+00	0.000E+00	0.000E+00	0.000E+00	0.000E+00
Cm-250	3.752E-09	3.752E-09	3.752E-09	3.752E-09	3.751E-09	3.737E-09	3.606E-09	2.519E-09	6.984E-11	0.000E+00
Bk-249	7.710E+00	7.710E+00	7.218E+00	3.444E+00	2.428E-03	0.000E+00	0.000E+00	0.000E+00	0.000E+00	0.000E+00
Bk-250	4.124E-01	3.326E-01	3.752E-09	3.752E-09	3.751E-09	3.737E-09	3.606E-09	2.519E-09	6.984E-11	0.000E+00
Cf-249	9.558E-03	9.558E-03	1.076E-02	1.996E-02	2.788E-02	2.336E-02	3.969E-03	7.962E-11	0.000E+00	0.000E+00
Cf-250	2.686E-02	2.686E-02	2.675E-02	2.548E-02	1.582E-02	1.341E-04	3.606E-09	2.519E-09	6.984E-11	0.000E+00
Cf-251	8.784E-06	8.784E-06	8.784E-06	8.778E-06	8.718E-06	8.136E-06	4.067E-06	3.968E-09	0.000E+00	0.000E+00
Cf-252	5.428E-05	5.428E-05	5.312E-05	4.177E-05	3.952E-06	2.273E-16	0.000E+00	0.000E+00	0.000E+00	0.000E+00
Cf-253	9.468E-06	9.456E-06	2.948E-06	6.342E-12	0.000E+00	0.000E+00	0.000E+00	0.000E+00	0.000E+00	0.000E+00
Cf-254	1.924E-08	1.923E-08	1.364E-08	2.929E-10	0.000E+00	0.000E+00	0.000E+00	0.000E+00	0.000E+00	0.000E+00
Es-253	9.444E-06	9.444E-06	6.666E-06	2.951E-10	0.000E+00	0.000E+00	0.000E+00	0.000E+00	0.000E+00	0.000E+00
Total	3.567E+08	3.535E+08	5.959E+06	4.149E+06	2.545E+06	4.091E+05	1.501E+05	6.743E+04	3.446E+03	3.958E+01

Table A.46

Breeder core heavy TRU elements nuclide radioactivity (all regions), Curies

Basis = (EQULIBRIUM DISCHARGE FUEL - EQULIBRIUM FRESH FUEL) BREEDER CORE

	Initial	1 hour	30 days	1 year	10 y	100 y	1000 y	10000 y	100000 y	1000000 y
Np-236	2.751E+02	2.666E+02	3.901E-08	0.000E+00	0.000E+00	0.000E+00	0.000E+00	0.000E+00	0.000E+00	0.000E+00
Np-237	-6.600E-03	-6.600E-03	1.260E-02	1.020E-02	-6.000E-03	0.000E+00	2.580E-01	3.360E-01	3.300E-01	2.400E-01
Np-238	1.650E+06	1.627E+06	8.292E+01	0.000E+00	0.000E+00	0.000E+00	0.000E+00	0.000E+00	0.000E+00	0.000E+00
Np-239	3.436E+08	3.414E+08	4.984E+04	7.200E+00	7.200E+00	7.200E+00	6.600E+00	2.880E+00	1.260E-03	4.352E-04
Np-240m	4.293E+02	4.123E+02	4.045E-07	4.045E-07	4.046E-07	4.053E-07	4.125E-07	4.840E-07	1.132E-06	3.889E-06
Np-240	1.967E+05	1.016E+05	0.000E+00	0.000E+00	0.000E+00	0.000E+00	0.000E+00	0.000E+00	0.000E+00	0.000E+00
Pu-236	4.568E+01	4.568E+01	4.493E+01	3.593E+01	4.027E+00	1.252E-09	0.000E+00	0.000E+00	0.000E+00	0.000E+00
Pu-238	-5.340E+03	-5.340E+03	-3.900E+03	3.960E+03	5.940E+03	2.640E+03	-5.580E+00	-2.898E-17	0.000E+00	0.000E+00
Pu-239	3.540E+02	3.540E+02	4.440E+02	4.380E+02	4.380E+02	4.440E+02	4.260E+02	3.360E+02	2.640E+01	4.353E-04
Pu-240	5.400E+02	5.400E+02	5.400E+02	5.400E+02	5.400E+02	5.400E+02	5.220E+02	2.100E+02	2.040E-02	3.893E-06
Pu-241	2.826E+05	2.826E+05	2.814E+05	2.688E+05	1.758E+05	2.442E+03	4.200E-01	1.980E-01	1.020E-04	0.000E+00
Pu-242	7.200E-01	7.200E-01	7.200E-01	7.200E-01	7.800E-01	7.200E-01	7.200E-01	7.200E-01	5.400E-01	1.140E-01
Pu-243	1.060E+06	9.228E+05	4.540E-04	4.540E-04	4.540E-04	4.540E-04	4.540E-04	4.538E-04	4.521E-04	4.352E-04
Pu-244	4.051E-07	4.051E-07	4.051E-07	4.051E-07	4.051E-07	4.058E-07	4.130E-07	4.846E-07	1.134E-06	3.893E-06
Pu-245	4.849E-01	4.543E-01	0.000E+00	0.000E+00	0.000E+00	0.000E+00	0.000E+00	0.000E+00	0.000E+00	0.000E+00
Am-241	-7.908E+03	-7.908E+03	-7.914E+03	-7.482E+03	-4.260E+03	1.560E+03	3.900E+02	1.980E-01	1.080E-04	0.000E+00
Am-242m	-9.360E+02	-9.360E+02	-9.360E+02	-9.300E+02	-8.940E+02	-5.880E+02	-9.840E+00	-1.458E-17	0.000E+00	0.000E+00
Am-242	3.191E+06	3.055E+06	-9.360E+02	-9.300E+02	-8.940E+02	-5.880E+02	-9.840E+00	-1.458E-17	0.000E+00	0.000E+00
Am-243	7.200E+00	7.200E+00	7.200E+00	7.200E+00	7.200E+00	7.200E+00	6.600E+00	2.880E+00	1.260E-03	4.352E-04
Am-244	7.932E+05	1.604E+05	5.266E-10	5.266E-10	5.266E-10	5.276E-10	5.370E-10	6.300E-10	1.474E-09	5.062E-09
Am-245	2.832E+00	2.160E+00	1.083E-04	5.167E-05	3.643E-08	0.000E+00	0.000E+00	0.000E+00	0.000E+00	0.000E+00
Cm-242	2.317E+06	2.317E+06	2.049E+06	4.918E+05	-7.320E+02	-4.866E+02	-8.040E+00	-1.200E-17	0.000E+00	0.000E+00
Cm-243	5.280E+01	5.280E+01	5.220E+01	5.160E+01	4.200E+01	6.000E+00	2.040E-08	0.000E+00	0.000E+00	0.000E+00
Cm-244	1.266E+04	1.266E+04	1.254E+04	1.212E+04	8.640E+03	2.748E+02	5.373E-10	6.300E-10	1.474E-09	5.062E-09
Cm-245	4.800E-01	4.800E-01	4.800E-01	4.200E-01	4.200E-01	4.800E-01	4.200E-01	1.980E-01	1.080E-04	0.000E+00
Cm-246	4.800E-01	4.800E-01	4.800E-01	4.800E-01	4.800E-01	4.620E-01	4.020E-01	1.080E-01	1.920E-07	0.000E+00
Cm-247	4.540E-04	4.540E-04	4.540E-04	4.540E-04	4.540E-04	4.540E-04	4.540E-04	4.538E-04	4.521E-04	4.352E-04
Cm-248	1.047E-03	1.047E-03	1.047E-03	1.047E-03	1.047E-03	1.047E-03	1.045E-03	1.027E-03	8.598E-04	1.461E-04
Cm-249	1.752E+01	9.150E+00	9.162E-09	1.971E-14	0.000E+00	0.000E+00	0.000E+00	0.000E+00	0.000E+00	0.000E+00
Cm-250	3.752E-09	3.752E-09	3.752E-09	3.752E-09	3.751E-09	3.737E-09	3.606E-09	2.519E-09	6.984E-11	0.000E+00
Bk-249	7.710E+00	7.710E+00	7.218E+00	3.444E+00	2.428E-03	0.000E+00	0.000E+00	0.000E+00	0.000E+00	0.000E+00
Bk-250	4.124E-01	3.326E-01	3.752E-09	3.752E-09	3.751E-09	3.737E-09	3.606E-09	2.519E-09	6.984E-11	0.000E+00
Cf-249	9.558E-03	9.558E-03	1.076E-02	1.996E-02	2.788E-02	2.336E-02	3.969E-03	7.962E-11	0.000E+00	0.000E+00
Cf-250	2.686E-02	2.686E-02	2.675E-02	2.548E-02	1.582E-02	1.341E-04	3.606E-09	2.519E-09	6.984E-11	0.000E+00
Cf-251	8.784E-06	8.784E-06	8.784E-06	8.778E-06	8.718E-06	8.136E-06	4.067E-06	3.968E-09	0.000E+00	0.000E+00
Cf-252	5.428E-05	5.428E-05	5.312E-05	4.177E-05	3.952E-06	2.273E-16	0.000E+00	0.000E+00	0.000E+00	0.000E+00
Cf-253	9.468E-06	9.456E-06	2.948E-06	6.342E-12	0.000E+00	0.000E+00	0.000E+00	0.000E+00	0.000E+00	0.000E+00
Cf-254	1.924E-08	1.923E-08	1.364E-08	2.929E-10	0.000E+00	0.000E+00	0.000E+00	0.000E+00	0.000E+00	0.000E+00
Es-253	9.444E-06	9.444E-06	6.666E-06	2.951E-10	0.000E+00	0.000E+00	0.000E+00	0.000E+00	0.000E+00	0.000E+00
Total	3.531E+08	3.499E+08	2.380E+06	7.684E+05	1.846E+05	6.260E+03	1.320E+03	5.535E+02	2.730E+01	3.563E-01
Total/MWd	7.232E+02	7.166E+02	4.875E+00	1.574E+00	3.782E-01	1.282E-02	2.704E-03	1.134E-03	5.591E-05	7.298E-07

Table A.47

Converter core heavy TRU elements nuclide radioactivity (core), Curies Basis = EQULIBRIUM FRESH FUEL CONVERTER CORE

	Initial	1 hour	30 days	1 year	10 y	100 y	1000 y	10000 y	100000 y	1000000 y
Np-236	0.000E+00	0.000E+00	0.000E+00	0.000E+00	0.000E+00	0.000E+00	0.000E+00	0.000E+00	0.000E+00	0.000E+00
Np-237	2.354E+01	2.354E+01	2.355E+01	2.360E+01	2.428E+01	3.494E+01	9.300E+01	1.120E+02	1.100E+02	8.220E+01
Np-238	0.000E+00	0.000E+00	0.000E+00	0.000E+00	0.000E+00	0.000E+00	0.000E+00	0.000E+00	0.000E+00	0.000E+00
Np-239	0.000E+00	8.316E+01	6.810E+03	6.810E+03	6.804E+03	6.750E+03	6.222E+03	2.753E+03	7.902E-01	0.000E+00
Np-240m	0.000E+00	0.000E+00	0.000E+00	0.000E+00	0.000E+00	0.000E+00	0.000E+00	0.000E+00	0.000E+00	0.000E+00
Np-240	0.000E+00	0.000E+00	0.000E+00	0.000E+00	0.000E+00	0.000E+00	0.000E+00	0.000E+00	0.000E+00	0.000E+00
Pu-236	1.227E+02	1.227E+02	1.202E+02	9.618E+01	1.078E+01	3.352E-09	0.000E+00	0.000E+00	0.000E+00	0.000E+00
Pu-238	6.864E+05	6.864E+05	6.864E+05	6.828E+05	6.384E+05	3.268E+05	5.672E+02	9.828E-16	0.000E+00	0.000E+00
Pu-239	4.855E+04	4.855E+04	4.855E+04	4.855E+04	4.853E+04	4.843E+04	4.738E+04	3.763E+04	3.015E+03	2.369E-08
Pu-240	9.396E+04	9.396E+04	9.396E+04	9.408E+04	9.504E+04	9.654E+04	8.808E+04	3.501E+04	3.436E+00	0.000E+00
Pu-241	7.614E+06	7.614E+06	7.584E+06	7.260E+06	4.738E+06	6.672E+04	6.588E+02	3.096E+02	1.631E-01	0.000E+00
Pu-242	3.307E+02	3.307E+02	3.307E+02	3.307E+02	3.308E+02	3.315E+02	3.332E+02	3.322E+02	2.831E+02	5.458E+01
Pu-243	0.000E+00	0.000E+00	0.000E+00	0.000E+00	0.000E+00	0.000E+00	0.000E+00	0.000E+00	0.000E+00	0.000E+00
Pu-244	0.000E+00	0.000E+00	0.000E+00	0.000E+00	0.000E+00	0.000E+00	0.000E+00	0.000E+00	0.000E+00	0.000E+00
Pu-245	0.000E+00	0.000E+00	0.000E+00	0.000E+00	0.000E+00	0.000E+00	0.000E+00	0.000E+00	0.000E+00	0.000E+00
Am241	1.748E+05	1.748E+05	1.758E+05	1.864E+05	2.683E+05	3.732E+05	8.940E+04	3.096E+02	1.721E-01	0.000E+00
Am-242m	3.173E+04	3.173E+04	3.172E+04	3.159E+04	3.032E+04	2.011E+04	3.318E+02	4.954E-16	0.000E+00	0.000E+00
Am-242	0.000E+00	1.345E+03	3.172E+04	3.159E+04	3.032E+04	2.011E+04	3.318E+02	4.954E-16	0.000E+00	0.000E+00
Am-243	6.810E+03	6.810E+03	6.810E+03	6.810E+03	6.804E+03	6.750E+03	6.222E+03	2.753E+03	7.902E-01	0.000E+00
Am-244	0.000E+00	0.000E+00	0.000E+00	0.000E+00	0.000E+00	0.000E+00	0.000E+00	0.000E+00	0.000E+00	0.000E+00
Am-245	0.000E+00	0.000E+00	0.000E+00	0.000E+00	0.000E+00	0.000E+00	0.000E+00	0.000E+00	0.000E+00	0.000E+00
Cm-242	3.323E+05	3.322E+05	2.955E+05	9.072E+04	2.486E+04	1.649E+04	2.721E+02	4.075E-16	0.000E+00	0.000E+00
Cm-243	5.846E+03	5.846E+03	5.836E+03	5.720E+03	4.707E+03	6.702E+02	2.284E-06	0.000E+00	0.000E+00	0.000E+00
Cm-244	1.368E+06	1.368E+06	1.364E+06	1.316E+06	9.330E+05	2.970E+04	3.184E-11	0.000E+00	0.000E+00	0.000E+00
Cm-245	7.152E+02	7.152E+02	7.152E+02	7.146E+02	7.146E+02	7.092E+02	6.576E+02	3.091E+02	1.628E-01	0.000E+00
Cm-246	5.527E+02	5.527E+02	5.527E+02	5.526E+02	5.519E+02	5.446E+02	4.771E+02	1.269E+02	2.249E-04	0.000E+00
Cm-247	0.000E+00	0.000E+00	0.000E+00	0.000E+00	0.000E+00	0.000E+00	0.000E+00	0.000E+00	0.000E+00	0.000E+00
Cm-248	0.000E+00	0.000E+00	0.000E+00	0.000E+00	0.000E+00	0.000E+00	0.000E+00	0.000E+00	0.000E+00	0.000E+00
Cm-249	0.000E+00	0.000E+00	0.000E+00	0.000E+00	0.000E+00	0.000E+00	0.000E+00	0.000E+00	0.000E+00	0.000E+00
Cm-250	0.000E+00	0.000E+00	0.000E+00	0.000E+00	0.000E+00	0.000E+00	0.000E+00	0.000E+00	0.000E+00	0.000E+00
Bk-249	0.000E+00	0.000E+00	0.000E+00	0.000E+00	0.000E+00	0.000E+00	0.000E+00	0.000E+00	0.000E+00	0.000E+00
Bk-250	0.000E+00	0.000E+00	0.000E+00	0.000E+00	0.000E+00	0.000E+00	0.000E+00	0.000E+00	0.000E+00	0.000E+00
Cf-249	0.000E+00	0.000E+00	0.000E+00	0.000E+00	0.000E+00	0.000E+00	0.000E+00	0.000E+00	0.000E+00	0.000E+00
Cf-250	0.000E+00	0.000E+00	0.000E+00	0.000E+00	0.000E+00	0.000E+00	0.000E+00	0.000E+00	0.000E+00	0.000E+00
Cf-251	0.000E+00	0.000E+00	0.000E+00	0.000E+00	0.000E+00	0.000E+00	0.000E+00	0.000E+00	0.000E+00	0.000E+00
Cf-252	0.000E+00	0.000E+00	0.000E+00	0.000E+00	0.000E+00	0.000E+00	0.000E+00	0.000E+00	0.000E+00	0.000E+00
Cf-253	0.000E+00	0.000E+00	0.000E+00	0.000E+00	0.000E+00	0.000E+00	0.000E+00	0.000E+00	0.000E+00	0.000E+00
Cf-254	0.000E+00	0.000E+00	0.000E+00	0.000E+00	0.000E+00	0.000E+00	0.000E+00	0.000E+00	0.000E+00	0.000E+00
Es-253	0.000E+00	0.000E+00	0.000E+00	0.000E+00	0.000E+00	0.000E+00	0.000E+00	0.000E+00	0.000E+00	0.000E+00
Total	1.036E+07	1.037E+07	1.033E+07	9.763E+06	6.826E+06	1.014E+06	2.410E+05	7.965E+04	3.414E+03	1.368E+02

Table A.48

Converter core heavy TRU elements nuclide radioactivity (core), Curies Basis = EQULIBRIUM DISCHARGE FUEL CONVERTER CORE

	Initial	1 hour	30 days	1 year	10 y	100 y	1000 y	10000 y	100000 y	1000000 y
Np-236	9.732E+02	9.426E+02	1.379E-07	0.000E+00	0.000E+00	0.000E+00	0.000E+00	0.000E+00	0.000E+00	0.000E+00
Np-237	1.769E+01	1.769E+01	1.771E+01	1.775E+01	1.830E+01	2.698E+01	7.422E+01	9.000E+01	8.862E+01	6.618E+01
Np-238	5.426E+06	5.352E+06	2.726E+02	0.000E+00	0.000E+00	0.000E+00	0.000E+00	0.000E+00	0.000E+00	0.000E+00
Np-239	1.676E+08	1.665E+08	3.040E+04	6.102E+03	6.096E+03	6.048E+03	5.574E+03	2.466E+03	7.122E-01	3.983E-03
Np-240m	2.975E+02	2.857E+02	1.795E-06	1.795E-06	1.796E-06	1.802E-06	1.870E-06	2.537E-06	8.592E-06	3.433E-05
Np-240	1.363E+05	7.044E+04	0.000E+00	0.000E+00	0.000E+00	0.000E+00	0.000E+00	0.000E+00	0.000E+00	0.000E+00
Pu-236	2.921E+02	2.921E+02	2.869E+02	2.294E+02	2.570E+01	7.998E-09	0.000E+00	0.000E+00	0.000E+00	0.000E+00
Pu-238	6.438E+05	6.438E+05	6.480E+05	6.660E+05	6.294E+05	3.212E+05	5.358E+02	8.880E-16	0.000E+00	0.000E+00
Pu-239	3.982E+04	3.982E+04	3.986E+04	3.986E+04	3.986E+04	3.977E+04	3.892E+04	3.098E+04	2.489E+03	3.983E-03
Pu-240	8.676E+04	8.676E+04	8.676E+04	8.688E+04	8.784E+04	8.952E+04	8.166E+04	3.246E+04	3.185E+00	3.437E-05
Pu-241	6.168E+06	6.168E+06	6.144E+06	5.884E+06	3.838E+06	5.417E+04	6.450E+02	3.032E+02	1.598E-01	0.000E+00
Pu-242	3.088E+02	3.088E+02	3.088E+02	3.088E+02	3.089E+02	3.096E+02	3.112E+02	3.104E+02	2.647E+02	5.103E+01
Pu-243	4.508E+06	3.923E+06	4.155E-03	4.155E-03	4.155E-03	4.155E-03	4.155E-03	4.154E-03	4.138E-03	3.983E-03
Pu-244	1.798E-06	1.798E-06	1.798E-06	1.798E-06	1.798E-06	1.805E-06	1.873E-06	2.540E-06	8.598E-06	3.437E-05
Pu-245	2.149E+00	2.013E+00	0.000E+00	0.000E+00	0.000E+00	0.000E+00	0.000E+00	0.000E+00	0.000E+00	0.000E+00
Am-241	1.432E+05	1.432E+05	1.440E+05	1.526E+05	2.189E+05	3.037E+05	7.284E+04	3.033E+02	1.686E-01	0.000E+00
Am-242m	2.866E+04	2.866E+04	2.865E+04	2.853E+04	2.738E+04	1.816E+04	2.996E+02	4.474E-16	0.000E+00	0.000E+00
Am-242	8.730E+06	8.364E+06	2.865E+04	2.853E+04	2.738E+04	1.816E+04	2.996E+02	4.474E-16	0.000E+00	0.000E+00
Am-243	6.102E+03	6.102E+03	6.102E+03	6.102E+03	6.096E+03	6.048E+03	5.574E+03	2.466E+03	7.122E-01	3.983E-03
Am-244	5.459E+06	1.104E+06	2.337E-09	2.337E-09	2.338E-09	2.347E-09	2.434E-09	3.302E-09	1.118E-08	4.468E-08
Am-245	1.852E+01	1.384E+01	1.035E-03	4.937E-04	3.481E-07	0.000E+00	0.000E+00	0.000E+00	0.000E+00	0.000E+00
Cm-242	6.768E+06	6.768E+06	5.985E+06	1.456E+06	2.246E+04	1.490E+04	2.457E+02	3.680E-16	0.000E+00	0.000E+00
Cm-243	5.459E+03	5.459E+03	5.449E+03	5.342E+03	4.396E+03	6.258E+02	2.133E-06	0.000E+00	0.000E+00	0.000E+00
Cm-244	1.395E+06	1.395E+06	1.390E+06	1.342E+06	9.510E+05	3.028E+04	2.467E-09	3.302E-09	1.118E-08	4.468E-08
Cm-245	7.002E+02	7.002E+02	7.002E+02	7.002E+02	6.996E+02	6.942E+02	6.438E+02	3.027E+02	1.595E-01	0.000E+00
Cm-246	5.504E+02	5.504E+02	5.504E+02	5.503E+02	5.496E+02	5.424E+02	4.751E+02	1.264E+02	2.239E-04	0.000E+00
Cm-247	4.155E-03	4.155E-03	4.155E-03	4.155E-03	4.155E-03	4.155E-03	4.155E-03	4.154E-03	4.138E-03	3.983E-03
Cm-248	9.774E-03	9.774E-03	9.774E-03	9.774E-03	9.774E-03	9.768E-03	9.756E-03	9.582E-03	8.028E-03	1.364E-03
Cm-249	1.672E+02	8.730E+01	9.810E-08	2.111E-13	0.000E+00	0.000E+00	0.000E+00	0.000E+00	0.000E+00	0.000E+00
Cm-250	3.687E-08	3.687E-08	3.687E-08	3.687E-08	3.686E-08	3.673E-08	3.543E-08	2.476E-08	6.864E-10	0.000E+00
Bk-249	7.368E+01	7.368E+01	6.900E+01	3.290E+01	2.320E-02	0.000E+00	0.000E+00	0.000E+00	0.000E+00	0.000E+00
Bk-250	4.045E+00	3.262E+00	3.687E-08	3.687E-08	3.686E-08	3.673E-08	3.543E-08	2.476E-08	6.864E-10	0.000E+00
Cf-249	9.108E-02	9.108E-02	1.026E-01	1.904E-01	2.662E-01	2.230E-01	3.788E-02	7.602E-10	0.000E+00	0.000E+00
Cf-250	2.630E-01	2.630E-01	2.620E-01	2.495E-01	1.549E-01	1.313E-03	3.543E-08	2.476E-08	6.864E-10	0.000E+00
Cf-251	8.838E-05	8.838E-05	8.838E-05	8.832E-05	8.772E-05	8.184E-05	4.093E-05	3.994E-08	0.000E+00	0.000E+00
Cf-252	5.632E-04	5.632E-04	5.512E-04	4.334E-04	4.101E-05	2.359E-15	0.000E+00	0.000E+00	0.000E+00	0.000E+00
Cf-253	1.014E-04	1.012E-04	3.156E-05	6.792E-11	0.000E+00	0.000E+00	0.000E+00	0.000E+00	0.000E+00	0.000E+00
Cf-254	2.132E-07	2.131E-07	1.512E-07	3.246E-09	0.000E+00	0.000E+00	0.000E+00	0.000E+00	0.000E+00	0.000E+00
Es-253	1.011E-04	1.011E-04	7.140E-05	3.160E-09	0.000E+00	0.000E+00	0.000E+00	0.000E+00	0.000E+00	0.000E+00
Total	2.071E+08	2.006E+08	1.454E+07	9.704E+06	5.861E+06	9.042E+05	2.081E+05	6.981E+04	2.848E+03	1.172E+02

Table A.49

Converter core heavy TRU elements nuclide radioactivity (core), Curies Basis = (EQULIBRIUM DISCHARGE FUEL - EQULIBRIUM FRESH FUEL) CONVERTER CORE

	Initial	1 hour	30 days	1 year	10 y	100 y	1000 y	10000 y	100000 y	1000000 y
Np-236	9.732E+02	9.426E+02	1.379E-07	0.000E+00	0.000E+00	0.000E+00	0.000E+00	0.000E+00	0.000E+00	0.000E+00
Np-237	-5.856E+00	-5.856E+00	-5.844E+00	-5.856E+00	-5.976E+00	-7.956E+00	-1.878E+01	-2.202E+01	-2.142E+01	-1.602E+01
Np-238	5.426E+06	5.352E+06	2.726E+02	0.000E+00	0.000E+00	0.000E+00	0.000E+00	0.000E+00	0.000E+00	0.000E+00
Np-239	1.676E+08	1.665E+08	2.359E+04	-7.080E+02	-7.080E+02	-7.020E+02	-6.480E+02	-2.868E+02	-7.800E-02	3.983E-03
Np-240m	2.975E+02	2.857E+02	1.795E-06	1.795E-06	1.796E-06	1.802E-06	1.870E-06	2.537E-06	8.592E-06	3.433E-05
Np-240	1.363E+05	7.044E+04	0.000E+00	0.000E+00	0.000E+00	0.000E+00	0.000E+00	0.000E+00	0.000E+00	0.000E+00
Pu-236	1.694E+02	1.694E+02	1.666E+02	1.333E+02	1.493E+01	4.646E-09	0.000E+00	0.000E+00	0.000E+00	0.000E+00
Pu-238	-4.260E+04	-4.260E+04	-3.840E+04	-1.680E+04	-9.000E+03	-5.580E+03	-3.138E+01	-9.480E-17	0.000E+00	0.000E+00
Pu-239	-8.724E+03	-8.724E+03	-8.682E+03	-8.682E+03	-8.676E+03	-8.658E+03	-8.460E+03	-6.648E+03	-5.256E+02	3.983E-03
Pu-240	-7.200E+03	-7.200E+03	-7.200E+03	-7.200E+03	-7.200E+03	-7.020E+03	-6.420E+03	-2.550E+03	-2.502E-01	3.437E-05
Pu-241	-1.446E+06	-1.446E+06	-1.440E+06	-1.376E+06	-8.994E+05	-1.255E+04	-1.380E+01	-6.360E+00	-3.300E-03	0.000E+00
Pu-242	-2.184E+01	-2.184E+01	-2.184E+01	-2.184E+01	-2.184E+01	-2.190E+01	-2.202E+01	-2.172E+01	-1.842E+01	-3.546E+00
Pu-243	4.508E+06	3.923E+06	4.155E-03	4.155E-03	4.155E-03	4.155E-03	4.155E-03	4.154E-03	4.138E-03	3.983E-03
Pu-244	1.798E-06	1.798E-06	1.798E-06	1.798E-06	1.798E-06	1.805E-06	1.873E-06	2.540E-06	8.598E-06	3.437E-05
Pu-245	2.149E+00	2.013E+00	0.000E+00	0.000E+00	0.000E+00	0.000E+00	0.000E+00	0.000E+00	0.000E+00	0.000E+00
Am-241	-3.162E+04	-3.162E+04	-3.180E+04	-3.384E+04	-4.938E+04	-6.954E+04	-1.656E+04	-6.300E+00	-3.540E-03	0.000E+00
Am-242m	-3.072E+03	-3.072E+03	-3.072E+03	-3.060E+03	-2.934E+03	-1.950E+03	-3.216E+01	-4.800E-17	0.000E+00	0.000E+00
Am-242	8.730E+06	8.363E+06	-3.072E+03	-3.060E+03	-2.934E+03	-1.950E+03	-3.216E+01	-4.800E-17	0.000E+00	0.000E+00
Am-243	-7.080E+02	-7.080E+02	-7.080E+02	-7.080E+02	-7.080E+02	-7.020E+02	-6.480E+02	-2.868E+02	-7.800E-02	3.983E-03
Am-244	5.459E+06	1.104E+06	2.337E-09	2.337E-09	2.338E-09	2.347E-09	2.434E-09	3.302E-09	1.118E-08	4.468E-08
Am-245	1.852E+01	1.384E+01	1.035E-03	4.937E-04	3.481E-07	0.000E+00	0.000E+00	0.000E+00	0.000E+00	0.000E+00
Cm-242	6.436E+06	6.436E+06	5.690E+06	1.365E+06	-2.406E+03	-1.596E+03	-2.640E+01	-3.948E-17	0.000E+00	0.000E+00
Cm-243	-3.870E+02	-3.870E+02	-3.864E+02	-3.780E+02	-3.114E+02	-4.440E+01	-1.512E-07	0.000E+00	0.000E+00	0.000E+00
Cm-244	2.700E+04	2.700E+04	2.640E+04	2.580E+04	1.800E+04	5.820E+02	2.435E-09	3.302E-09	1.118E-08	4.468E-08
Cm-245	-1.500E+01	-1.500E+01	-1.500E+01	-1.440E+01	-1.500E+01	-1.500E+01	-1.380E+01	-6.360E+00	-3.360E-03	0.000E+00
Cm-246	-2.280E+00	-2.280E+00	-2.340E+00	-2.280E+00	-2.280E+00	-2.220E+00	-1.980E+00	-5.400E-01	-9.600E-07	0.000E+00
Cm-247	4.155E-03	4.155E-03	4.155E-03	4.155E-03	4.155E-03	4.155E-03	4.155E-03	4.154E-03	4.138E-03	3.983E-03
Cm-248	9.774E-03	9.774E-03	9.774E-03	9.774E-03	9.774E-03	9.768E-03	9.756E-03	9.582E-03	8.028E-03	1.364E-03
Cm-249	1.672E+02	8.730E+01	9.810E-08	2.111E-13	0.000E+00	0.000E+00	0.000E+00	0.000E+00	0.000E+00	0.000E+00
Cm-250	3.687E-08	3.687E-08	3.687E-08	3.687E-08	3.686E-08	3.673E-08	3.543E-08	2.476E-08	6.864E-10	0.000E+00
Bk-249	7.368E+01	7.368E+01	6.900E+01	3.290E+01	2.320E-02	0.000E+00	0.000E+00	0.000E+00	0.000E+00	0.000E+00
Bk-250	4.045E+00	3.262E+00	3.687E-08	3.687E-08	3.686E-08	3.673E-08	3.543E-08	2.476E-08	6.864E-10	0.000E+00
Cf-249	9.108E-02	9.108E-02	1.026E-01	1.904E-01	2.662E-01	2.230E-01	3.788E-02	7.602E-10	0.000E+00	0.000E+00
Cf-250	2.630E-01	2.630E-01	2.620E-01	2.495E-01	1.549E-01	1.313E-03	3.543E-08	2.476E-08	6.864E-10	0.000E+00
Cf-251	8.838E-05	8.838E-05	8.838E-05	8.832E-05	8.772E-05	8.184E-05	4.093E-05	3.994E-08	0.000E+00	0.000E+00
Cf-252	5.632E-04	5.632E-04	5.512E-04	4.334E-04	4.101E-05	2.359E-15	0.000E+00	0.000E+00	0.000E+00	0.000E+00
Cf-253	1.014E-04	1.012E-04	3.156E-05	6.792E-11	0.000E+00	0.000E+00	0.000E+00	0.000E+00	0.000E+00	0.000E+00
Cf-254	2.132E-07	2.131E-07	1.512E-07	3.246E-09	0.000E+00	0.000E+00	0.000E+00	0.000E+00	0.000E+00	0.000E+00
Es-253	1.011E-04	1.011E-04	7.140E-05	3.160E-09	0.000E+00	0.000E+00	0.000E+00	0.000E+00	0.000E+00	0.000E+00
Total	1.968E+08	1.902E+08	4.207E+06	-5.884E+04	-9.657E+05	-1.098E+05	-3.293E+04	-9.835E+03	-5.658E+02	-1.954E+01
Total/MWd	4.030E+02	3.896E+02	8.616E+00	-1.205E-01	-1.978E+00	-2.248E-01	-6.744E-02	-2.014E-02	-1.159E-03	-4.003E-05

Table A.50

Converter core heavy TRU elements nuclide radioactivity (axial blanket), Curies Basis = EQULIBRIUM FRESH FUEL CONVERTER CORE

	Initial	1 hour	30 days	1 year	10 y	100 y	1000 y	10000 y	100000 y	1000000 y
Np-236	0.000E+00	0.000E+00	0.000E+00	0.000E+00	0.000E+00	0.000E+00	0.000E+00	0.000E+00	0.000E+00	0.000E+00
Np-237	0.000E+00	0.000E+00	0.000E+00	0.000E+00	0.000E+00	0.000E+00	0.000E+00	0.000E+00	0.000E+00	0.000E+00
Np-238	0.000E+00	0.000E+00	0.000E+00	0.000E+00	0.000E+00	0.000E+00	0.000E+00	0.000E+00	0.000E+00	0.000E+00
Np-239	0.000E+00	0.000E+00	0.000E+00	0.000E+00	0.000E+00	0.000E+00	0.000E+00	0.000E+00	0.000E+00	0.000E+00
Np-240m	0.000E+00	0.000E+00	0.000E+00	0.000E+00	0.000E+00	0.000E+00	0.000E+00	0.000E+00	0.000E+00	0.000E+00
Np-240	0.000E+00	0.000E+00	0.000E+00	0.000E+00	0.000E+00	0.000E+00	0.000E+00	0.000E+00	0.000E+00	0.000E+00
Pu-236	0.000E+00	0.000E+00	0.000E+00	0.000E+00	0.000E+00	0.000E+00	0.000E+00	0.000E+00	0.000E+00	0.000E+00
Pu-238	0.000E+00	0.000E+00	0.000E+00	0.000E+00	0.000E+00	0.000E+00	0.000E+00	0.000E+00	0.000E+00	0.000E+00
Pu-239	0.000E+00	0.000E+00	0.000E+00	0.000E+00	0.000E+00	0.000E+00	0.000E+00	0.000E+00	0.000E+00	0.000E+00
Pu-240	0.000E+00	0.000E+00	0.000E+00	0.000E+00	0.000E+00	0.000E+00	0.000E+00	0.000E+00	0.000E+00	0.000E+00
Pu-241	0.000E+00	0.000E+00	0.000E+00	0.000E+00	0.000E+00	0.000E+00	0.000E+00	0.000E+00	0.000E+00	0.000E+00
Pu-242	0.000E+00	0.000E+00	0.000E+00	0.000E+00	0.000E+00	0.000E+00	0.000E+00	0.000E+00	0.000E+00	0.000E+00
Pu-243	0.000E+00	0.000E+00	0.000E+00	0.000E+00	0.000E+00	0.000E+00	0.000E+00	0.000E+00	0.000E+00	0.000E+00
Pu-244	0.000E+00	0.000E+00	0.000E+00	0.000E+00	0.000E+00	0.000E+00	0.000E+00	0.000E+00	0.000E+00	0.000E+00
Pu-245	0.000E+00	0.000E+00	0.000E+00	0.000E+00	0.000E+00	0.000E+00	0.000E+00	0.000E+00	0.000E+00	0.000E+00
Am-241	0.000E+00	0.000E+00	0.000E+00	0.000E+00	0.000E+00	0.000E+00	0.000E+00	0.000E+00	0.000E+00	0.000E+00
Am-242m	0.000E+00	0.000E+00	0.000E+00	0.000E+00	0.000E+00	0.000E+00	0.000E+00	0.000E+00	0.000E+00	0.000E+00
Am-242	0.000E+00	0.000E+00	0.000E+00	0.000E+00	0.000E+00	0.000E+00	0.000E+00	0.000E+00	0.000E+00	0.000E+00
Am-243	0.000E+00	0.000E+00	0.000E+00	0.000E+00	0.000E+00	0.000E+00	0.000E+00	0.000E+00	0.000E+00	0.000E+00
Am-244	0.000E+00	0.000E+00	0.000E+00	0.000E+00	0.000E+00	0.000E+00	0.000E+00	0.000E+00	0.000E+00	0.000E+00
Am-245	0.000E+00	0.000E+00	0.000E+00	0.000E+00	0.000E+00	0.000E+00	0.000E+00	0.000E+00	0.000E+00	0.000E+00
Cm-242	0.000E+00	0.000E+00	0.000E+00	0.000E+00	0.000E+00	0.000E+00	0.000E+00	0.000E+00	0.000E+00	0.000E+00
Cm-243	0.000E+00	0.000E+00	0.000E+00	0.000E+00	0.000E+00	0.000E+00	0.000E+00	0.000E+00	0.000E+00	0.000E+00
Cm-244	0.000E+00	0.000E+00	0.000E+00	0.000E+00	0.000E+00	0.000E+00	0.000E+00	0.000E+00	0.000E+00	0.000E+00
Cm-245	0.000E+00	0.000E+00	0.000E+00	0.000E+00	0.000E+00	0.000E+00	0.000E+00	0.000E+00	0.000E+00	0.000E+00
Cm-246	0.000E+00	0.000E+00	0.000E+00	0.000E+00	0.000E+00	0.000E+00	0.000E+00	0.000E+00	0.000E+00	0.000E+00
Cm-247	0.000E+00	0.000E+00	0.000E+00	0.000E+00	0.000E+00	0.000E+00	0.000E+00	0.000E+00	0.000E+00	0.000E+00
Cm-248	0.000E+00	0.000E+00	0.000E+00	0.000E+00	0.000E+00	0.000E+00	0.000E+00	0.000E+00	0.000E+00	0.000E+00
Cm-249	0.000E+00	0.000E+00	0.000E+00	0.000E+00	0.000E+00	0.000E+00	0.000E+00	0.000E+00	0.000E+00	0.000E+00
Cm-250	0.000E+00	0.000E+00	0.000E+00	0.000E+00	0.000E+00	0.000E+00	0.000E+00	0.000E+00	0.000E+00	0.000E+00
Bk-249	0.000E+00	0.000E+00	0.000E+00	0.000E+00	0.000E+00	0.000E+00	0.000E+00	0.000E+00	0.000E+00	0.000E+00
Bk-250	0.000E+00	0.000E+00	0.000E+00	0.000E+00	0.000E+00	0.000E+00	0.000E+00	0.000E+00	0.000E+00	0.000E+00
Cf-249	0.000E+00	0.000E+00	0.000E+00	0.000E+00	0.000E+00	0.000E+00	0.000E+00	0.000E+00	0.000E+00	0.000E+00
Cf-250	0.000E+00	0.000E+00	0.000E+00	0.000E+00	0.000E+00	0.000E+00	0.000E+00	0.000E+00	0.000E+00	0.000E+00
Cf-251	0.000E+00	0.000E+00	0.000E+00	0.000E+00	0.000E+00	0.000E+00	0.000E+00	0.000E+00	0.000E+00	0.000E+00
Cf-252	0.000E+00	0.000E+00	0.000E+00	0.000E+00	0.000E+00	0.000E+00	0.000E+00	0.000E+00	0.000E+00	0.000E+00
Cf-253	0.000E+00	0.000E+00	0.000E+00	0.000E+00	0.000E+00	0.000E+00	0.000E+00	0.000E+00	0.000E+00	0.000E+00
Cf-254	0.000E+00	0.000E+00	0.000E+00	0.000E+00	0.000E+00	0.000E+00	0.000E+00	0.000E+00	0.000E+00	0.000E+00
Es-253	0.000E+00	0.000E+00	0.000E+00	0.000E+00	0.000E+00	0.000E+00	0.000E+00	0.000E+00	0.000E+00	0.000E+00
Total	0.000E+00	0.000E+00	0.000E+00	0.000E+00	0.000E+00	0.000E+00	0.000E+00	0.000E+00	0.000E+00	0.000E+00

Table A.51

Converter core heavy TRU elements nuclide radioactivity (axial blanket), Curies Basis = EQULIBRIUM DISCHARGE FUEL CONVERTER CORE

	Initial	**1 hour**	**30 days**	**1 year**	**10 y**	**100 y**	**1000 y**	**10000 y**	**100000 y**	**1000000 y**
Np-236	3.367E+00	3.263E+00	4.774E-10	0.000E+00	0.000E+00	0.000E+00	0.000E+00	0.000E+00	0.000E+00	0.000E+00
Np-237	1.923E-01	1.923E-01	1.946E-01	1.947E-01	1.949E-01	2.020E-01	2.428E-01	2.548E-01	2.474E-01	1.849E-01
Np-238	5.885E+04	5.805E+04	2.957E+00	0.000E+00	0.000E+00	0.000E+00	0.000E+00	0.000E+00	0.000E+00	0.000E+00
Np-239	6.702E+07	6.660E+07	9.720E+03	5.131E-03	5.126E-03	5.085E-03	4.687E-03	2.074E-03	5.954E-07	6.030E-14
Np-240m	7.428E+01	7.134E+01	9.828E-12	9.792E-12	9.792E-12	9.792E-12	9.792E-12	9.792E-12	9.786E-12	9.714E-12
Np-240	3.404E+04	1.760E+04	0.000E+00	0.000E+00	0.000E+00	0.000E+00	0.000E+00	0.000E+00	0.000E+00	0.000E+00
Pu-236	4.582E-01	4.582E-01	4.508E-01	3.606E-01	4.039E-02	1.256E-11	0.000E+00	0.000E+00	0.000E+00	0.000E+00
Pu-238	5.643E+02	5.644E+02	5.678E+02	5.646E+02	5.266E+02	2.613E+02	2.381E-01	7.896E-21	0.000E+00	0.000E+00
Pu-239	4.111E+03	4.112E+03	4.129E+03	4.129E+03	4.128E+03	4.118E+03	4.013E+03	3.108E+03	2.410E+02	1.894E-09
Pu-240	5.786E+02	5.786E+02	5.786E+02	5.786E+02	5.781E+02	5.728E+02	5.222E+02	2.075E+02	2.036E-02	9.726E-12
Pu-241	8.646E+03	8.646E+03	8.610E+03	8.244E+03	5.378E+03	7.494E+01	3.587E-06	1.686E-06	8.880E-10	0.000E+00
Pu-242	6.618E-03	6.618E-03	6.618E-03	6.618E-03	6.618E-03	6.624E-03	6.624E-03	6.516E-03	5.528E-03	1.066E-03
Pu-243	1.094E+02	9.516E+01	6.288E-14	6.288E-14	6.288E-14	6.288E-14	6.288E-14	6.288E-14	6.264E-14	6.030E-14
Pu-244	9.810E-12	9.810E-12	9.810E-12	9.810E-12	9.810E-12	9.810E-12	9.810E-12	9.804E-12	9.798E-12	9.726E-12
Pu-245	1.681E-05	1.575E-05	0.000E+00	0.000E+00	0.000E+00	0.000E+00	0.000E+00	0.000E+00	0.000E+00	0.000E+00
Am-241	9.474E+00	9.474E+00	1.061E+01	2.297E+01	1.186E+02	2.626E+02	6.276E+01	3.637E-05	9.372E-10	0.000E+00
Am-242m	2.549E-01	2.549E-01	2.548E-01	2.537E-01	2.435E-01	1.615E-01	2.665E-03	3.979E-21	0.000E+00	0.000E+00
Am-242	5.714E+02	5.473E+02	2.548E-01	2.537E-01	2.435E-01	1.615E-01	2.665E-03	0.000E+00	0.000E+00	0.000E+00
Am-243	5.123E-03	5.124E-03	5.131E-03	5.131E-03	5.126E-03	5.085E-03	4.687E-03	2.074E-03	5.954E-07	6.030E-14
Am-244	5.116E+00	1.034E+00	1.275E-14	1.275E-14	1.275E-14	1.275E-14	1.275E-14	1.275E-14	1.274E-14	1.264E-14
Am-245	2.775E-05	2.448E-05	0.000E+00	0.000E+00	0.000E+00	0.000E+00	0.000E+00	0.000E+00	0.000E+00	0.000E+00
Cm-242	2.317E+02	2.317E+02	2.057E+02	4.958E+01	1.997E-01	1.325E-01	2.185E-03	0.000E+00	0.000E+00	0.000E+00
Cm-243	2.999E-02	2.999E-02	2.993E-02	2.935E-02	2.415E-02	3.437E-03	1.172E-11	0.000E+00	0.000E+00	0.000E+00
Cm-244	8.316E-02	8.316E-02	8.292E-02	8.004E-02	5.671E-02	1.806E-03	1.275E-14	1.275E-14	1.274E-14	1.264E-14
Cm-245	3.893E-06	3.893E-06	3.893E-06	3.893E-06	3.890E-06	3.861E-06	3.580E-06	1.683E-06	8.868E-10	0.000E+00
Cm-246	4.815E-08	4.815E-08	4.815E-08	4.814E-08	4.808E-08	4.745E-08	4.156E-08	1.105E-08	1.959E-14	0.000E+00
Cm-247	6.288E-14	6.288E-14	6.288E-14	6.288E-14	6.288E-14	6.288E-14	6.288E-14	6.288E-14	6.264E-14	6.030E-14
Cm-248	2.125E-14	2.125E-14	2.125E-14	2.125E-14	2.125E-14	2.125E-14	2.121E-14	2.084E-14	1.745E-14	2.965E-15
Cm-249	2.552E-10	1.333E-10	0.000E+00	0.000E+00	0.000E+00	0.000E+00	0.000E+00	0.000E+00	0.000E+00	0.000E+00
Cm-250	9.804E-21	9.804E-21	9.804E-21	9.804E-21	9.804E-21	9.768E-21	9.420E-21	6.576E-21	1.787E-22	0.000E+00
Bk-249	4.134E-11	4.135E-11	3.872E-11	1.847E-11	1.303E-14	0.000E+00	0.000E+00	0.000E+00	0.000E+00	0.000E+00
Bk-250	1.532E-12	1.235E-12	0.000E+00	0.000E+00	0.000E+00	0.000E+00	0.000E+00	0.000E+00	0.000E+00	0.000E+00
Cf-249	1.712E-14	1.714E-14	2.360E-14	7.296E-14	1.161E-13	9.726E-14	1.652E-14	0.000E+00	0.000E+00	0.000E+00
Cf-250	2.320E-14	2.321E-14	2.315E-14	2.205E-14	1.369E-14	1.161E-16	0.000E+00	0.000E+00	0.000E+00	0.000E+00
Cf-251	2.116E-18	2.116E-18	2.116E-18	2.114E-18	2.099E-18	1.958E-18	9.792E-19	9.558E-22	0.000E+00	0.000E+00
Cf-252	3.631E-18	3.631E-18	3.631E-18	2.863E-18	2.174E-19	0.000E+00	0.000E+00	0.000E+00	0.000E+00	0.000E+00
Cf-253	0.000E+00	0.000E+00	0.000E+00	0.000E+00	0.000E+00	0.000E+00	0.000E+00	0.000E+00	0.000E+00	0.000E+00
Cf-254	0.000E+00	0.000E+00	0.000E+00	0.000E+00	0.000E+00	0.000E+00	0.000E+00	0.000E+00	0.000E+00	0.000E+00
Es-253	0.000E+00	0.000E+00	0.000E+00	0.000E+00	0.000E+00	0.000E+00	0.000E+00	0.000E+00	0.000E+00	0.000E+00
Total	6.713E+07	6.669E+07	2.383E+04	1.359E+04	1.073E+04	5.290E+03	4.599E+03	3.316E+03	2.413E+02	1.859E-01

Table A.52

Converter core heavy TRU elements nuclide radioactivity (axial blanket), Curies

Basis = (EQULIBRIUM DISCHARGE FUEL - EQULIBRIUM FRESH FUEL) CONVERTER CORE

	Initial	1 hour	30 days	1 year	10 y	100 y	1000 y	10000 y	100000 y	1000000 y
Np-236	3.367E+00	3.263E+00	4.774E-10	0.000E+00	0.000E+00	0.000E+00	0.000E+00	0.000E+00	0.000E+00	0.000E+00
Np-237	1.923E-01	1.923E-01	1.946E-01	1.947E-01	1.949E-01	2.020E-01	2.428E-01	2.548E-01	2.474E-01	1.849E-01
Np-238	5.885E+04	5.805E+04	2.957E+00	0.000E+00	0.000E+00	0.000E+00	0.000E+00	0.000E+00	0.000E+00	0.000E+00
Np-239	6.702E+07	6.660E+07	9.720E+03	5.131E-03	5.126E-03	5.085E-03	4.687E-03	2.074E-03	5.954E-07	6.030E-14
Np-240m	7.428E+01	7.134E+01	9.828E-12	9.792E-12	9.792E-12	9.792E-12	9.792E-12	9.792E-12	9.786E-12	9.714E-12
Np-240	3.404E+04	1.760E+04	0.000E+00	0.000E+00	0.000E+00	0.000E+00	0.000E+00	0.000E+00	0.000E+00	0.000E+00
Pu-236	4.582E-01	4.582E-01	4.508E-01	3.606E-01	4.039E-02	1.256E-11	0.000E+00	0.000E+00	0.000E+00	0.000E+00
Pu-238	5.643E+02	5.644E+02	5.678E+02	5.646E+02	5.266E+02	2.613E+02	2.381E-01	7.896E-21	0.000E+00	0.000E+00
Pu-239	4.111E+03	4.112E+03	4.129E+03	4.129E+03	4.128E+03	4.118E+03	4.013E+03	3.108E+03	2.410E+02	1.894E-09
Pu-240	5.786E+02	5.786E+02	5.786E+02	5.786E+02	5.781E+02	5.728E+02	5.222E+02	2.075E+02	2.036E-02	9.726E-12
Pu-241	8.646E+03	8.646E+03	8.610E+03	8.244E+03	5.378E+03	7.494E+01	3.587E-06	1.686E-06	8.880E-10	0.000E+00
Pu-242	6.618E-03	6.618E-03	6.618E-03	6.618E-03	6.618E-03	6.624E-03	6.624E-03	6.516E-03	5.528E-03	1.066E-03
Pu-243	1.094E+02	9.516E+01	6.288E-14	6.288E-14	6.288E-14	6.288E-14	6.288E-14	6.288E-14	6.264E-14	6.030E-14
Pu-244	9.810E-12	9.810E-12	9.810E-12	9.810E-12	9.810E-12	9.810E-12	9.810E-12	9.804E-12	9.798E-12	9.726E-12
Pu-245	1.681E-05	1.575E-05	0.000E+00	0.000E+00	0.000E+00	0.000E+00	0.000E+00	0.000E+00	0.000E+00	0.000E+00
Am-241	9.474E+00	9.474E+00	1.061E+01	2.297E+01	1.186E+02	2.626E+02	6.276E+01	3.637E-05	9.372E-10	0.000E+00
Am-242m	2.549E-01	2.549E-01	2.548E-01	2.537E-01	2.435E-01	1.615E-01	2.665E-03	3.979E-21	0.000E+00	0.000E+00
Am-242	5.714E+02	5.473E+02	2.548E-01	2.537E-01	2.435E-01	1.615E-01	2.665E-03	0.000E+00	0.000E+00	0.000E+00
Am-243	5.123E-03	5.124E-03	5.131E-03	5.131E-03	5.126E-03	5.085E-03	4.687E-03	2.074E-03	5.954E-07	6.030E-14
Am-244	5.116E+00	1.034E+00	1.275E-14	1.275E-14	1.275E-14	1.275E-14	1.275E-14	1.275E-14	1.274E-14	1.264E-14
Am-245	2.775E-05	2.448E-05	0.000E+00	0.000E+00	0.000E+00	0.000E+00	0.000E+00	0.000E+00	0.000E+00	0.000E+00
Cm-242	2.317E+02	2.317E+02	2.057E+02	4.958E+01	1.997E-01	1.325E-01	2.185E-03	0.000E+00	0.000E+00	0.000E+00
Cm-243	2.999E-02	2.999E-02	2.993E-02	2.935E-02	2.415E-02	3.437E-03	1.172E-11	0.000E+00	0.000E+00	0.000E+00
Cm-244	8.316E-02	8.316E-02	8.292E-02	8.004E-02	5.671E-02	1.806E-03	1.275E-14	1.275E-14	1.274E-14	1.264E-14
Cm-245	3.893E-06	3.893E-06	3.893E-06	3.893E-06	3.890E-06	3.861E-06	3.580E-06	1.683E-06	8.868E-10	0.000E+00
Cm-246	4.815E-08	4.815E-08	4.815E-08	4.814E-08	4.808E-08	4.745E-08	4.156E-08	1.105E-08	1.959E-14	0.000E+00
Cm-247	6.288E-14	6.288E-14	6.288E-14	6.288E-14	6.288E-14	6.288E-14	6.288E-14	6.288E-14	6.264E-14	6.030E-14
Cm-248	2.125E-14	2.125E-14	2.125E-14	2.125E-14	2.125E-14	2.125E-14	2.121E-14	2.084E-14	1.745E-14	2.965E-15
Cm-249	2.552E-10	1.333E-10	0.000E+00	0.000E+00	0.000E+00	0.000E+00	0.000E+00	0.000E+00	0.000E+00	0.000E+00
Cm-250	9.804E-21	9.804E-21	9.804E-21	9.804E-21	9.804E-21	9.768E-21	9.420E-21	6.576E-21	1.787E-22	0.000E+00
Bk-249	4.134E-11	4.135E-11	3.872E-11	1.847E-11	1.303E-14	0.000E+00	0.000E+00	0.000E+00	0.000E+00	0.000E+00
Bk-250	1.532E-12	1.235E-12	0.000E+00	0.000E+00	0.000E+00	0.000E+00	0.000E+00	0.000E+00	0.000E+00	0.000E+00
Cf-249	1.712E-14	1.714E-14	2.360E-14	7.296E-14	1.161E-13	9.726E-14	1.652E-14	0.000E+00	0.000E+00	0.000E+00
Cf-250	2.320E-14	2.321E-14	2.315E-14	2.205E-14	1.369E-14	1.161E-16	0.000E+00	0.000E+00	0.000E+00	0.000E+00
Cf-251	2.116E-18	2.116E-18	2.116E-18	2.114E-18	2.099E-18	1.958E-18	9.792E-19	9.558E-22	0.000E+00	0.000E+00
Cf-252	3.631E-18	3.631E-18	3.631E-18	2.863E-18	2.174E-19	0.000E+00	0.000E+00	0.000E+00	0.000E+00	0.000E+00
Cf-253	0.000E+00	0.000E+00	0.000E+00	0.000E+00	0.000E+00	0.000E+00	0.000E+00	0.000E+00	0.000E+00	0.000E+00
Cf-254	0.000E+00	0.000E+00	0.000E+00	0.000E+00	0.000E+00	0.000E+00	0.000E+00	0.000E+00	0.000E+00	0.000E+00
Es-253	0.000E+00	0.000E+00	0.000E+00	0.000E+00	0.000E+00	0.000E+00	0.000E+00	0.000E+00	0.000E+00	0.000E+00
Total	6.713E+07	6.669E+07	2.383E+04	1.359E+04	1.073E+04	5.290E+03	4.599E+03	3.316E+03	2.413E+02	1.859E-01
Total/MWd	1.375E+02	1.366E+02	4.880E-02	2.783E-02	2.198E-02	1.083E-02	9.419E-03	6.791E-03	4.942E-04	3.808E-07

Table A.53

Breeder core heavy TRU elements nuclide radioactivity (Core), Curies Basis = EQULIBRIUM FRESH FUEL BREEDER CORE

	Initial	1 hour	30 days	1 year	10 y	100 y	1000 y	10000 y	100000 y	1000000 y
Np-236	0.000E+00	0.000E+00	0.000E+00	0.000E+00	0.000E+00	0.000E+00	0.000E+00	0.000E+00	0.000E+00	0.000E+00
Np-237	5.446E+00	5.446E+00	5.447E+00	5.467E+00	5.705E+00	9.618E+00	3.098E+01	3.766E+01	3.672E+01	2.744E+01
Np-238	0.000E+00	0.000E+00	0.000E+00	0.000E+00	0.000E+00	0.000E+00	0.000E+00	0.000E+00	0.000E+00	0.000E+00
Np-239	0.000E+00	1.072E+01	8.778E+02	8.778E+02	8.772E+02	8.700E+02	8.016E+02	3.547E+02	1.019E-01	0.000E+00
Np-240m	0.000E+00	0.000E+00	0.000E+00	0.000E+00	0.000E+00	0.000E+00	0.000E+00	0.000E+00	0.000E+00	0.000E+00
Np-240	0.000E+00	0.000E+00	0.000E+00	0.000E+00	0.000E+00	0.000E+00	0.000E+00	0.000E+00	0.000E+00	0.000E+00
Pu-236	2.896E+01	2.896E+01	2.839E+01	2.271E+01	2.543E+00	7.914E-10	0.000E+00	0.000E+00	0.000E+00	0.000E+00
Pu-238	1.958E+05	1.958E+05	1.958E+05	1.948E+05	1.823E+05	9.396E+04	1.796E+02	3.427E-16	0.000E+00	0.000E+00
Pu-239	5.646E+04	5.646E+04	5.646E+04	5.646E+04	5.645E+04	5.630E+04	5.491E+04	4.264E+04	3.319E+03	2.608E-08
Pu-240	6.456E+04	6.456E+04	6.456E+04	6.456E+04	6.462E+04	6.432E+04	5.866E+04	2.330E+04	2.287E+00	0.000E+00
Pu-241	2.921E+06	2.921E+06	2.910E+06	2.786E+06	1.817E+06	2.541E+04	7.638E+01	3.590E+01	1.892E-02	0.000E+00
Pu-242	7.188E+01	7.188E+01	7.188E+01	7.188E+01	7.188E+01	7.212E+01	7.260E+01	7.188E+01	6.114E+01	1.178E+01
Pu-243	0.000E+00	0.000E+00	0.000E+00	0.000E+00	0.000E+00	0.000E+00	0.000E+00	0.000E+00	0.000E+00	0.000E+00
Pu-244	0.000E+00	0.000E+00	0.000E+00	0.000E+00	0.000E+00	0.000E+00	0.000E+00	0.000E+00	0.000E+00	0.000E+00
Pu-245	0.000E+00	0.000E+00	0.000E+00	0.000E+00	0.000E+00	0.000E+00	0.000E+00	0.000E+00	0.000E+00	0.000E+00
Am-241	6.036E+04	6.036E+04	6.078E+04	6.486E+04	9.636E+04	1.375E+05	3.280E+04	3.592E+01	1.996E-02	0.000E+00
Am-242m	1.106E+04	1.106E+04	1.106E+04	1.101E+04	1.057E+04	7.008E+03	1.157E+02	1.727E-16	0.000E+00	0.000E+00
Am-242	0.000E+00	4.687E+02	1.106E+04	1.101E+04	1.057E+04	7.008E+03	1.157E+02	1.727E-16	0.000E+00	0.000E+00
Am-243	8.778E+02	8.778E+02	8.778E+02	8.778E+02	8.772E+02	8.700E+02	8.016E+02	3.547E+02	1.019E-01	0.000E+00
Am-244	0.000E+00	0.000E+00	0.000E+00	0.000E+00	0.000E+00	0.000E+00	0.000E+00	0.000E+00	0.000E+00	0.000E+00
Am-245	0.000E+00	0.000E+00	0.000E+00	0.000E+00	0.000E+00	0.000E+00	0.000E+00	0.000E+00	0.000E+00	0.000E+00
Cm-242	1.174E+05	1.174E+05	1.044E+05	3.194E+04	8.664E+03	5.749E+03	9.486E+01	1.420E-16	0.000E+00	0.000E+00
Cm-243	1.630E+03	1.630E+03	1.628E+03	1.595E+03	1.313E+03	1.868E+02	6.372E-07	0.000E+00	0.000E+00	0.000E+00
Cm-244	1.618E+05	1.618E+05	1.613E+05	1.557E+05	1.103E+05	3.512E+03	3.765E-12	0.000E+00	0.000E+00	0.000E+00
Cm-245	8.292E+01	8.292E+01	8.292E+01	8.292E+01	8.286E+01	8.220E+01	7.626E+01	3.584E+01	1.888E-02	0.000E+00
Cm-246	6.000E+01	6.000E+01	6.000E+01	6.000E+01	5.994E+01	5.915E+01	5.182E+01	1.378E+01	2.442E-05	0.000E+00
Cm-247	0.000E+00	0.000E+00	0.000E+00	0.000E+00	0.000E+00	0.000E+00	0.000E+00	0.000E+00	0.000E+00	0.000E+00
Cm-248	0.000E+00	0.000E+00	0.000E+00	0.000E+00	0.000E+00	0.000E+00	0.000E+00	0.000E+00	0.000E+00	0.000E+00
Cm-249	0.000E+00	0.000E+00	0.000E+00	0.000E+00	0.000E+00	0.000E+00	0.000E+00	0.000E+00	0.000E+00	0.000E+00
Cm-250	0.000E+00	0.000E+00	0.000E+00	0.000E+00	0.000E+00	0.000E+00	0.000E+00	0.000E+00	0.000E+00	0.000E+00
Bk-249	0.000E+00	0.000E+00	0.000E+00	0.000E+00	0.000E+00	0.000E+00	0.000E+00	0.000E+00	0.000E+00	0.000E+00
Bk-250	0.000E+00	0.000E+00	0.000E+00	0.000E+00	0.000E+00	0.000E+00	0.000E+00	0.000E+00	0.000E+00	0.000E+00
Cf-249	0.000E+00	0.000E+00	0.000E+00	0.000E+00	0.000E+00	0.000E+00	0.000E+00	0.000E+00	0.000E+00	0.000E+00
Cf-250	0.000E+00	0.000E+00	0.000E+00	0.000E+00	0.000E+00	0.000E+00	0.000E+00	0.000E+00	0.000E+00	0.000E+00
Cf-251	0.000E+00	0.000E+00	0.000E+00	0.000E+00	0.000E+00	0.000E+00	0.000E+00	0.000E+00	0.000E+00	0.000E+00
Cf-252	0.000E+00	0.000E+00	0.000E+00	0.000E+00	0.000E+00	0.000E+00	0.000E+00	0.000E+00	0.000E+00	0.000E+00
Cf-253	0.000E+00	0.000E+00	0.000E+00	0.000E+00	0.000E+00	0.000E+00	0.000E+00	0.000E+00	0.000E+00	0.000E+00
Cf-254	0.000E+00	0.000E+00	0.000E+00	0.000E+00	0.000E+00	0.000E+00	0.000E+00	0.000E+00	0.000E+00	0.000E+00
Es-253	0.000E+00	0.000E+00	0.000E+00	0.000E+00	0.000E+00	0.000E+00	0.000E+00	0.000E+00	0.000E+00	0.000E+00
Total	3.592E+06	3.592E+06	3.579E+06	3.380E+06	2.360E+06	4.029E+05	1.488E+05	6.688E+04	3.419E+03	3.922E+01

Table A.54

Breeder core heavy TRU elements nuclide radioactivity (Core), Curies Basis = EQULIBRIUM DISCHARGE FUEL BREEDER CORE

	Initial	**1 hour**	**30 days**	**1 year**	**10 y**	**100 y**	**1000 y**	**10000 y**	**100000 y**	**1000000 y**
Np-236	2.710E+02	2.626E+02	3.843E-08	0.000E+00	0.000E+00	0.000E+00	0.000E+00	0.000E+00	0.000E+00	0.000E+00
Np-237	5.070E+00	5.070E+00	5.086E+00	5.103E+00	5.324E+00	9.234E+00	3.080E+01	3.754E+01	3.661E+01	2.735E+01
Np-238	1.569E+06	1.547E+06	7.884E+01	0.000E+00	0.000E+00	0.000E+00	0.000E+00	0.000E+00	0.000E+00	0.000E+00
Np-239	1.769E+08	1.758E+08	2.655E+04	8.850E+02	8.844E+02	8.772E+02	8.082E+02	3.576E+02	1.031E-01	4.352E-04
Np-240m	3.082E+02	2.960E+02	4.045E-07	4.045E-07	4.046E-07	4.053E-07	4.125E-07	4.840E-07	1.132E-06	3.889E-06
Np-240	1.412E+05	7.296E+04	0.000E+00	0.000E+00	0.000E+00	0.000E+00	0.000E+00	0.000E+00	0.000E+00	0.000E+00
Pu-236	7.404E+01	7.404E+01	7.272E+01	5.818E+01	6.516E+00	2.027E-09	0.000E+00	0.000E+00	0.000E+00	0.000E+00
Pu-238	1.896E+05	1.896E+05	1.910E+05	1.979E+05	1.874E+05	9.618E+04	1.737E+02	3.137E-16	0.000E+00	0.000E+00
Pu-239	4.532E+04	4.532E+04	4.536E+04	4.536E+04	4.535E+04	4.524E+04	4.412E+04	3.428E+04	2.671E+03	4.352E-04
Pu-240	6.402E+04	6.402E+04	6.402E+04	6.402E+04	6.408E+04	6.384E+04	5.821E+04	2.313E+04	2.270E+00	3.893E-06
Pu-241	3.193E+06	3.192E+06	3.180E+06	3.044E+06	1.985E+06	2.776E+04	7.680E+01	3.610E+01	1.902E-02	0.000E+00
Pu-242	7.260E+01	7.260E+01	7.260E+01	7.260E+01	7.260E+01	7.284E+01	7.326E+01	7.260E+01	6.168E+01	1.190E+01
Pu-243	1.060E+06	9.228E+05	4.540E-04	4.540E-04	4.540E-04	4.540E-04	4.540E-04	4.538E-04	4.521E-04	4.352E-04
Pu-244	4.051E-07	4.051E-07	4.051E-07	4.051E-07	4.051E-07	4.058E-07	4.130E-07	4.846E-07	1.134E-06	3.893E-06
Pu-245	4.849E-01	4.542E-01	0.000E+00	0.000E+00	0.000E+00	0.000E+00	0.000E+00	0.000E+00	0.000E+00	0.000E+00
Am-241	5.244E+04	5.244E+04	5.285E+04	5.734E+04	9.192E+04	1.387E+05	3.311E+04	3.612E+01	2.007E-02	0.000E+00
Am-242m	1.013E+04	1.013E+04	1.012E+04	1.008E+04	9.672E+03	6.414E+03	1.058E+02	1.580E-16	0.000E+00	0.000E+00
Am-242	3.190E+06	3.055E+06	1.012E+04	1.008E+04	9.672E+03	6.420E+03	1.058E+02	1.581E-16	0.000E+00	0.000E+00
Am-243	8.850E+02	8.850E+02	8.850E+02	8.850E+02	8.844E+02	8.772E+02	8.082E+02	3.576E+02	1.031E-01	4.352E-04
Am-244	7.932E+05	1.604E+05	5.266E-10	5.266E-10	5.266E-10	5.276E-10	5.369E-10	6.300E-10	1.474E-09	5.062E-09
Am-245	2.832E+00	2.160E+00	1.083E-04	5.167E-05	3.643E-08	0.000E+00	0.000E+00	0.000E+00	0.000E+00	0.000E+00
Cm-242	2.434E+06	2.434E+06	2.153E+06	5.237E+05	7.932E+03	5.262E+03	8.682E+01	1.300E-16	0.000E+00	0.000E+00
Cm-243	1.683E+03	1.683E+03	1.680E+03	1.647E+03	1.355E+03	1.928E+02	6.576E-07	0.000E+00	0.000E+00	0.000E+00
Cm-244	1.744E+05	1.744E+05	1.738E+05	1.678E+05	1.189E+05	3.787E+03	5.410E-10	6.300E-10	1.474E-09	5.062E-09
Cm-245	8.340E+01	8.340E+01	8.340E+01	8.334E+01	8.328E+01	8.268E+01	7.668E+01	3.604E+01	1.899E-02	0.000E+00
Cm-246	6.048E+01	6.048E+01	6.048E+01	6.048E+01	6.042E+01	5.962E+01	5.222E+01	1.389E+01	2.461E-05	0.000E+00
Cm-247	4.540E-04	4.540E-04	4.540E-04	4.540E-04	4.540E-04	4.540E-04	4.540E-04	4.538E-04	4.521E-04	4.352E-04
Cm-248	1.047E-03	1.047E-03	1.047E-03	1.047E-03	1.047E-03	1.047E-03	1.045E-03	1.027E-03	8.598E-04	1.461E-04
Cm-249	1.752E+01	9.150E+00	9.162E-09	1.971E-14	0.000E+00	0.000E+00	0.000E+00	0.000E+00	0.000E+00	0.000E+00
Cm-250	3.752E-09	3.752E-09	3.752E-09	3.752E-09	3.751E-09	3.737E-09	3.606E-09	2.519E-09	6.984E-11	0.000E+00
Bk-249	7.710E+00	7.710E+00	7.218E+00	3.444E+00	2.428E-03	0.000E+00	0.000E+00	0.000E+00	0.000E+00	0.000E+00
Bk-250	4.124E-01	3.326E-01	3.752E-09	3.752E-09	3.751E-09	3.737E-09	3.606E-09	2.519E-09	6.984E-11	0.000E+00
Cf-249	9.558E-03	9.558E-03	1.076E-02	1.996E-02	2.788E-02	2.336E-02	3.969E-03	7.962E-11	0.000E+00	0.000E+00
Cf-250	2.686E-02	2.686E-02	2.675E-02	2.548E-02	1.582E-02	1.341E-04	3.606E-09	2.519E-09	6.984E-11	0.000E+00
Cf-251	8.784E-06	8.784E-06	8.784E-06	8.778E-06	8.718E-06	8.136E-06	4.067E-06	3.968E-09	0.000E+00	0.000E+00
Cf-252	5.428E-05	5.428E-05	5.312E-05	4.177E-05	3.952E-06	2.273E-16	0.000E+00	0.000E+00	0.000E+00	0.000E+00
Cf-253	9.468E-06	9.456E-06	2.948E-06	6.342E-12	0.000E+00	0.000E+00	0.000E+00	0.000E+00	0.000E+00	0.000E+00
Cf-254	1.924E-08	1.923E-08	1.364E-08	2.929E-10	0.000E+00	0.000E+00	0.000E+00	0.000E+00	0.000E+00	0.000E+00
Es-253	9.444E-06	9.444E-06	6.666E-06	2.951E-10	0.000E+00	0.000E+00	0.000E+00	0.000E+00	0.000E+00	0.000E+00
Total	1.899E+08	1.877E+08	5.910E+06	4.124E+06	2.524E+06	3.957E+05	1.378E+05	5.836E+04	2.772E+03	3.925E+01

Table A.55

Breeder core heavy TRU elements nuclide radioactivity (Core), Curies Basis = (EQULIBRIUM DISCHARGE FUEL - EQULIBRIUM FRESH FUEL) BREEDER CORE

	Initial	**1 hour**	**30 days**	**1 year**	**10 y**	**100 y**	**1000 y**	**10000 y**	**100000 y**	**1000000 y**
Np-236	2.710E+02	2.626E+02	3.843E-08	0.000E+00	0.000E+00	0.000E+00	0.000E+00	0.000E+00	0.000E+00	0.000E+00
Np-237	-3.762E-01	-3.762E-01	-3.618E-01	-3.636E-01	-3.804E-01	-3.840E-01	-1.800E-01	-1.200E-01	-1.140E-01	-9.000E-02
Np-238	1.569E+06	1.547E+06	7.884E+01	0.000E+00	0.000E+00	0.000E+00	0.000E+00	0.000E+00	0.000E+00	0.000E+00
Np-239	1.769E+08	1.758E+08	2.567E+04	7.200E+00	7.200E+00	7.200E+00	6.600E+00	2.880E+00	1.260E-03	4.352E-04
Np-240m	3.082E+02	2.960E+02	4.045E-07	4.045E-07	4.046E-07	4.053E-07	4.125E-07	4.840E-07	1.132E-06	3.889E-06
Np-240	1.412E+05	7.296E+04	0.000E+00	0.000E+00	0.000E+00	0.000E+00	0.000E+00	0.000E+00	0.000E+00	0.000E+00
Pu-236	4.508E+01	4.508E+01	4.433E+01	3.547E+01	3.973E+00	1.236E-09	0.000E+00	0.000E+00	0.000E+00	0.000E+00
Pu-238	-6.180E+03	-6.180E+03	-4.740E+03	3.120E+03	5.160E+03	2.220E+03	-5.940E+00	-2.898E-17	0.000E+00	0.000E+00
Pu-239	-1.114E+04	-1.114E+04	-1.110E+04	-1.110E+04	-1.110E+04	-1.106E+04	-1.079E+04	-8.352E+03	-6.474E+02	4.352E-04
Pu-240	-5.400E+02	-5.400E+02	-5.400E+02	-5.400E+02	-5.400E+02	-4.800E+02	-4.500E+02	-1.740E+02	-1.740E-02	3.893E-06
Pu-241	2.712E+05	2.706E+05	2.700E+05	2.580E+05	1.680E+05	2.346E+03	4.200E-01	1.980E-01	1.020E-04	0.000E+00
Pu-242	7.200E-01	7.200E-01	7.200E-01	7.200E-01	7.200E-01	7.200E-01	6.600E-01	7.200E-01	5.400E-01	1.140E-01
Pu-243	1.060E+06	9.228E+05	4.540E-04	4.540E-04	4.540E-04	4.540E-04	4.540E-04	4.538E-04	4.521E-04	4.352E-04
Pu-244	4.051E-07	4.051E-07	4.051E-07	4.051E-07	4.051E-07	4.058E-07	4.130E-07	4.846E-07	1.134E-06	3.893E-06
Pu-245	4.849E-01	4.542E-01	0.000E+00	0.000E+00	0.000E+00	0.000E+00	0.000E+00	0.000E+00	0.000E+00	0.000E+00
Am-241	-7.920E+03	-7.920E+03	-7.926E+03	-7.518E+03	-4.440E+03	1.200E+03	3.060E+02	1.980E-01	1.080E-04	0.000E+00
Am-242m	-9.360E+02	-9.360E+02	-9.360E+02	-9.300E+02	-8.940E+02	-5.940E+02	-9.840E+00	-1.464E-17	0.000E+00	0.000E+00
Am-242	3.190E+06	3.055E+06	-9.360E+02	-9.300E+02	-8.940E+02	-5.880E+02	-9.840E+00	-1.458E-17	0.000E+00	0.000E+00
Am-243	7.200E+00	7.200E+00	7.200E+00	7.200E+00	7.200E+00	7.200E+00	6.600E+00	2.880E+00	1.260E-03	4.352E-04
Am-244	7.932E+05	1.604E+05	5.266E-10	5.266E-10	5.266E-10	5.276E-10	5.369E-10	6.300E-10	1.474E-09	5.062E-09
Am-245	2.832E+00	2.160E+00	1.083E-04	5.167E-05	3.643E-08	0.000E+00	0.000E+00	0.000E+00	0.000E+00	0.000E+00
Cm-242	2.317E+06	2.317E+06	2.048E+06	4.917E+05	-7.320E+02	-4.866E+02	-8.040E+00	-1.200E-17	0.000E+00	0.000E+00
Cm-243	5.280E+01	5.280E+01	5.220E+01	5.160E+01	4.200E+01	6.000E+00	2.040E-08	0.000E+00	0.000E+00	0.000E+00
Cm-244	1.266E+04	1.266E+04	1.254E+04	1.212E+04	8.640E+03	2.748E+02	5.373E-10	6.300E-10	1.474E-09	5.062E-09
Cm-245	4.800E-01	4.800E-01	4.800E-01	4.200E-01	4.200E-01	4.800E-01	4.200E-01	1.980E-01	1.080E-04	0.000E+00
Cm-246	4.800E-01	4.800E-01	4.800E-01	4.800E-01	4.800E-01	4.620E-01	4.020E-01	1.080E-01	1.920E-07	0.000E+00
Cm-247	4.540E-04	4.540E-04	4.540E-04	4.540E-04	4.540E-04	4.540E-04	4.540E-04	4.538E-04	4.521E-04	4.352E-04
Cm-248	1.047E-03	1.047E-03	1.047E-03	1.047E-03	1.047E-03	1.047E-03	1.045E-03	1.027E-03	8.598E-04	1.461E-04
Cm-249	1.752E+01	9.150E+00	9.162E-09	1.971E-14	0.000E+00	0.000E+00	0.000E+00	0.000E+00	0.000E+00	0.000E+00
Cm-250	3.752E-09	3.752E-09	3.752E-09	3.752E-09	3.751E-09	3.737E-09	3.606E-09	2.519E-09	6.984E-11	0.000E+00
Bk-249	7.710E+00	7.710E+00	7.218E+00	3.444E+00	2.428E-03	0.000E+00	0.000E+00	0.000E+00	0.000E+00	0.000E+00
Bk-250	4.124E-01	3.326E-01	3.752E-09	3.752E-09	3.751E-09	3.737E-09	3.606E-09	2.519E-09	6.984E-11	0.000E+00
Cf-249	9.558E-03	9.558E-03	1.076E-02	1.996E-02	2.788E-02	2.336E-02	3.969E-03	7.962E-11	0.000E+00	0.000E+00
Cf-250	2.686E-02	2.686E-02	2.675E-02	2.548E-02	1.582E-02	1.341E-04	3.606E-09	2.519E-09	6.984E-11	0.000E+00
Cf-251	8.784E-06	8.784E-06	8.784E-06	8.778E-06	8.718E-06	8.136E-06	4.067E-06	3.968E-09	0.000E+00	0.000E+00
Cf-252	5.428E-05	5.428E-05	5.312E-05	4.177E-05	3.952E-06	2.273E-16	0.000E+00	0.000E+00	0.000E+00	0.000E+00
Cf-253	9.468E-06	9.456E-06	2.948E-06	6.342E-12	0.000E+00	0.000E+00	0.000E+00	0.000E+00	0.000E+00	0.000E+00
Cf-254	1.924E-08	1.923E-08	1.364E-08	2.929E-10	0.000E+00	0.000E+00	0.000E+00	0.000E+00	0.000E+00	0.000E+00
Es-253	9.444E-06	9.444E-06	6.666E-06	2.951E-10	0.000E+00	0.000E+00	0.000E+00	0.000E+00	0.000E+00	0.000E+00
Total	1.863E+08	1.841E+08	2.331E+06	7.441E+05	1.633E+05	-7.150E+03	-1.095E+04	-8.519E+03	-6.470E+02	2.633E-02
Total/MWd	3.815E+02	3.771E+02	4.773E+00	1.524E+00	3.344E-01	-1.464E-02	-2.243E-02	-1.745E-02	-1.325E-03	5.393E-08

Table A.56

Breeder core heavy TRU elements nuclide radioactivity (axial blanket), Curies Basis = EQULIBRIUM FRESH FUEL BREEDER CORE

	Initial	1 hour	30 days	1 year	10 y	100 y	1000 y	10000 y	100000 y	1000000 y
Np-236	0.000E+00	0.000E+00	0.000E+00	0.000E+00	0.000E+00	0.000E+00	0.000E+00	0.000E+00	0.000E+00	0.000E+00
Np-237	0.000E+00	0.000E+00	0.000E+00	0.000E+00	0.000E+00	0.000E+00	0.000E+00	0.000E+00	0.000E+00	0.000E+00
Np-238	0.000E+00	0.000E+00	0.000E+00	0.000E+00	0.000E+00	0.000E+00	0.000E+00	0.000E+00	0.000E+00	0.000E+00
Np-239	0.000E+00	0.000E+00	0.000E+00	0.000E+00	0.000E+00	0.000E+00	0.000E+00	0.000E+00	0.000E+00	0.000E+00
Np-240m	0.000E+00	0.000E+00	0.000E+00	0.000E+00	0.000E+00	0.000E+00	0.000E+00	0.000E+00	0.000E+00	0.000E+00
Np-240	0.000E+00	0.000E+00	0.000E+00	0.000E+00	0.000E+00	0.000E+00	0.000E+00	0.000E+00	0.000E+00	0.000E+00
Pu-236	0.000E+00	0.000E+00	0.000E+00	0.000E+00	0.000E+00	0.000E+00	0.000E+00	0.000E+00	0.000E+00	0.000E+00
Pu-238	0.000E+00	0.000E+00	0.000E+00	0.000E+00	0.000E+00	0.000E+00	0.000E+00	0.000E+00	0.000E+00	0.000E+00
Pu-239	0.000E+00	0.000E+00	0.000E+00	0.000E+00	0.000E+00	0.000E+00	0.000E+00	0.000E+00	0.000E+00	0.000E+00
Pu-240	0.000E+00	0.000E+00	0.000E+00	0.000E+00	0.000E+00	0.000E+00	0.000E+00	0.000E+00	0.000E+00	0.000E+00
Pu-241	0.000E+00	0.000E+00	0.000E+00	0.000E+00	0.000E+00	0.000E+00	0.000E+00	0.000E+00	0.000E+00	0.000E+00
Pu-242	0.000E+00	0.000E+00	0.000E+00	0.000E+00	0.000E+00	0.000E+00	0.000E+00	0.000E+00	0.000E+00	0.000E+00
Pu-243	0.000E+00	0.000E+00	0.000E+00	0.000E+00	0.000E+00	0.000E+00	0.000E+00	0.000E+00	0.000E+00	0.000E+00
Pu-244	0.000E+00	0.000E+00	0.000E+00	0.000E+00	0.000E+00	0.000E+00	0.000E+00	0.000E+00	0.000E+00	0.000E+00
Pu-245	0.000E+00	0.000E+00	0.000E+00	0.000E+00	0.000E+00	0.000E+00	0.000E+00	0.000E+00	0.000E+00	0.000E+00
Am-241	0.000E+00	0.000E+00	0.000E+00	0.000E+00	0.000E+00	0.000E+00	0.000E+00	0.000E+00	0.000E+00	0.000E+00
Am-242m	0.000E+00	0.000E+00	0.000E+00	0.000E+00	0.000E+00	0.000E+00	0.000E+00	0.000E+00	0.000E+00	0.000E+00
Am-242	0.000E+00	0.000E+00	0.000E+00	0.000E+00	0.000E+00	0.000E+00	0.000E+00	0.000E+00	0.000E+00	0.000E+00
Am-243	0.000E+00	0.000E+00	0.000E+00	0.000E+00	0.000E+00	0.000E+00	0.000E+00	0.000E+00	0.000E+00	0.000E+00
Am-244	0.000E+00	0.000E+00	0.000E+00	0.000E+00	0.000E+00	0.000E+00	0.000E+00	0.000E+00	0.000E+00	0.000E+00
Am-245	0.000E+00	0.000E+00	0.000E+00	0.000E+00	0.000E+00	0.000E+00	0.000E+00	0.000E+00	0.000E+00	0.000E+00
Cm-242	0.000E+00	0.000E+00	0.000E+00	0.000E+00	0.000E+00	0.000E+00	0.000E+00	0.000E+00	0.000E+00	0.000E+00
Cm-243	0.000E+00	0.000E+00	0.000E+00	0.000E+00	0.000E+00	0.000E+00	0.000E+00	0.000E+00	0.000E+00	0.000E+00
Cm-244	0.000E+00	0.000E+00	0.000E+00	0.000E+00	0.000E+00	0.000E+00	0.000E+00	0.000E+00	0.000E+00	0.000E+00
Cm-245	0.000E+00	0.000E+00	0.000E+00	0.000E+00	0.000E+00	0.000E+00	0.000E+00	0.000E+00	0.000E+00	0.000E+00
Cm-246	0.000E+00	0.000E+00	0.000E+00	0.000E+00	0.000E+00	0.000E+00	0.000E+00	0.000E+00	0.000E+00	0.000E+00
Cm-247	0.000E+00	0.000E+00	0.000E+00	0.000E+00	0.000E+00	0.000E+00	0.000E+00	0.000E+00	0.000E+00	0.000E+00
Cm-248	0.000E+00	0.000E+00	0.000E+00	0.000E+00	0.000E+00	0.000E+00	0.000E+00	0.000E+00	0.000E+00	0.000E+00
Cm-249	0.000E+00	0.000E+00	0.000E+00	0.000E+00	0.000E+00	0.000E+00	0.000E+00	0.000E+00	0.000E+00	0.000E+00
Cm-250	0.000E+00	0.000E+00	0.000E+00	0.000E+00	0.000E+00	0.000E+00	0.000E+00	0.000E+00	0.000E+00	0.000E+00
Bk-249	0.000E+00	0.000E+00	0.000E+00	0.000E+00	0.000E+00	0.000E+00	0.000E+00	0.000E+00	0.000E+00	0.000E+00
Bk-250	0.000E+00	0.000E+00	0.000E+00	0.000E+00	0.000E+00	0.000E+00	0.000E+00	0.000E+00	0.000E+00	0.000E+00
Cf-249	0.000E+00	0.000E+00	0.000E+00	0.000E+00	0.000E+00	0.000E+00	0.000E+00	0.000E+00	0.000E+00	0.000E+00
Cf-250	0.000E+00	0.000E+00	0.000E+00	0.000E+00	0.000E+00	0.000E+00	0.000E+00	0.000E+00	0.000E+00	0.000E+00
Cf-251	0.000E+00	0.000E+00	0.000E+00	0.000E+00	0.000E+00	0.000E+00	0.000E+00	0.000E+00	0.000E+00	0.000E+00
Cf-252	0.000E+00	0.000E+00	0.000E+00	0.000E+00	0.000E+00	0.000E+00	0.000E+00	0.000E+00	0.000E+00	0.000E+00
Cf-253	0.000E+00	0.000E+00	0.000E+00	0.000E+00	0.000E+00	0.000E+00	0.000E+00	0.000E+00	0.000E+00	0.000E+00
Cf-254	0.000E+00	0.000E+00	0.000E+00	0.000E+00	0.000E+00	0.000E+00	0.000E+00	0.000E+00	0.000E+00	0.000E+00
Es-253	0.000E+00	0.000E+00	0.000E+00	0.000E+00	0.000E+00	0.000E+00	0.000E+00	0.000E+00	0.000E+00	0.000E+00
Total	0.000E+00	0.000E+00	0.000E+00	0.000E+00	0.000E+00	0.000E+00	0.000E+00	0.000E+00	0.000E+00	0.000E+00

Table A.57

Breeder core heavy TRU elements nuclide radioactivity (axial blanket), Curies Basis = EQULIBRIUM DISCHARGE FUEL BREEDER CORE

	Initial	1 hour	30 days	1 year	10 y	100 y	1000 y	10000 y	100000 y	1000000 y
Np-236	3.209E+00	3.110E+00	4.550E-10	0.000E+00	0.000E+00	0.000E+00	0.000E+00	0.000E+00	0.000E+00	0.000E+00
Np-237	2.657E-01	2.657E-01	2.688E-01	2.690E-01	2.692E-01	2.768E-01	3.210E-01	3.337E-01	3.241E-01	2.422E-01
Np-238	6.564E+04	6.474E+04	3.298E+00	0.000E+00	0.000E+00	0.000E+00	0.000E+00	0.000E+00	0.000E+00	0.000E+00
Np-239	1.328E+08	1.320E+08	1.927E+04	3.628E-03	3.625E-03	3.595E-03	3.314E-03	1.466E-03	4.210E-07	2.259E-14
Np-240m	1.054E+02	1.012E+02	5.815E-12	5.769E-12	5.769E-12	5.769E-12	5.769E-12	5.768E-12	5.764E-12	5.721E-12
Np-240	4.832E+04	2.498E+04	0.000E+00	0.000E+00	0.000E+00	0.000E+00	0.000E+00	0.000E+00	0.000E+00	0.000E+00
Pu-236	4.393E-01	4.394E-01	4.322E-01	3.457E-01	3.873E-02	1.205E-11	0.000E+00	0.000E+00	0.000E+00	0.000E+00
Pu-238	6.348E+02	6.348E+02	6.384E+02	6.348E+02	5.920E+02	2.938E+02	2.671E-01	7.014E-21	0.000E+00	0.000E+00
Pu-239	8.550E+03	8.550E+03	8.586E+03	8.586E+03	8.586E+03	8.562E+03	8.346E+03	6.462E+03	5.011E+02	3.938E-09
Pu-240	8.412E+02	8.412E+02	8.412E+02	8.412E+02	8.406E+02	8.328E+02	7.590E+02	3.017E+02	2.960E-02	5.728E-12
Pu-241	9.342E+03	9.342E+03	9.306E+03	8.910E+03	5.812E+03	8.100E+01	1.824E-06	8.574E-07	4.517E-10	0.000E+00
Pu-242	5.705E-03	5.705E-03	5.706E-03	5.706E-03	5.707E-03	5.711E-03	5.712E-03	5.619E-03	4.766E-03	9.186E-04
Pu-243	7.722E+01	6.720E+01	2.356E-14	2.356E-14	2.356E-14	2.356E-14	2.356E-14	2.356E-14	2.347E-14	2.259E-14
Pu-244	5.777E-12	5.777E-12	5.777E-12	5.777E-12	5.777E-12	5.777E-12	5.777E-12	5.776E-12	5.772E-12	5.728E-12
Pu-245	7.668E-06	7.182E-06	0.000E+00	0.000E+00	0.000E+00	0.000E+00	0.000E+00	0.000E+00	0.000E+00	0.000E+00
Am-241	1.027E+01	1.027E+01	1.149E+01	2.485E+01	1.282E+02	2.837E+02	6.786E+01	3.834E-05	4.767E-10	0.000E+00
Am-242m	2.265E-01	2.265E-01	2.264E-01	2.255E-01	2.164E-01	1.435E-01	2.368E-03	3.536E-21	0.000E+00	0.000E+00
Am-242	5.012E+02	4.799E+02	2.264E-01	2.255E-01	2.164E-01	1.435E-01	2.368E-03	0.000E+00	0.000E+00	0.000E+00
Am-243	3.622E-03	3.623E-03	3.628E-03	3.628E-03	3.625E-03	3.595E-03	3.314E-03	1.466E-03	4.210E-07	2.259E-14
Am-244	3.040E+00	6.144E-01	7.512E-15	7.512E-15	7.512E-15	7.512E-15	7.512E-15	7.506E-15	7.506E-15	7.446E-15
Am-245	1.337E-05	1.168E-05	0.000E+00	0.000E+00	0.000E+00	0.000E+00	0.000E+00	0.000E+00	0.000E+00	0.000E+00
Cm-242	2.024E+02	2.025E+02	1.797E+02	4.333E+01	1.775E-01	1.177E-01	1.942E-03	0.000E+00	0.000E+00	0.000E+00
Cm-243	2.033E-02	2.033E-02	2.030E-02	1.990E-02	1.637E-02	2.330E-03	7.944E-12	0.000E+00	0.000E+00	0.000E+00
Cm-244	4.938E-02	4.939E-02	4.924E-02	4.753E-02	3.368E-02	1.072E-03	7.512E-15	7.506E-15	7.506E-15	7.446E-15
Cm-245	1.980E-06	1.980E-06	1.980E-06	1.980E-06	1.979E-06	1.964E-06	1.821E-06	8.562E-07	4.510E-10	0.000E+00
Cm-246	2.164E-08	2.164E-08	2.164E-08	2.164E-08	2.161E-08	2.132E-08	1.868E-08	4.967E-09	8.802E-15	0.000E+00
Cm-247	2.356E-14	2.356E-14	2.356E-14	2.356E-14	2.356E-14	2.356E-14	2.356E-14	2.356E-14	2.347E-14	2.259E-14
Cm-248	7.038E-15	7.038E-15	7.038E-15	7.038E-15	7.038E-15	7.038E-15	7.026E-15	6.900E-15	5.779E-15	9.822E-16
Cm-249	7.494E-11	3.914E-11	0.000E+00	0.000E+00	0.000E+00	0.000E+00	0.000E+00	0.000E+00	0.000E+00	0.000E+00
Cm-250	2.487E-21	2.487E-21	2.487E-21	2.487E-21	2.487E-21	2.487E-21	2.398E-21	1.673E-21	4.985E-23	0.000E+00
Bk-249	1.212E-11	1.212E-11	1.135E-11	5.416E-12	3.818E-15	0.000E+00	0.000E+00	0.000E+00	0.000E+00	0.000E+00
Bk-250	3.984E-13	3.213E-13	0.000E+00	0.000E+00	0.000E+00	0.000E+00	0.000E+00	0.000E+00	0.000E+00	0.000E+00
Cf-249	5.038E-15	5.041E-15	6.936E-15	2.141E-14	3.404E-14	2.852E-14	4.847E-15	0.000E+00	0.000E+00	0.000E+00
Cf-250	6.042E-15	6.042E-15	6.024E-15	5.740E-15	3.563E-15	3.022E-17	0.000E+00	0.000E+00	0.000E+00	0.000E+00
Cf-251	4.880E-19	4.880E-19	4.880E-19	4.876E-19	4.843E-19	4.519E-19	2.258E-19	0.000E+00	0.000E+00	0.000E+00
Cf-252	5.155E-19	5.155E-19	5.155E-19	3.915E-19	0.000E+00	0.000E+00	0.000E+00	0.000E+00	0.000E+00	0.000E+00
Cf-253	0.000E+00	0.000E+00	0.000E+00	0.000E+00	0.000E+00	0.000E+00	0.000E+00	0.000E+00	0.000E+00	0.000E+00
Cf-254	0.000E+00	0.000E+00	0.000E+00	0.000E+00	0.000E+00	0.000E+00	0.000E+00	0.000E+00	0.000E+00	0.000E+00
Es-253	0.000E+00	0.000E+00	0.000E+00	0.000E+00	0.000E+00	0.000E+00	0.000E+00	0.000E+00	0.000E+00	0.000E+00
Total	1.330E+08	1.321E+08	3.883E+04	1.904E+04	1.596E+04	1.005E+04	9.173E+03	6.764E+03	5.015E+02	2.431E-01

Table A.58

Breeder core heavy TRU elements nuclide radioactivity (axial blanket), Curies

Basis = (EQULIBRIUM DISCHARGE FUEL - EQULIBRIUM FRESH FUEL) BREEDER CORE

	Initial	1 hour	30 days	1 year	10 y	100 y	1000 y	10000 y	100000 y	1000000 y
Np-236	3.209E+00	3.110E+00	4.550E-10	0.000E+00	0.000E+00	0.000E+00	0.000E+00	0.000E+00	0.000E+00	0.000E+00
Np-237	2.657E-01	2.657E-01	2.688E-01	2.690E-01	2.692E-01	2.768E-01	3.210E-01	3.337E-01	3.241E-01	2.422E-01
Np-238	6.564E+04	6.474E+04	3.298E+00	0.000E+00	0.000E+00	0.000E+00	0.000E+00	0.000E+00	0.000E+00	0.000E+00
Np-239	1.328E+08	1.320E+08	1.927E+04	3.628E-03	3.625E-03	3.595E-03	3.314E-03	1.466E-03	4.210E-07	2.259E-14
Np-240m	1.054E+02	1.012E+02	5.815E-12	5.769E-12	5.769E-12	5.769E-12	5.769E-12	5.768E-12	5.764E-12	5.721E-12
Np-240	4.832E+04	2.498E+04	0.000E+00	0.000E+00	0.000E+00	0.000E+00	0.000E+00	0.000E+00	0.000E+00	0.000E+00
Pu-236	4.393E-01	4.394E-01	4.322E-01	3.457E-01	3.873E-02	1.205E-11	0.000E+00	0.000E+00	0.000E+00	0.000E+00
Pu-238	6.348E+02	6.348E+02	6.384E+02	6.348E+02	5.920E+02	2.938E+02	2.671E-01	7.014E-21	0.000E+00	0.000E+00
Pu-239	8.550E+03	8.550E+03	8.586E+03	8.586E+03	8.586E+03	8.562E+03	8.346E+03	6.462E+03	5.011E+02	3.938E-09
Pu-240	8.412E+02	8.412E+02	8.412E+02	8.412E+02	8.406E+02	8.328E+02	7.590E+02	3.017E+02	2.960E-02	5.728E-12
Pu-241	9.342E+03	9.342E+03	9.306E+03	8.910E+03	5.812E+03	8.100E+01	1.824E-06	8.574E-07	4.517E-10	0.000E+00
Pu-242	5.705E-03	5.705E-03	5.706E-03	5.706E-03	5.707E-03	5.711E-03	5.712E-03	5.619E-03	4.766E-03	9.186E-04
Pu-243	7.722E+01	6.720E+01	2.356E-14	2.356E-14	2.356E-14	2.356E-14	2.356E-14	2.356E-14	2.347E-14	2.259E-14
Pu-244	5.777E-12	5.777E-12	5.777E-12	5.777E-12	5.777E-12	5.777E-12	5.777E-12	5.776E-12	5.772E-12	5.728E-12
Pu-245	7.668E-06	7.182E-06	0.000E+00	0.000E+00	0.000E+00	0.000E+00	0.000E+00	0.000E+00	0.000E+00	0.000E+00
Am-241	1.027E+01	1.027E+01	1.149E+01	2.485E+01	1.282E+02	2.837E+02	6.786E+01	3.834E-05	4.767E-10	0.000E+00
Am-242m	2.265E-01	2.265E-01	2.264E-01	2.255E-01	2.164E-01	1.435E-01	2.368E-03	3.536E-21	0.000E+00	0.000E+00
Am-242	5.012E+02	4.799E+02	2.264E-01	2.255E-01	2.164E-01	1.435E-01	2.368E-03	0.000E+00	0.000E+00	0.000E+00
Am-243	3.622E-03	3.623E-03	3.628E-03	3.628E-03	3.625E-03	3.595E-03	3.314E-03	1.466E-03	4.210E-07	2.259E-14
Am-244	3.040E+00	6.144E-01	7.512E-15	7.512E-15	7.512E-15	7.512E-15	7.512E-15	7.506E-15	7.506E-15	7.446E-15
Am-245	1.337E-05	1.168E-05	0.000E+00	0.000E+00	0.000E+00	0.000E+00	0.000E+00	0.000E+00	0.000E+00	0.000E+00
Cm-242	2.024E+02	2.025E+02	1.797E+02	4.333E+01	1.775E-01	1.177E-01	1.942E-03	0.000E+00	0.000E+00	0.000E+00
Cm-243	2.033E-02	2.033E-02	2.030E-02	1.990E-02	1.637E-02	2.330E-03	7.944E-12	0.000E+00	0.000E+00	0.000E+00
Cm-244	4.938E-02	4.939E-02	4.924E-02	4.753E-02	3.368E-02	1.072E-03	7.512E-15	7.506E-15	7.506E-15	7.446E-15
Cm-245	1.980E-06	1.980E-06	1.980E-06	1.980E-06	1.979E-06	1.964E-06	1.821E-06	8.562E-07	4.510E-10	0.000E+00
Cm-246	2.164E-08	2.164E-08	2.164E-08	2.164E-08	2.161E-08	2.132E-08	1.868E-08	4.967E-09	8.802E-15	0.000E+00
Cm-247	2.356E-14	2.356E-14	2.356E-14	2.356E-14	2.356E-14	2.356E-14	2.356E-14	2.356E-14	2.347E-14	2.259E-14
Cm-248	7.038E-15	7.038E-15	7.038E-15	7.038E-15	7.038E-15	7.038E-15	7.026E-15	6.900E-15	5.779E-15	9.822E-16
Cm-249	7.494E-11	3.914E-11	0.000E+00	0.000E+00	0.000E+00	0.000E+00	0.000E+00	0.000E+00	0.000E+00	0.000E+00
Cm-250	2.487E-21	2.487E-21	2.487E-21	2.487E-21	2.487E-21	2.487E-21	2.398E-21	1.673E-21	4.985E-23	0.000E+00
Bk-249	1.212E-11	1.212E-11	1.135E-11	5.416E-12	3.818E-15	0.000E+00	0.000E+00	0.000E+00	0.000E+00	0.000E+00
Bk-250	3.984E-13	3.213E-13	0.000E+00	0.000E+00	0.000E+00	0.000E+00	0.000E+00	0.000E+00	0.000E+00	0.000E+00
Cf-249	5.038E-15	5.041E-15	6.936E-15	2.141E-14	3.404E-14	2.852E-14	4.847E-15	0.000E+00	0.000E+00	0.000E+00
Cf-250	6.042E-15	6.042E-15	6.024E-15	5.740E-15	3.563E-15	3.022E-17	0.000E+00	0.000E+00	0.000E+00	0.000E+00
Cf-251	4.880E-19	4.880E-19	4.880E-19	4.876E-19	4.843E-19	4.519E-19	2.258E-19	0.000E+00	0.000E+00	0.000E+00
Cf-252	5.155E-19	5.155E-19	5.155E-19	3.915E-19	0.000E+00	0.000E+00	0.000E+00	0.000E+00	0.000E+00	0.000E+00
Cf-253	0.000E+00	0.000E+00	0.000E+00	0.000E+00	0.000E+00	0.000E+00	0.000E+00	0.000E+00	0.000E+00	0.000E+00
Cf-254	0.000E+00	0.000E+00	0.000E+00	0.000E+00	0.000E+00	0.000E+00	0.000E+00	0.000E+00	0.000E+00	0.000E+00
Es-253	0.000E+00	0.000E+00	0.000E+00	0.000E+00	0.000E+00	0.000E+00	0.000E+00	0.000E+00	0.000E+00	0.000E+00
Total	1.330E+08	1.321E+08	3.883E+04	1.904E+04	1.596E+04	1.005E+04	9.173E+03	6.764E+03	5.015E+02	2.431E-01
Total/MWd	2.723E+02	2.706E+02	7.954E-02	3.900E-02	3.269E-02	2.059E-02	1.879E-02	1.385E-02	1.027E-03	4.979E-07

Table A.59

Breeder core heavy TRU elements nuclide radioactivity (inner blanket), Curies Basis = EQULIBRIUM FRESH FUEL BREEDER CORE

	Initial	1 hour	30 days	1 year	10 y	100 y	1000 y	10000 y	100000 y	1000000 y
Np-236	0.000E+00	0.000E+00	0.000E+00	0.000E+00	0.000E+00	0.000E+00	0.000E+00	0.000E+00	0.000E+00	0.000E+00
Np-237	0.000E+00	0.000E+00	0.000E+00	0.000E+00	0.000E+00	0.000E+00	0.000E+00	0.000E+00	0.000E+00	0.000E+00
Np-238	0.000E+00	0.000E+00	0.000E+00	0.000E+00	0.000E+00	0.000E+00	0.000E+00	0.000E+00	0.000E+00	0.000E+00
Np-239	0.000E+00	0.000E+00	0.000E+00	0.000E+00	0.000E+00	0.000E+00	0.000E+00	0.000E+00	0.000E+00	0.000E+00
Np-240m	0.000E+00	0.000E+00	0.000E+00	0.000E+00	0.000E+00	0.000E+00	0.000E+00	0.000E+00	0.000E+00	0.000E+00
Np-240	0.000E+00	0.000E+00	0.000E+00	0.000E+00	0.000E+00	0.000E+00	0.000E+00	0.000E+00	0.000E+00	0.000E+00
Pu-236	0.000E+00	0.000E+00	0.000E+00	0.000E+00	0.000E+00	0.000E+00	0.000E+00	0.000E+00	0.000E+00	0.000E+00
Pu-238	0.000E+00	0.000E+00	0.000E+00	0.000E+00	0.000E+00	0.000E+00	0.000E+00	0.000E+00	0.000E+00	0.000E+00
Pu-239	0.000E+00	0.000E+00	0.000E+00	0.000E+00	0.000E+00	0.000E+00	0.000E+00	0.000E+00	0.000E+00	0.000E+00
Pu-240	0.000E+00	0.000E+00	0.000E+00	0.000E+00	0.000E+00	0.000E+00	0.000E+00	0.000E+00	0.000E+00	0.000E+00
Pu-241	0.000E+00	0.000E+00	0.000E+00	0.000E+00	0.000E+00	0.000E+00	0.000E+00	0.000E+00	0.000E+00	0.000E+00
Pu-242	0.000E+00	0.000E+00	0.000E+00	0.000E+00	0.000E+00	0.000E+00	0.000E+00	0.000E+00	0.000E+00	0.000E+00
Pu-243	0.000E+00	0.000E+00	0.000E+00	0.000E+00	0.000E+00	0.000E+00	0.000E+00	0.000E+00	0.000E+00	0.000E+00
Pu-244	0.000E+00	0.000E+00	0.000E+00	0.000E+00	0.000E+00	0.000E+00	0.000E+00	0.000E+00	0.000E+00	0.000E+00
Pu-245	0.000E+00	0.000E+00	0.000E+00	0.000E+00	0.000E+00	0.000E+00	0.000E+00	0.000E+00	0.000E+00	0.000E+00
Am-241	0.000E+00	0.000E+00	0.000E+00	0.000E+00	0.000E+00	0.000E+00	0.000E+00	0.000E+00	0.000E+00	0.000E+00
Am-242m	0.000E+00	0.000E+00	0.000E+00	0.000E+00	0.000E+00	0.000E+00	0.000E+00	0.000E+00	0.000E+00	0.000E+00
Am-242	0.000E+00	0.000E+00	0.000E+00	0.000E+00	0.000E+00	0.000E+00	0.000E+00	0.000E+00	0.000E+00	0.000E+00
Am-243	0.000E+00	0.000E+00	0.000E+00	0.000E+00	0.000E+00	0.000E+00	0.000E+00	0.000E+00	0.000E+00	0.000E+00
Am-244	0.000E+00	0.000E+00	0.000E+00	0.000E+00	0.000E+00	0.000E+00	0.000E+00	0.000E+00	0.000E+00	0.000E+00
Am-245	0.000E+00	0.000E+00	0.000E+00	0.000E+00	0.000E+00	0.000E+00	0.000E+00	0.000E+00	0.000E+00	0.000E+00
Cm-242	0.000E+00	0.000E+00	0.000E+00	0.000E+00	0.000E+00	0.000E+00	0.000E+00	0.000E+00	0.000E+00	0.000E+00
Cm-243	0.000E+00	0.000E+00	0.000E+00	0.000E+00	0.000E+00	0.000E+00	0.000E+00	0.000E+00	0.000E+00	0.000E+00
Cm-244	0.000E+00	0.000E+00	0.000E+00	0.000E+00	0.000E+00	0.000E+00	0.000E+00	0.000E+00	0.000E+00	0.000E+00
Cm-245	0.000E+00	0.000E+00	0.000E+00	0.000E+00	0.000E+00	0.000E+00	0.000E+00	0.000E+00	0.000E+00	0.000E+00
Cm-246	0.000E+00	0.000E+00	0.000E+00	0.000E+00	0.000E+00	0.000E+00	0.000E+00	0.000E+00	0.000E+00	0.000E+00
Cm-247	0.000E+00	0.000E+00	0.000E+00	0.000E+00	0.000E+00	0.000E+00	0.000E+00	0.000E+00	0.000E+00	0.000E+00
Cm-248	0.000E+00	0.000E+00	0.000E+00	0.000E+00	0.000E+00	0.000E+00	0.000E+00	0.000E+00	0.000E+00	0.000E+00
Cm-249	0.000E+00	0.000E+00	0.000E+00	0.000E+00	0.000E+00	0.000E+00	0.000E+00	0.000E+00	0.000E+00	0.000E+00
Cm-250	0.000E+00	0.000E+00	0.000E+00	0.000E+00	0.000E+00	0.000E+00	0.000E+00	0.000E+00	0.000E+00	0.000E+00
Bk-249	0.000E+00	0.000E+00	0.000E+00	0.000E+00	0.000E+00	0.000E+00	0.000E+00	0.000E+00	0.000E+00	0.000E+00
Bk-250	0.000E+00	0.000E+00	0.000E+00	0.000E+00	0.000E+00	0.000E+00	0.000E+00	0.000E+00	0.000E+00	0.000E+00
Cf-249	0.000E+00	0.000E+00	0.000E+00	0.000E+00	0.000E+00	0.000E+00	0.000E+00	0.000E+00	0.000E+00	0.000E+00
Cf-250	0.000E+00	0.000E+00	0.000E+00	0.000E+00	0.000E+00	0.000E+00	0.000E+00	0.000E+00	0.000E+00	0.000E+00
Cf-251	0.000E+00	0.000E+00	0.000E+00	0.000E+00	0.000E+00	0.000E+00	0.000E+00	0.000E+00	0.000E+00	0.000E+00
Cf-252	0.000E+00	0.000E+00	0.000E+00	0.000E+00	0.000E+00	0.000E+00	0.000E+00	0.000E+00	0.000E+00	0.000E+00
Cf-253	0.000E+00	0.000E+00	0.000E+00	0.000E+00	0.000E+00	0.000E+00	0.000E+00	0.000E+00	0.000E+00	0.000E+00
Cf-254	0.000E+00	0.000E+00	0.000E+00	0.000E+00	0.000E+00	0.000E+00	0.000E+00	0.000E+00	0.000E+00	0.000E+00
Es-253	0.000E+00	0.000E+00	0.000E+00	0.000E+00	0.000E+00	0.000E+00	0.000E+00	0.000E+00	0.000E+00	0.000E+00
Total	0.000E+00	0.000E+00	0.000E+00	0.000E+00	0.000E+00	0.000E+00	0.000E+00	0.000E+00	0.000E+00	0.000E+00

Table A.60

Breeder core heavy TRU hlements nuclide radioactivity (inner blanket), Curies Basis = EQULIBRIUM DISCHARGE FUEL BREEDER CORE

	Initial	1 hour	30 days	1 year	10 y	100 y	1000 y	10000 y	100000 y	1000000 y
Np-236	4.482E-03	4.343E-03	6.354E-13	0.000E+00	0.000E+00	0.000E+00	0.000E+00	0.000E+00	0.000E+00	0.000E+00
Np-237	3.811E-03	3.811E-03	3.841E-03	3.842E-03	3.843E-03	3.851E-03	3.895E-03	3.897E-03	3.785E-03	2.828E-03
Np-238	1.592E+02	1.570E+02	7.998E-03	0.000E+00	0.000E+00	0.000E+00	0.000E+00	0.000E+00	0.000E+00	0.000E+00
Np-239	2.103E+06	2.089E+06	3.050E+02	1.629E-07	1.628E-07	1.615E-07	1.488E-07	6.582E-08	1.891E-11	0.000E+00
Np-240m	2.596E-01	2.492E-01	0.000E+00	0.000E+00	0.000E+00	0.000E+00	0.000E+00	0.000E+00	0.000E+00	0.000E+00
Np-240	1.190E+02	6.156E+01	0.000E+00	0.000E+00	0.000E+00	0.000E+00	0.000E+00	0.000E+00	0.000E+00	0.000E+00
Pu-236	7.818E-04	7.818E-04	7.686E-04	6.150E-04	6.888E-05	2.142E-14	0.000E+00	0.000E+00	0.000E+00	0.000E+00
Pu-238	2.110E+00	2.110E+00	2.119E+00	2.104E+00	1.962E+00	9.732E-01	8.790E-04	0.000E+00	0.000E+00	0.000E+00
Pu-239	1.960E+02	1.960E+02	1.965E+02	1.965E+02	1.964E+02	1.960E+02	1.910E+02	1.479E+02	1.147E+01	9.012E-11
Pu-240	4.285E+00	4.285E+00	4.285E+00	4.285E+00	4.280E+00	4.241E+00	3.867E+00	1.537E+00	1.508E-04	4.025E-17
Pu-241	9.384E+00	9.384E+00	9.348E+00	8.946E+00	5.837E+00	8.136E-02	3.500E-12	1.645E-12	8.670E-16	0.000E+00
Pu-242	1.240E-06	1.240E-06	1.240E-06	1.240E-06	1.240E-06	1.241E-06	1.243E-06	1.222E-06	1.037E-06	1.999E-07
Pu-243	2.608E-03	2.270E-03	0.000E+00	0.000E+00	0.000E+00	0.000E+00	0.000E+00	0.000E+00	0.000E+00	0.000E+00
Pu-244	4.059E-17	4.059E-17	4.059E-17	4.059E-17	4.059E-17	4.059E-17	4.059E-17	4.058E-17	4.055E-17	4.025E-17
Pu-245	8.238E-12	7.716E-12	0.000E+00	0.000E+00	0.000E+00	0.000E+00	0.000E+00	0.000E+00	0.000E+00	0.000E+00
Am-241	1.414E-02	1.414E-02	1.537E-02	2.878E-02	1.325E-01	2.882E-01	6.888E-02	3.808E-08	9.144E-16	0.000E+00
Am-242m	7.170E-05	7.170E-05	7.170E-05	7.140E-05	6.852E-05	4.545E-05	7.500E-07	0.000E+00	0.000E+00	0.000E+00
Am-242	1.154E-01	1.105E-01	7.170E-05	7.140E-05	6.852E-05	4.545E-05	7.500E-07	0.000E+00	0.000E+00	0.000E+00
Am-243	1.628E-07	1.628E-07	1.630E-07	1.629E-07	1.628E-07	1.615E-07	1.488E-07	6.582E-08	1.891E-11	1.796E-21
Am-244	2.177E-05	4.403E-06	0.000E+00	0.000E+00	0.000E+00	0.000E+00	0.000E+00	0.000E+00	0.000E+00	0.000E+00
Am-245	1.365E-11	1.203E-11	0.000E+00	0.000E+00	0.000E+00	0.000E+00	0.000E+00	0.000E+00	0.000E+00	0.000E+00
Cm-242	5.241E-02	5.242E-02	4.648E-02	1.121E-02	5.619E-05	3.727E-05	6.150E-07	0.000E+00	0.000E+00	0.000E+00
Cm-243	1.090E-06	1.090E-06	1.088E-06	1.066E-06	8.778E-07	1.249E-07	4.258E-16	0.000E+00	0.000E+00	0.000E+00
Cm-244	4.631E-07	4.631E-07	4.616E-07	4.457E-07	3.158E-07	1.006E-08	5.278E-20	5.276E-20	5.272E-20	5.232E-20
Cm-245	3.800E-12	3.800E-12	3.799E-12	3.799E-12	3.796E-12	3.768E-12	3.494E-12	1.642E-12	8.652E-16	0.000E+00
Cm-246	8.598E-15	8.598E-15	8.598E-15	8.592E-15	8.586E-15	8.472E-15	7.422E-15	1.973E-15	3.497E-21	0.000E+00
Cm-247	1.874E-21	1.874E-21	1.874E-21	1.874E-21	1.874E-21	1.874E-21	1.873E-21	1.873E-21	1.865E-21	1.796E-21
Cm-248	8.580E-23	8.580E-23	8.580E-23	8.580E-23	8.580E-23	8.580E-23	8.580E-23	8.430E-23	7.068E-23	1.187E-23
Cm-249	0.000E+00	0.000E+00	0.000E+00	0.000E+00	0.000E+00	0.000E+00	0.000E+00	0.000E+00	0.000E+00	0.000E+00
Cm-250	0.000E+00	0.000E+00	0.000E+00	0.000E+00	0.000E+00	0.000E+00	0.000E+00	0.000E+00	0.000E+00	0.000E+00
Bk-249	0.000E+00	0.000E+00	0.000E+00	0.000E+00	0.000E+00	0.000E+00	0.000E+00	0.000E+00	0.000E+00	0.000E+00
Bk-250	0.000E+00	0.000E+00	0.000E+00	0.000E+00	0.000E+00	0.000E+00	0.000E+00	0.000E+00	0.000E+00	0.000E+00
Cf-249	0.000E+00	0.000E+00	0.000E+00	0.000E+00	0.000E+00	0.000E+00	0.000E+00	0.000E+00	0.000E+00	0.000E+00
Cf-250	0.000E+00	0.000E+00	0.000E+00	0.000E+00	0.000E+00	0.000E+00	0.000E+00	0.000E+00	0.000E+00	0.000E+00
Cf-251	0.000E+00	0.000E+00	0.000E+00	0.000E+00	0.000E+00	0.000E+00	0.000E+00	0.000E+00	0.000E+00	0.000E+00
Cf-252	0.000E+00	0.000E+00	0.000E+00	0.000E+00	0.000E+00	0.000E+00	0.000E+00	0.000E+00	0.000E+00	0.000E+00
Cf-253	0.000E+00	0.000E+00	0.000E+00	0.000E+00	0.000E+00	0.000E+00	0.000E+00	0.000E+00	0.000E+00	0.000E+00
Cf-254	0.000E+00	0.000E+00	0.000E+00	0.000E+00	0.000E+00	0.000E+00	0.000E+00	0.000E+00	0.000E+00	0.000E+00
Es-253	0.000E+00	0.000E+00	0.000E+00	0.000E+00	0.000E+00	0.000E+00	0.000E+00	0.000E+00	0.000E+00	0.000E+00
Total	2.103E+06	2.090E+06	5.173E+02	2.119E+02	2.087E+02	2.015E+02	1.950E+02	1.494E+02	1.148E+01	2.828E-03

Table A.61

Breeder core heavy TRU elements nuclide radioactivity (inner blanket), Curies

Basis = (EQULIBRIUM DISCHARGE FUEL - EQULIBRIUM FRESH FUEL) BREEDER CORE

	Initial	1 hour	30 days	1 year	10 y	100 y	1000 y	10000 y	100000 y	1000000 y
Np-236	4.482E-03	4.343E-03	6.354E-13	0.000E+00	0.000E+00	0.000E+00	0.000E+00	0.000E+00	0.000E+00	0.000E+00
Np-237	3.811E-03	3.811E-03	3.841E-03	3.842E-03	3.843E-03	3.851E-03	3.895E-03	3.897E-03	3.785E-03	2.828E-03
Np-238	1.592E+02	1.570E+02	7.998E-03	0.000E+00	0.000E+00	0.000E+00	0.000E+00	0.000E+00	0.000E+00	0.000E+00
Np-239	2.103E+06	2.089E+06	3.050E+02	1.629E-07	1.628E-07	1.615E-07	1.488E-07	6.582E-08	1.891E-11	0.000E+00
Np-240m	2.596E-01	2.492E-01	0.000E+00	0.000E+00	0.000E+00	0.000E+00	0.000E+00	0.000E+00	0.000E+00	0.000E+00
Np-240	1.190E+02	6.156E+01	0.000E+00	0.000E+00	0.000E+00	0.000E+00	0.000E+00	0.000E+00	0.000E+00	0.000E+00
Pu-236	7.818E-04	7.818E-04	7.686E-04	6.150E-04	6.888E-05	2.142E-14	0.000E+00	0.000E+00	0.000E+00	0.000E+00
Pu-238	2.110E+00	2.110E+00	2.119E+00	2.104E+00	1.962E+00	9.732E-01	8.790E-04	0.000E+00	0.000E+00	0.000E+00
Pu-239	1.960E+02	1.960E+02	1.965E+02	1.965E+02	1.964E+02	1.960E+02	1.910E+02	1.479E+02	1.147E+01	9.012E-11
Pu-240	4.285E+00	4.285E+00	4.285E+00	4.285E+00	4.280E+00	4.241E+00	3.867E+00	1.537E+00	1.508E-04	4.025E-17
Pu-241	9.384E+00	9.384E+00	9.348E+00	8.946E+00	5.837E+00	8.136E-02	3.500E-12	1.645E-12	8.670E-16	0.000E+00
Pu-242	1.240E-06	1.240E-06	1.240E-06	1.240E-06	1.240E-06	1.241E-06	1.243E-06	1.222E-06	1.037E-06	1.999E-07
Pu-243	2.608E-03	2.270E-03	0.000E+00	0.000E+00	0.000E+00	0.000E+00	0.000E+00	0.000E+00	0.000E+00	0.000E+00
Pu-244	4.059E-17	4.059E-17	4.059E-17	4.059E-17	4.059E-17	4.059E-17	4.059E-17	4.058E-17	4.055E-17	4.025E-17
Pu-245	8.238E-12	7.716E-12	0.000E+00	0.000E+00	0.000E+00	0.000E+00	0.000E+00	0.000E+00	0.000E+00	0.000E+00
Am-241	1.414E-02	1.414E-02	1.537E-02	2.878E-02	1.325E-01	2.882E-01	6.888E-02	3.808E-08	9.144E-16	0.000E+00
Am-242m	7.170E-05	7.170E-05	7.170E-05	7.140E-05	6.852E-05	4.545E-05	7.500E-07	0.000E+00	0.000E+00	0.000E+00
Am-242	1.154E-01	1.105E-01	7.170E-05	7.140E-05	6.852E-05	4.545E-05	7.500E-07	0.000E+00	0.000E+00	0.000E+00
Am-243	1.628E-07	1.628E-07	1.630E-07	1.629E-07	1.628E-07	1.615E-07	1.488E-07	6.582E-08	1.891E-11	1.796E-21
Am-244	2.177E-05	4.403E-06	0.000E+00	0.000E+00	0.000E+00	0.000E+00	0.000E+00	0.000E+00	0.000E+00	0.000E+00
Am-245	1.365E-11	1.203E-11	0.000E+00	0.000E+00	0.000E+00	0.000E+00	0.000E+00	0.000E+00	0.000E+00	0.000E+00
Cm-242	5.241E-02	5.242E-02	4.648E-02	1.121E-02	5.619E-05	3.727E-05	6.150E-07	0.000E+00	0.000E+00	0.000E+00
Cm-243	1.090E-06	1.090E-06	1.088E-06	1.066E-06	8.778E-07	1.249E-07	4.258E-16	0.000E+00	0.000E+00	0.000E+00
Cm-244	4.631E-07	4.631E-07	4.616E-07	4.457E-07	3.158E-07	1.006E-08	5.278E-20	5.276E-20	5.272E-20	5.232E-20
Cm-245	3.800E-12	3.800E-12	3.799E-12	3.799E-12	3.796E-12	3.768E-12	3.494E-12	1.642E-12	8.652E-16	0.000E+00
Cm-246	8.598E-15	8.598E-15	8.598E-15	8.592E-15	8.586E-15	8.472E-15	7.422E-15	1.973E-15	3.497E-21	0.000E+00
Cm-247	1.874E-21	1.874E-21	1.874E-21	1.874E-21	1.874E-21	1.874E-21	1.873E-21	1.873E-21	1.865E-21	1.796E-21
Cm-248	8.580E-23	8.580E-23	8.580E-23	8.580E-23	8.580E-23	8.580E-23	8.580E-23	8.430E-23	7.068E-23	1.187E-23
Cm-249	0.000E+00	0.000E+00	0.000E+00	0.000E+00	0.000E+00	0.000E+00	0.000E+00	0.000E+00	0.000E+00	0.000E+00
Cm-250	0.000E+00	0.000E+00	0.000E+00	0.000E+00	0.000E+00	0.000E+00	0.000E+00	0.000E+00	0.000E+00	0.000E+00
Bk-249	0.000E+00	0.000E+00	0.000E+00	0.000E+00	0.000E+00	0.000E+00	0.000E+00	0.000E+00	0.000E+00	0.000E+00
Bk-250	0.000E+00	0.000E+00	0.000E+00	0.000E+00	0.000E+00	0.000E+00	0.000E+00	0.000E+00	0.000E+00	0.000E+00
Cf-249	0.000E+00	0.000E+00	0.000E+00	0.000E+00	0.000E+00	0.000E+00	0.000E+00	0.000E+00	0.000E+00	0.000E+00
Cf-250	0.000E+00	0.000E+00	0.000E+00	0.000E+00	0.000E+00	0.000E+00	0.000E+00	0.000E+00	0.000E+00	0.000E+00
Cf-251	0.000E+00	0.000E+00	0.000E+00	0.000E+00	0.000E+00	0.000E+00	0.000E+00	0.000E+00	0.000E+00	0.000E+00
Cf-252	0.000E+00	0.000E+00	0.000E+00	0.000E+00	0.000E+00	0.000E+00	0.000E+00	0.000E+00	0.000E+00	0.000E+00
Cf-253	0.000E+00	0.000E+00	0.000E+00	0.000E+00	0.000E+00	0.000E+00	0.000E+00	0.000E+00	0.000E+00	0.000E+00
Cf-254	0.000E+00	0.000E+00	0.000E+00	0.000E+00	0.000E+00	0.000E+00	0.000E+00	0.000E+00	0.000E+00	0.000E+00
Es-253	0.000E+00	0.000E+00	0.000E+00	0.000E+00	0.000E+00	0.000E+00	0.000E+00	0.000E+00	0.000E+00	0.000E+00
Total	2.103E+06	2.090E+06	5.173E+02	2.119E+02	2.087E+02	2.015E+02	1.950E+02	1.494E+02	1.148E+01	2.828E-03
Total/MWd	4.308E+00	4.280E+00	1.060E-03	4.340E-04	4.274E-04	4.128E-04	3.993E-04	3.061E-04	2.350E-05	5.792E-09

Table A.62

Breeder core heavy TRU elements nuclide radioactivity (outer blanket), Curies Basis = EQULIBRIUM FRESH FUEL BREEDER CORE

	Initial	1 hour	30 days	1 year	10 y	100 y	1000 y	10000 y	100000 y	1000000 y
Np-236	0.000E+00	0.000E+00	0.000E+00	0.000E+00	0.000E+00	0.000E+00	0.000E+00	0.000E+00	0.000E+00	0.000E+00
Np-237	0.000E+00	0.000E+00	0.000E+00	0.000E+00	0.000E+00	0.000E+00	0.000E+00	0.000E+00	0.000E+00	0.000E+00
Np-238	0.000E+00	0.000E+00	0.000E+00	0.000E+00	0.000E+00	0.000E+00	0.000E+00	0.000E+00	0.000E+00	0.000E+00
Np-239	0.000E+00	0.000E+00	0.000E+00	0.000E+00	0.000E+00	0.000E+00	0.000E+00	0.000E+00	0.000E+00	0.000E+00
Np-240m	0.000E+00	0.000E+00	0.000E+00	0.000E+00	0.000E+00	0.000E+00	0.000E+00	0.000E+00	0.000E+00	0.000E+00
Np-240	0.000E+00	0.000E+00	0.000E+00	0.000E+00	0.000E+00	0.000E+00	0.000E+00	0.000E+00	0.000E+00	0.000E+00
Pu-236	0.000E+00	0.000E+00	0.000E+00	0.000E+00	0.000E+00	0.000E+00	0.000E+00	0.000E+00	0.000E+00	0.000E+00
Pu-238	0.000E+00	0.000E+00	0.000E+00	0.000E+00	0.000E+00	0.000E+00	0.000E+00	0.000E+00	0.000E+00	0.000E+00
Pu-239	0.000E+00	0.000E+00	0.000E+00	0.000E+00	0.000E+00	0.000E+00	0.000E+00	0.000E+00	0.000E+00	0.000E+00
Pu-240	0.000E+00	0.000E+00	0.000E+00	0.000E+00	0.000E+00	0.000E+00	0.000E+00	0.000E+00	0.000E+00	0.000E+00
Pu-241	0.000E+00	0.000E+00	0.000E+00	0.000E+00	0.000E+00	0.000E+00	0.000E+00	0.000E+00	0.000E+00	0.000E+00
Pu-242	0.000E+00	0.000E+00	0.000E+00	0.000E+00	0.000E+00	0.000E+00	0.000E+00	0.000E+00	0.000E+00	0.000E+00
Pu-243	0.000E+00	0.000E+00	0.000E+00	0.000E+00	0.000E+00	0.000E+00	0.000E+00	0.000E+00	0.000E+00	0.000E+00
Pu244	0.000E+00	0.000E+00	0.000E+00	0.000E+00	0.000E+00	0.000E+00	0.000E+00	0.000E+00	0.000E+00	0.000E+00
Pu-245	0.000E+00	0.000E+00	0.000E+00	0.000E+00	0.000E+00	0.000E+00	0.000E+00	0.000E+00	0.000E+00	0.000E+00
Am-241	0.000E+00	0.000E+00	0.000E+00	0.000E+00	0.000E+00	0.000E+00	0.000E+00	0.000E+00	0.000E+00	0.000E+00
Am-242m	0.000E+00	0.000E+00	0.000E+00	0.000E+00	0.000E+00	0.000E+00	0.000E+00	0.000E+00	0.000E+00	0.000E+00
Am-242	0.000E+00	0.000E+00	0.000E+00	0.000E+00	0.000E+00	0.000E+00	0.000E+00	0.000E+00	0.000E+00	0.000E+00
Am-243	0.000E+00	0.000E+00	0.000E+00	0.000E+00	0.000E+00	0.000E+00	0.000E+00	0.000E+00	0.000E+00	0.000E+00
Am-244	0.000E+00	0.000E+00	0.000E+00	0.000E+00	0.000E+00	0.000E+00	0.000E+00	0.000E+00	0.000E+00	0.000E+00
Am-245	0.000E+00	0.000E+00	0.000E+00	0.000E+00	0.000E+00	0.000E+00	0.000E+00	0.000E+00	0.000E+00	0.000E+00
Cm-242	0.000E+00	0.000E+00	0.000E+00	0.000E+00	0.000E+00	0.000E+00	0.000E+00	0.000E+00	0.000E+00	0.000E+00
Cm-243	0.000E+00	0.000E+00	0.000E+00	0.000E+00	0.000E+00	0.000E+00	0.000E+00	0.000E+00	0.000E+00	0.000E+00
Cm-244	0.000E+00	0.000E+00	0.000E+00	0.000E+00	0.000E+00	0.000E+00	0.000E+00	0.000E+00	0.000E+00	0.000E+00
Cm-245	0.000E+00	0.000E+00	0.000E+00	0.000E+00	0.000E+00	0.000E+00	0.000E+00	0.000E+00	0.000E+00	0.000E+00
Cm-246	0.000E+00	0.000E+00	0.000E+00	0.000E+00	0.000E+00	0.000E+00	0.000E+00	0.000E+00	0.000E+00	0.000E+00
Cm-247	0.000E+00	0.000E+00	0.000E+00	0.000E+00	0.000E+00	0.000E+00	0.000E+00	0.000E+00	0.000E+00	0.000E+00
Cm-248	0.000E+00	0.000E+00	0.000E+00	0.000E+00	0.000E+00	0.000E+00	0.000E+00	0.000E+00	0.000E+00	0.000E+00
Cm-249	0.000E+00	0.000E+00	0.000E+00	0.000E+00	0.000E+00	0.000E+00	0.000E+00	0.000E+00	0.000E+00	0.000E+00
Cm-250	0.000E+00	0.000E+00	0.000E+00	0.000E+00	0.000E+00	0.000E+00	0.000E+00	0.000E+00	0.000E+00	0.000E+00
Bk-249	0.000E+00	0.000E+00	0.000E+00	0.000E+00	0.000E+00	0.000E+00	0.000E+00	0.000E+00	0.000E+00	0.000E+00
Bk-250	0.000E+00	0.000E+00	0.000E+00	0.000E+00	0.000E+00	0.000E+00	0.000E+00	0.000E+00	0.000E+00	0.000E+00
Cf-249	0.000E+00	0.000E+00	0.000E+00	0.000E+00	0.000E+00	0.000E+00	0.000E+00	0.000E+00	0.000E+00	0.000E+00
Cf-250	0.000E+00	0.000E+00	0.000E+00	0.000E+00	0.000E+00	0.000E+00	0.000E+00	0.000E+00	0.000E+00	0.000E+00
Cf-251	0.000E+00	0.000E+00	0.000E+00	0.000E+00	0.000E+00	0.000E+00	0.000E+00	0.000E+00	0.000E+00	0.000E+00
Cf-252	0.000E+00	0.000E+00	0.000E+00	0.000E+00	0.000E+00	0.000E+00	0.000E+00	0.000E+00	0.000E+00	0.000E+00
Cf-253	0.000E+00	0.000E+00	0.000E+00	0.000E+00	0.000E+00	0.000E+00	0.000E+00	0.000E+00	0.000E+00	0.000E+00
Cf-254	0.000E+00	0.000E+00	0.000E+00	0.000E+00	0.000E+00	0.000E+00	0.000E+00	0.000E+00	0.000E+00	0.000E+00
Es-253	0.000E+00	0.000E+00	0.000E+00	0.000E+00	0.000E+00	0.000E+00	0.000E+00	0.000E+00	0.000E+00	0.000E+00
Total	0.000E+00	0.000E+00	0.000E+00	0.000E+00	0.000E+00	0.000E+00	0.000E+00	0.000E+00	0.000E+00	0.000E+00

Table A.63

Breeder core heavy TRU elements nuclide radioactivity (outer blanket), Curies Basis = EQULIBRIUM DISCHARGE FUEL BREEDER CORE

	Initial	1 hour	30 days	1 year	10 y	100 y	1000 y	10000 y	100000 y	1000000 y
Np-236	8.244E-01	7.986E-01	1.169E-10	0.000E+00	0.000E+00	0.000E+00	0.000E+00	0.000E+00	0.000E+00	0.000E+00
Np-237	1.004E-01	1.004E-01	1.013E-01	1.013E-01	1.013E-01	1.033E-01	1.144E-01	1.175E-01	1.141E-01	8.526E-02
Np-238	1.522E+04	1.501E+04	7.650E-01	0.000E+00	0.000E+00	0.000E+00	0.000E+00	0.000E+00	0.000E+00	0.000E+00
Np-239	3.170E+07	3.150E+07	4.598E+03	6.330E-04	6.330E-04	6.276E-04	5.784E-04	2.559E-04	7.350E-08	1.711E-15
Np-240m	1.538E+01	1.477E+01	6.372E-13	6.306E-13	6.306E-13	6.306E-13	6.306E-13	6.300E-13	6.300E-13	6.252E-13
Np-240	7.050E+03	3.645E+03	0.000E+00	0.000E+00	0.000E+00	0.000E+00	0.000E+00	0.000E+00	0.000E+00	0.000E+00
Pu-236	1.407E-01	1.407E-01	1.383E-01	1.106E-01	1.240E-02	3.855E-12	0.000E+00	0.000E+00	0.000E+00	0.000E+00
Pu-238	1.962E+02	1.962E+02	1.971E+02	1.958E+02	1.826E+02	9.060E+01	8.232E-02	0.000E+00	0.000E+00	0.000E+00
Pu-239	2.749E+03	2.749E+03	2.757E+03	2.757E+03	2.756E+03	2.749E+03	2.680E+03	2.075E+03	1.609E+02	1.265E-09
Pu-240	2.326E+02	2.326E+02	2.326E+02	2.325E+02	2.323E+02	2.302E+02	2.099E+02	8.340E+01	8.184E-03	6.258E-13
Pu-241	2.320E+03	2.320E+03	2.311E+03	2.213E+03	1.443E+03	2.011E+01	2.132E-07	1.002E-07	5.279E-11	0.000E+00
Pu-242	1.178E-03	1.178E-03	1.178E-03	1.178E-03	1.178E-03	1.180E-03	1.181E-03	1.162E-03	9.852E-04	1.899E-04
Pu-243	1.018E+01	8.862E+00	1.786E-15	1.786E-15	1.786E-15	1.786E-15	1.785E-15	1.784E-15	1.778E-15	1.711E-15
Pu-244	6.312E-13	6.312E-13	6.312E-13	6.312E-13	6.312E-13	6.312E-13	6.312E-13	6.312E-13	6.306E-13	6.258E-13
Pu-245	6.624E-07	6.210E-07	0.000E+00	0.000E+00	0.000E+00	0.000E+00	0.000E+00	0.000E+00	0.000E+00	0.000E+00
Am-241	3.475E+00	3.475E+00	3.779E+00	7.092E+00	3.274E+01	7.122E+01	1.703E+01	9.510E-06	5.571E-11	0.000E+00
Am-242m	6.276E-02	6.276E-02	6.276E-02	6.252E-02	5.999E-02	3.979E-02	6.564E-04	0.000E+00	0.000E+00	0.000E+00
Am-242	1.047E+02	1.003E+02	6.276E-02	6.252E-02	5.999E-02	3.980E-02	6.564E-04	0.000E+00	0.000E+00	0.000E+00
Am-243	6.324E-04	6.324E-04	6.330E-04	6.330E-04	6.330E-04	6.276E-04	5.784E-04	2.559E-04	7.350E-08	1.711E-15
Am-244	3.320E-01	6.714E-02	0.000E+00	0.000E+00	0.000E+00	0.000E+00	0.000E+00	0.000E+00	0.000E+00	0.000E+00
Am-245	9.798E-07	8.832E-07	0.000E+00	0.000E+00	0.000E+00	0.000E+00	0.000E+00	0.000E+00	0.000E+00	0.000E+00
Cm-242	4.806E+01	4.807E+01	4.262E+01	1.028E+01	4.920E-02	3.263E-02	5.383E-04	0.000E+00	0.000E+00	0.000E+00
Cm-243	4.565E-03	4.565E-03	4.557E-03	4.467E-03	3.676E-03	5.231E-04	1.784E-12	0.000E+00	0.000E+00	0.000E+00
Cm-244	7.086E-03	7.086E-03	7.062E-03	6.816E-03	4.830E-03	1.538E-04	8.208E-16	8.202E-16	8.196E-16	8.136E-16
Cm-245	2.314E-07	2.314E-07	2.314E-07	2.314E-07	2.312E-07	2.295E-07	2.128E-07	1.000E-07	5.270E-11	0.000E+00
Cm-246	1.856E-09	1.856E-09	1.856E-09	1.856E-09	1.853E-09	1.829E-09	1.602E-09	4.262E-10	7.554E-16	0.000E+00
Cm-247	1.786E-15	1.786E-15	1.786E-15	1.786E-15	1.786E-15	1.786E-15	1.785E-15	1.784E-15	1.778E-15	1.711E-15
Cm-248	3.412E-16	3.412E-16	3.412E-16	3.412E-16	3.412E-16	3.412E-16	3.406E-16	3.346E-16	2.802E-16	4.762E-17
Cm-249	1.764E-12	9.228E-13	0.000E+00	0.000E+00	0.000E+00	0.000E+00	0.000E+00	0.000E+00	0.000E+00	0.000E+00
Cm-250	0.000E+00	0.000E+00	0.000E+00	0.000E+00	0.000E+00	0.000E+00	0.000E+00	0.000E+00	0.000E+00	0.000E+00
Bk-249	3.574E-13	3.575E-13	3.348E-13	1.597E-13	1.126E-16	0.000E+00	0.000E+00	0.000E+00	0.000E+00	0.000E+00
Bk-250	5.153E-15	4.045E-15	0.000E+00	0.000E+00	0.000E+00	0.000E+00	0.000E+00	0.000E+00	0.000E+00	0.000E+00
Cf-249	2.012E-16	2.012E-16	2.571E-16	6.840E-16	1.055E-15	8.844E-16	1.503E-16	0.000E+00	0.000E+00	0.000E+00
Cf-250	1.252E-16	1.252E-16	1.248E-16	1.189E-16	7.380E-17	6.294E-19	0.000E+00	0.000E+00	0.000E+00	0.000E+00
Cf-251	4.502E-21	4.502E-21	4.502E-21	4.502E-21	4.502E-21	4.190E-21	2.293E-21	0.000E+00	0.000E+00	0.000E+00
Cf-252	0.000E+00	0.000E+00	0.000E+00	0.000E+00	0.000E+00	0.000E+00	0.000E+00	0.000E+00	0.000E+00	0.000E+00
Cf-253	0.000E+00	0.000E+00	0.000E+00	0.000E+00	0.000E+00	0.000E+00	0.000E+00	0.000E+00	0.000E+00	0.000E+00
Cf-254	0.000E+00	0.000E+00	0.000E+00	0.000E+00	0.000E+00	0.000E+00	0.000E+00	0.000E+00	0.000E+00	0.000E+00
Es-253	0.000E+00	0.000E+00	0.000E+00	0.000E+00	0.000E+00	0.000E+00	0.000E+00	0.000E+00	0.000E+00	0.000E+00
Total	3.173E+07	3.152E+07	1.014E+04	5.416E+03	4.647E+03	3.162E+03	2.907E+03	2.159E+03	1.610E+02	8.545E-02

Table A.64

Breeder core heavy TRU elements nuclide radioactivity (outer blanket), Curies

Basis = (EQULIBRIUM DISCHARGE FUEL - EQULIBRIUM FRESH FUEL) BREEDER CORE

	Initial	**1 hour**	**30 days**	**1 year**	**10 y**	**100 y**	**1000 y**	**10000 y**	**100000 y**	**1000000 y**
Np-236	8.244E-01	7.986E-01	1.169E-10	0.000E+00	0.000E+00	0.000E+00	0.000E+00	0.000E+00	0.000E+00	0.000E+00
Np-237	1.004E-01	1.004E-01	1.013E-01	1.013E-01	1.013E-01	1.033E-01	1.144E-01	1.175E-01	1.141E-01	8.526E-02
Np-238	1.522E+04	1.501E+04	7.650E-01	0.000E+00	0.000E+00	0.000E+00	0.000E+00	0.000E+00	0.000E+00	0.000E+00
Np-239	3.170E+07	3.150E+07	4.598E+03	6.330E-04	6.330E-04	6.276E-04	5.784E-04	2.559E-04	7.350E-08	1.711E-15
Np-240m	1.538E+01	1.477E+01	6.372E-13	6.306E-13	6.306E-13	6.306E-13	6.306E-13	6.300E-13	6.300E-13	6.252E-13
Np-240	7.050E+03	3.645E+03	0.000E+00	0.000E+00	0.000E+00	0.000E+00	0.000E+00	0.000E+00	0.000E+00	0.000E+00
Pu-236	1.407E-01	1.407E-01	1.383E-01	1.106E-01	1.240E-02	3.855E-12	0.000E+00	0.000E+00	0.000E+00	0.000E+00
Pu-238	1.962E+02	1.962E+02	1.971E+02	1.958E+02	1.826E+02	9.060E+01	8.232E-02	0.000E+00	0.000E+00	0.000E+00
Pu-239	2.749E+03	2.749E+03	2.757E+03	2.757E+03	2.756E+03	2.749E+03	2.680E+03	2.075E+03	1.609E+02	1.265E-09
Pu-240	2.326E+02	2.326E+02	2.326E+02	2.325E+02	2.323E+02	2.302E+02	2.099E+02	8.340E+01	8.184E-03	6.258E-13
Pu-241	2.320E+03	2.320E+03	2.311E+03	2.213E+03	1.443E+03	2.011E+01	2.132E-07	1.002E-07	5.279E-11	0.000E+00
Pu-242	1.178E-03	1.178E-03	1.178E-03	1.178E-03	1.178E-03	1.180E-03	1.181E-03	1.162E-03	9.852E-04	1.899E-04
Pu-243	1.018E+01	8.862E+00	1.786E-15	1.786E-15	1.786E-15	1.786E-15	1.785E-15	1.784E-15	1.778E-15	1.711E-15
Pu-244	6.312E-13	6.312E-13	6.312E-13	6.312E-13	6.312E-13	6.312E-13	6.312E-13	6.312E-13	6.306E-13	6.258E-13
Pu-245	6.624E-07	6.210E-07	0.000E+00	0.000E+00	0.000E+00	0.000E+00	0.000E+00	0.000E+00	0.000E+00	0.000E+00
Am-241	3.475E+00	3.475E+00	3.779E+00	7.092E+00	3.274E+01	7.122E+01	1.703E+01	9.510E-06	5.571E-11	0.000E+00
Am-242m	6.276E-02	6.276E-02	6.276E-02	6.252E-02	5.999E-02	3.979E-02	6.564E-04	0.000E+00	0.000E+00	0.000E+00
Am-242	1.047E+02	1.003E+02	6.276E-02	6.252E-02	5.999E-02	3.980E-02	6.564E-04	0.000E+00	0.000E+00	0.000E+00
Am-243	6.324E-04	6.324E-04	6.330E-04	6.330E-04	6.330E-04	6.276E-04	5.784E-04	2.559E-04	7.350E-08	1.711E-15
Am-244	3.320E-01	6.714E-02	0.000E+00	0.000E+00	0.000E+00	0.000E+00	0.000E+00	0.000E+00	0.000E+00	0.000E+00
Am-245	9.798E-07	8.832E-07	0.000E+00	0.000E+00	0.000E+00	0.000E+00	0.000E+00	0.000E+00	0.000E+00	0.000E+00
Cm-242	4.806E+01	4.807E+01	4.262E+01	1.028E+01	4.920E-02	3.263E-02	5.383E-04	0.000E+00	0.000E+00	0.000E+00
Cm-243	4.565E-03	4.565E-03	4.557E-03	4.467E-03	3.676E-03	5.231E-04	1.784E-12	0.000E+00	0.000E+00	0.000E+00
Cm-244	7.086E-03	7.086E-03	7.062E-03	6.816E-03	4.830E-03	1.538E-04	8.208E-16	8.202E-16	8.196E-16	8.136E-16
Cm-245	2.314E-07	2.314E-07	2.314E-07	2.314E-07	2.312E-07	2.295E-07	2.128E-07	1.000E-07	5.270E-11	0.000E+00
Cm-246	1.856E-09	1.856E-09	1.856E-09	1.856E-09	1.853E-09	1.829E-09	1.602E-09	4.262E-10	7.554E-16	0.000E+00
Cm-247	1.786E-15	1.786E-15	1.786E-15	1.786E-15	1.786E-15	1.786E-15	1.785E-15	1.784E-15	1.778E-15	1.711E-15
Cm-248	3.412E-16	3.412E-16	3.412E-16	3.412E-16	3.412E-16	3.412E-16	3.406E-16	3.346E-16	2.802E-16	4.762E-17
Cm-249	1.764E-12	9.228E-13	0.000E+00	0.000E+00	0.000E+00	0.000E+00	0.000E+00	0.000E+00	0.000E+00	0.000E+00
Cm-250	0.000E+00	0.000E+00	0.000E+00	0.000E+00	0.000E+00	0.000E+00	0.000E+00	0.000E+00	0.000E+00	0.000E+00
Bk-249	3.574E-13	3.575E-13	3.348E-13	1.597E-13	1.126E-16	0.000E+00	0.000E+00	0.000E+00	0.000E+00	0.000E+00
Bk-250	5.153E-15	4.045E-15	0.000E+00	0.000E+00	0.000E+00	0.000E+00	0.000E+00	0.000E+00	0.000E+00	0.000E+00
Cf-249	2.012E-16	2.012E-16	2.571E-16	6.840E-16	1.055E-15	8.844E-16	1.503E-16	0.000E+00	0.000E+00	0.000E+00
Cf-250	1.252E-16	1.252E-16	1.248E-16	1.189E-16	7.380E-17	6.294E-19	0.000E+00	0.000E+00	0.000E+00	0.000E+00
Cf-251	4.502E-21	4.502E-21	4.502E-21	4.502E-21	4.502E-21	4.190E-21	2.293E-21	0.000E+00	0.000E+00	0.000E+00
Cf-252	0.000E+00	0.000E+00	0.000E+00	0.000E+00	0.000E+00	0.000E+00	0.000E+00	0.000E+00	0.000E+00	0.000E+00
Cf-253	0.000E+00	0.000E+00	0.000E+00	0.000E+00	0.000E+00	0.000E+00	0.000E+00	0.000E+00	0.000E+00	0.000E+00
Cf-254	0.000E+00	0.000E+00	0.000E+00	0.000E+00	0.000E+00	0.000E+00	0.000E+00	0.000E+00	0.000E+00	0.000E+00
Es-253	0.000E+00	0.000E+00	0.000E+00	0.000E+00	0.000E+00	0.000E+00	0.000E+00	0.000E+00	0.000E+00	0.000E+00
Total	3.173E+07	3.152E+07	1.014E+04	5.416E+03	4.647E+03	3.162E+03	2.907E+03	2.159E+03	1.610E+02	8.545E-02
Total/MWd	6.499E+01	6.457E+01	2.078E-02	1.109E-02	9.518E-03	6.475E-03	5.955E-03	4.422E-03	3.298E-04	1.750E-07

Table A.65

Burner Core Heavy TRU Elements Nuclide Toxicity Hazard (Core) Basis = EQULIBRIUM FRESH FUEL BURNER CORE

	Initial	**1 hour**	**30 days**	**1 year**	**10 y**	**100 y**	**1000 y**	**10000 y**	**100000 y**	**1000000 y**	**Factor**
Np-236	0.000E+00	0.000E+00	0.000E+00	0.000E+00	0.000E+00	0.000E+00	0.000E+00	0.000E+00	0.000E+00	0.000E+00	0.000E+00
Np-237	7.202E+03	7.202E+03	7.203E+03	7.219E+03	7.410E+03	1.042E+04	2.679E+04	3.217E+04	3.160E+04	2.362E+04	1.972E+02
Np-238	0.000E+00	0.000E+00	0.000E+00	0.000E+00	0.000E+00	0.000E+00	0.000E+00	0.000E+00	0.000E+00	0.000E+00	0.000E+00
Np-239	0.000E+00	0.000E+00	0.000E+00	0.000E+00	0.000E+00	0.000E+00	0.000E+00	0.000E+00	0.000E+00	0.000E+00	0.000E+00
Np-240m	0.000E+00	0.000E+00	0.000E+00	0.000E+00	0.000E+00	0.000E+00	0.000E+00	0.000E+00	0.000E+00	0.000E+00	0.000E+00
Np-240	0.000E+00	0.000E+00	0.000E+00	0.000E+00	0.000E+00	0.000E+00	0.000E+00	0.000E+00	0.000E+00	0.000E+00	0.000E+00
Pu-236	0.000E+00	0.000E+00	0.000E+00	0.000E+00	0.000E+00	0.000E+00	0.000E+00	0.000E+00	0.000E+00	0.000E+00	0.000E+00
Pu-238	2.423E+08	2.423E+08	2.422E+08	2.408E+08	2.252E+08	1.152E+08	1.973E+05	3.373E-13	0.000E+00	0.000E+00	2.461E+02
Pu-239	1.212E+07	1.212E+07	1.212E+07	1.212E+07	1.211E+07	1.209E+07	1.186E+07	9.577E+06	7.834E+05	6.157E-06	2.675E+02
Pu-240	2.985E+07	2.985E+07	2.987E+07	2.992E+07	3.030E+07	3.101E+07	2.830E+07	1.125E+07	1.104E+03	0.000E+00	2.675E+02
Pu-241	0.000E+00	0.000E+00	0.000E+00	0.000E+00	0.000E+00	0.000E+00	0.000E+00	0.000E+00	0.000E+00	0.000E+00	0.000E+00
Pu-242	1.322E+05	1.322E+05	1.322E+05	1.322E+05	1.322E+05	1.325E+05	1.331E+05	1.326E+05	1.130E+05	2.178E+04	2.675E+02
Pu-243	0.000E+00	0.000E+00	0.000E+00	0.000E+00	0.000E+00	0.000E+00	0.000E+00	0.000E+00	0.000E+00	0.000E+00	0.000E+00
Pu-244	0.000E+00	0.000E+00	0.000E+00	0.000E+00	0.000E+00	0.000E+00	0.000E+00	0.000E+00	0.000E+00	0.000E+00	0.000E+00
Pu-245	0.000E+00	0.000E+00	0.000E+00	0.000E+00	0.000E+00	0.000E+00	0.000E+00	0.000E+00	0.000E+00	0.000E+00	0.000E+00
Am-241	6.984E+07	6.984E+07	7.021E+07	7.429E+07	1.057E+08	1.456E+08	3.488E+07	1.262E+05	7.016E+01	0.000E+00	2.729E+02
Am-242m	1.183E+07	1.183E+07	1.183E+07	1.178E+07	1.131E+07	7.500E+06	1.237E+05	1.847E-13	0.000E+00	0.000E+00	2.675E+02
Am-242	0.000E+00	0.000E+00	0.000E+00	0.000E+00	0.000E+00	0.000E+00	0.000E+00	0.000E+00	0.000E+00	0.000E+00	0.000E+00
Am-243	2.916E+06	2.916E+06	2.916E+06	2.916E+06	2.913E+06	2.890E+06	2.664E+06	1.178E+06	3.383E+02	0.000E+00	2.729E+02
Am-244	0.000E+00	0.000E+00	0.000E+00	0.000E+00	0.000E+00	0.000E+00	0.000E+00	0.000E+00	0.000E+00	0.000E+00	0.000E+00
Am-245	0.000E+00	0.000E+00	0.000E+00	0.000E+00	0.000E+00	0.000E+00	0.000E+00	0.000E+00	0.000E+00	0.000E+00	0.000E+00
Cm-242	3.178E+06	3.177E+06	2.826E+06	8.686E+05	2.392E+05	1.586E+05	2.617E+03	3.919E-15	0.000E+00	0.000E+00	6.900E+00
Cm-243	1.609E+06	1.609E+06	1.607E+06	1.575E+06	1.296E+06	1.844E+05	6.287E-04	0.000E+00	0.000E+00	0.000E+00	1.969E+02
Cm-244	3.409E+08	3.409E+08	3.399E+08	3.281E+08	2.325E+08	7.402E+06	7.935E-09	0.000E+00	0.000E+00	0.000E+00	1.630E+02
Cm-245	3.033E+05	3.033E+05	3.033E+05	3.033E+05	3.030E+05	3.008E+05	2.789E+05	1.311E+05	6.906E+01	0.000E+00	2.840E+02
Cm-246	2.231E+05	2.231E+05	2.231E+05	2.231E+05	2.228E+05	2.199E+05	1.925E+05	5.124E+04	9.080E-02	0.000E+00	2.845E+02
Cm-247	0.000E+00	0.000E+00	0.000E+00	0.000E+00	0.000E+00	0.000E+00	0.000E+00	0.000E+00	0.000E+00	0.000E+00	0.000E+00
Cm-248	0.000E+00	0.000E+00	0.000E+00	0.000E+00	0.000E+00	0.000E+00	0.000E+00	0.000E+00	0.000E+00	0.000E+00	0.000E+00
Cm-249	0.000E+00	0.000E+00	0.000E+00	0.000E+00	0.000E+00	0.000E+00	0.000E+00	0.000E+00	0.000E+00	0.000E+00	0.000E+00
Cm-250	0.000E+00	0.000E+00	0.000E+00	0.000E+00	0.000E+00	0.000E+00	0.000E+00	0.000E+00	0.000E+00	0.000E+00	0.000E+00
Bk-249	0.000E+00	0.000E+00	0.000E+00	0.000E+00	0.000E+00	0.000E+00	0.000E+00	0.000E+00	0.000E+00	0.000E+00	0.000E+00
Bk-250	0.000E+00	0.000E+00	0.000E+00	0.000E+00	0.000E+00	0.000E+00	0.000E+00	0.000E+00	0.000E+00	0.000E+00	0.000E+00
Cf-249	0.000E+00	0.000E+00	0.000E+00	0.000E+00	0.000E+00	0.000E+00	0.000E+00	0.000E+00	0.000E+00	0.000E+00	0.000E+00
Cf-250	0.000E+00	0.000E+00	0.000E+00	0.000E+00	0.000E+00	0.000E+00	0.000E+00	0.000E+00	0.000E+00	0.000E+00	0.000E+00
Cf-251	0.000E+00	0.000E+00	0.000E+00	0.000E+00	0.000E+00	0.000E+00	0.000E+00	0.000E+00	0.000E+00	0.000E+00	0.000E+00
Cf-252	0.000E+00	0.000E+00	0.000E+00	0.000E+00	0.000E+00	0.000E+00	0.000E+00	0.000E+00	0.000E+00	0.000E+00	0.000E+00
Cf-253	0.000E+00	0.000E+00	0.000E+00	0.000E+00	0.000E+00	0.000E+00	0.000E+00	0.000E+00	0.000E+00	0.000E+00	0.000E+00
Cf-254	0.000E+00	0.000E+00	0.000E+00	0.000E+00	0.000E+00	0.000E+00	0.000E+00	0.000E+00	0.000E+00	0.000E+00	0.000E+00
Es-253	0.000E+00	0.000E+00	0.000E+00	0.000E+00	0.000E+00	0.000E+00	0.000E+00	0.000E+00	0.000E+00	0.000E+00	0.000E+00
Total	7.152E+08	7.152E+08	7.141E+08	7.031E+08	6.221E+08	3.226E+08	7.865E+07	2.247E+07	9.296E+05	4.540E+04	0.000E+00

Table A.66

Burner core heavy TRU elements nuclide toxicity hazard (Core) Basis = EQULIBRIUM DISCHARGE FUEL BURNER CORE

	Initial	1 hour	30 days	1 year	10 y	100 y	1000 y	10000 y	100000 y	1000000 y	Factor
Np-236	0.000E+00	0.000E+00	0.000E+00	0.000E+00	0.000E+00	0.000E+00	0.000E+00	0.000E+00	0.000E+00	0.000E+00	0.000E+00
Np-237	5.282E+03	5.282E+03	5.285E+03	5.298E+03	5.452E+03	7.793E+03	2.048E+04	2.474E+04	2.439E+04	1.822E+04	1.972E+02
Np-238	0.000E+00	0.000E+00	0.000E+00	0.000E+00	0.000E+00	0.000E+00	0.000E+00	0.000E+00	0.000E+00	0.000E+00	0.000E+00
Np-239	0.000E+00	0.000E+00	0.000E+00	0.000E+00	0.000E+00	0.000E+00	0.000E+00	0.000E+00	0.000E+00	0.000E+00	0.000E+00
Np-240m	0.000E+00	0.000E+00	0.000E+00	0.000E+00	0.000E+00	0.000E+00	0.000E+00	0.000E+00	0.000E+00	0.000E+00	0.000E+00
Np-240	0.000E+00	0.000E+00	0.000E+00	0.000E+00	0.000E+00	0.000E+00	0.000E+00	0.000E+00	0.000E+00	0.000E+00	0.000E+00
Pu-236	0.000E+00	0.000E+00	0.000E+00	0.000E+00	0.000E+00	0.000E+00	0.000E+00	0.000E+00	0.000E+00	0.000E+00	0.000E+00
Pu-238	2.256E+08	2.256E+08	2.271E+08	2.332E+08	2.203E+08	1.124E+08	1.856E+05	3.040E-13	0.000E+00	0.000E+00	2.461E+02
Pu-239	9.993E+06	9.993E+06	1.001E+07	1.000E+07	1.000E+07	9.986E+06	9.795E+06	7.935E+06	6.515E+05	1.474E+00	2.675E+02
Pu-240	2.680E+07	2.680E+07	2.680E+07	2.685E+07	2.725E+07	2.799E+07	2.555E+07	1.016E+07	9.965E+02	1.410E-02	2.675E+02
Pu-241	0.000E+00	0.000E+00	0.000E+00	0.000E+00	0.000E+00	0.000E+00	0.000E+00	0.000E+00	0.000E+00	0.000E+00	0.000E+00
Pu-242	1.220E+05	1.220E+05	1.220E+05	1.220E+05	1.221E+05	1.223E+05	1.229E+05	1.226E+05	1.045E+05	2.014E+04	2.675E+02
Pu-243	0.000E+00	0.000E+00	0.000E+00	0.000E+00	0.000E+00	0.000E+00	0.000E+00	0.000E+00	0.000E+00	0.000E+00	0.000E+00
Pu-244	0.000E+00	0.000E+00	0.000E+00	0.000E+00	0.000E+00	0.000E+00	0.000E+00	0.000E+00	0.000E+00	0.000E+00	0.000E+00
Pu-245	0.000E+00	0.000E+00	0.000E+00	0.000E+00	0.000E+00	0.000E+00	0.000E+00	0.000E+00	0.000E+00	0.000E+00	0.000E+00
Am-241	5.683E+07	5.683E+07	5.711E+07	6.016E+07	8.359E+07	1.129E+08	2.708E+07	1.233E+05	6.853E+01	0.000E+00	2.729E+02
Am-242m	1.067E+07	1.067E+07	1.067E+07	1.062E+07	1.019E+07	6.762E+06	1.116E+05	1.666E-13	0.000E+00	0.000E+00	2.675E+02
Am-242	0.000E+00	0.000E+00	0.000E+00	0.000E+00	0.000E+00	0.000E+00	0.000E+00	0.000E+00	0.000E+00	0.000E+00	0.000E+00
Am-243	2.582E+06	2.582E+06	2.582E+06	2.582E+06	2.579E+06	2.559E+06	2.358E+06	1.043E+06	3.011E+02	1.504E+00	2.729E+02
Am-244	0.000E+00	0.000E+00	0.000E+00	0.000E+00	0.000E+00	0.000E+00	0.000E+00	0.000E+00	0.000E+00	0.000E+00	0.000E+00
Am-245	0.000E+00	0.000E+00	0.000E+00	0.000E+00	0.000E+00	0.000E+00	0.000E+00	0.000E+00	0.000E+00	0.000E+00	0.000E+00
Cm-242	6.479E+07	6.479E+07	5.730E+07	1.394E+07	2.157E+05	1.430E+05	2.360E+03	3.534E-15	0.000E+00	0.000E+00	6.900E+00
Cm-243	1.468E+06	1.468E+06	1.466E+06	1.438E+06	1.183E+06	1.683E+05	5.739E-04	0.000E+00	0.000E+00	0.000E+00	1.969E+02
Cm-244	3.454E+08	3.454E+08	3.444E+08	3.324E+08	2.355E+08	7.500E+06	1.440E-06	1.631E-06	3.448E-06	1.117E-05	1.630E+02
Cm-245	2.962E+05	2.962E+05	2.962E+05	2.962E+05	2.960E+05	2.938E+05	2.723E+05	1.280E+05	6.746E+01	0.000E+00	2.840E+02
Cm-246	2.219E+05	2.219E+05	2.219E+05	2.219E+05	2.217E+05	2.187E+05	1.917E+05	5.097E+04	9.032E-02	0.000E+00	2.845E+02
Cm-247	0.000E+00	0.000E+00	0.000E+00	0.000E+00	0.000E+00	0.000E+00	0.000E+00	0.000E+00	0.000E+00	0.000E+00	0.000E+00
Cm-248	0.000E+00	0.000E+00	0.000E+00	0.000E+00	0.000E+00	0.000E+00	0.000E+00	0.000E+00	0.000E+00	0.000E+00	0.000E+00
Cm-249	0.000E+00	0.000E+00	0.000E+00	0.000E+00	0.000E+00	0.000E+00	0.000E+00	0.000E+00	0.000E+00	0.000E+00	0.000E+00
Cm-250	0.000E+00	0.000E+00	0.000E+00	0.000E+00	0.000E+00	0.000E+00	0.000E+00	0.000E+00	0.000E+00	0.000E+00	0.000E+00
Bk-249	0.000E+00	0.000E+00	0.000E+00	0.000E+00	0.000E+00	0.000E+00	0.000E+00	0.000E+00	0.000E+00	0.000E+00	0.000E+00
Bk-250	0.000E+00	0.000E+00	0.000E+00	0.000E+00	0.000E+00	0.000E+00	0.000E+00	0.000E+00	0.000E+00	0.000E+00	0.000E+00
Cf-249	0.000E+00	0.000E+00	0.000E+00	0.000E+00	0.000E+00	0.000E+00	0.000E+00	0.000E+00	0.000E+00	0.000E+00	0.000E+00
Cf-250	0.000E+00	0.000E+00	0.000E+00	0.000E+00	0.000E+00	0.000E+00	0.000E+00	0.000E+00	0.000E+00	0.000E+00	0.000E+00
Cf-251	0.000E+00	0.000E+00	0.000E+00	0.000E+00	0.000E+00	0.000E+00	0.000E+00	0.000E+00	0.000E+00	0.000E+00	0.000E+00
Cf-252	0.000E+00	0.000E+00	0.000E+00	0.000E+00	0.000E+00	0.000E+00	0.000E+00	0.000E+00	0.000E+00	0.000E+00	0.000E+00
Cf-253	0.000E+00	0.000E+00	0.000E+00	0.000E+00	0.000E+00	0.000E+00	0.000E+00	0.000E+00	0.000E+00	0.000E+00	0.000E+00
Cf-254	0.000E+00	0.000E+00	0.000E+00	0.000E+00	0.000E+00	0.000E+00	0.000E+00	0.000E+00	0.000E+00	0.000E+00	0.000E+00
Es-253	0.000E+00	0.000E+00	0.000E+00	0.000E+00	0.000E+00	0.000E+00	0.000E+00	0.000E+00	0.000E+00	0.000E+00	0.000E+00
Total	7.448E+08	7.448E+08	7.380E+08	6.918E+08	5.915E+08	2.810E+08	6.569E+07	1.958E+07	7.818E+05	3.837E+04	0.000E+00

Table A.67

Burner core heavy TRU elements nuclide toxicity hazard (Core) Basis = (EQULIBRIUM DISCHARGE FUEL - EQULIBRIUM FRESH FUEL) BURNER CORE

	Initial	1 hour	30 days	1 year	10 y	100 y	1000 y	10000 y	100000 y	1000000 y	Factor
Np-236	0.000E+00	0.000E+00	0.000E+00	0.000E+00	0.000E+00	0.000E+00	0.000E+00	0.000E+00	0.000E+00	0.000E+00	0.000E+00
Np-237	-1.920E+03	-1.920E+03	-1.918E+03	-1.920E+03	-1.958E+03	-2.628E+03	-6.306E+03	-7.430E+03	-7.218E+03	-5.395E+03	1.972E+02
Np-238	0.000E+00	0.000E+00	0.000E+00	0.000E+00	0.000E+00	0.000E+00	0.000E+00	0.000E+00	0.000E+00	0.000E+00	0.000E+00
Np-239	0.000E+00	0.000E+00	0.000E+00	0.000E+00	0.000E+00	0.000E+00	0.000E+00	0.000E+00	0.000E+00	0.000E+00	0.000E+00
Np-240m	0.000E+00	0.000E+00	0.000E+00	0.000E+00	0.000E+00	0.000E+00	0.000E+00	0.000E+00	0.000E+00	0.000E+00	0.000E+00
Np-240	0.000E+00	0.000E+00	0.000E+00	0.000E+00	0.000E+00	0.000E+00	0.000E+00	0.000E+00	0.000E+00	0.000E+00	0.000E+00
Pu-236	0.000E+00	0.000E+00	0.000E+00	0.000E+00	0.000E+00	0.000E+00	0.000E+00	0.000E+00	0.000E+00	0.000E+00	0.000E+00
Pu-238	-1.669E+07	-1.669E+07	-1.506E+07	-7.678E+06	-4.873E+06	-2.806E+06	-1.167E+04	-3.322E-14	0.000E+00	0.000E+00	2.461E+02
Pu-239	-2.123E+06	-2.123E+06	-2.111E+06	-2.112E+06	-2.109E+06	-2.106E+06	-2.061E+06	-1.642E+06	-1.319E+05	1.474E+00	2.675E+02
Pu-240	-3.050E+06	-3.050E+06	-3.066E+06	-3.066E+06	-3.050E+06	-3.017E+06	-2.745E+06	-1.090E+06	-1.071E+02	1.410E-02	2.675E+02
Pu-241	0.000E+00	0.000E+00	0.000E+00	0.000E+00	0.000E+00	0.000E+00	0.000E+00	0.000E+00	0.000E+00	0.000E+00	0.000E+00
Pu-242	-1.013E+04	-1.013E+04	-1.013E+04	-1.014E+04	-1.014E+04	-1.016E+04	-1.019E+04	-1.005E+04	-8.523E+03	-1.637E+03	2.675E+02
Pu-243	0.000E+00	0.000E+00	0.000E+00	0.000E+00	0.000E+00	0.000E+00	0.000E+00	0.000E+00	0.000E+00	0.000E+00	0.000E+00
Pu-244	0.000E+00	0.000E+00	0.000E+00	0.000E+00	0.000E+00	0.000E+00	0.000E+00	0.000E+00	0.000E+00	0.000E+00	0.000E+00
Pu-245	0.000E+00	0.000E+00	0.000E+00	0.000E+00	0.000E+00	0.000E+00	0.000E+00	0.000E+00	0.000E+00	0.000E+00	0.000E+00
Am-241	-1.300E+07	-1.300E+07	-1.310E+07	-1.413E+07	-2.207E+07	-3.270E+07	-7.794E+06	-2.964E+03	-1.637E+00	0.000E+00	2.729E+02
Am-242m	-1.164E+06	-1.164E+06	-1.162E+06	-1.159E+06	-1.111E+06	-7.383E+05	-1.217E+04	-1.814E-14	0.000E+00	0.000E+00	2.675E+02
Am-242	0.000E+00	0.000E+00	0.000E+00	0.000E+00	0.000E+00	0.000E+00	0.000E+00	0.000E+00	0.000E+00	0.000E+00	0.000E+00
Am-243	-3.340E+05	-3.340E+05	-3.340E+05	-3.340E+05	-3.340E+05	-3.308E+05	-3.062E+05	-1.351E+05	-3.717E+01	1.504E+00	2.729E+02
Am-244	0.000E+00	0.000E+00	0.000E+00	0.000E+00	0.000E+00	0.000E+00	0.000E+00	0.000E+00	0.000E+00	0.000E+00	0.000E+00
Am-245	0.000E+00	0.000E+00	0.000E+00	0.000E+00	0.000E+00	0.000E+00	0.000E+00	0.000E+00	0.000E+00	0.000E+00	0.000E+00
Cm-242	6.161E+07	6.161E+07	5.447E+07	1.307E+07	-2.352E+04	-1.561E+04	-2.571E+02	-3.850E-16	0.000E+00	0.000E+00	6.900E+00
Cm-243	-1.406E+05	-1.406E+05	-1.406E+05	-1.370E+05	-1.134E+05	-1.607E+04	-5.482E-05	0.000E+00	0.000E+00	0.000E+00	1.969E+02
Cm-244	4.499E+06	4.499E+06	4.499E+06	4.303E+06	3.032E+06	9.878E+04	1.432E-06	1.631E-06	3.448E-06	1.117E-05	1.630E+02
Cm-245	-7.157E+03	-7.157E+03	-7.157E+03	-7.157E+03	-6.986E+03	-6.986E+03	-6.646E+03	-3.067E+03	-1.602E+00	0.000E+00	2.840E+02
Cm-246	-1.195E+03	-1.195E+03	-1.195E+03	-1.195E+03	-1.024E+03	-1.195E+03	-8.535E+02	-2.731E+02	-4.780E-04	0.000E+00	2.845E+02
Cm-247	0.000E+00	0.000E+00	0.000E+00	0.000E+00	0.000E+00	0.000E+00	0.000E+00	0.000E+00	0.000E+00	0.000E+00	0.000E+00
Cm-248	0.000E+00	0.000E+00	0.000E+00	0.000E+00	0.000E+00	0.000E+00	0.000E+00	0.000E+00	0.000E+00	0.000E+00	0.000E+00
Cm-249	0.000E+00	0.000E+00	0.000E+00	0.000E+00	0.000E+00	0.000E+00	0.000E+00	0.000E+00	0.000E+00	0.000E+00	0.000E+00
Cm-250	0.000E+00	0.000E+00	0.000E+00	0.000E+00	0.000E+00	0.000E+00	0.000E+00	0.000E+00	0.000E+00	0.000E+00	0.000E+00
Bk-249	0.000E+00	0.000E+00	0.000E+00	0.000E+00	0.000E+00	0.000E+00	0.000E+00	0.000E+00	0.000E+00	0.000E+00	0.000E+00
Bk-250	0.000E+00	0.000E+00	0.000E+00	0.000E+00	0.000E+00	0.000E+00	0.000E+00	0.000E+00	0.000E+00	0.000E+00	0.000E+00
Cf-249	0.000E+00	0.000E+00	0.000E+00	0.000E+00	0.000E+00	0.000E+00	0.000E+00	0.000E+00	0.000E+00	0.000E+00	0.000E+00
Cf-250	0.000E+00	0.000E+00	0.000E+00	0.000E+00	0.000E+00	0.000E+00	0.000E+00	0.000E+00	0.000E+00	0.000E+00	0.000E+00
Cf-251	0.000E+00	0.000E+00	0.000E+00	0.000E+00	0.000E+00	0.000E+00	0.000E+00	0.000E+00	0.000E+00	0.000E+00	0.000E+00
Cf-252	0.000E+00	0.000E+00	0.000E+00	0.000E+00	0.000E+00	0.000E+00	0.000E+00	0.000E+00	0.000E+00	0.000E+00	0.000E+00
Cf-253	0.000E+00	0.000E+00	0.000E+00	0.000E+00	0.000E+00	0.000E+00	0.000E+00	0.000E+00	0.000E+00	0.000E+00	0.000E+00
Cf-254	0.000E+00	0.000E+00	0.000E+00	0.000E+00	0.000E+00	0.000E+00	0.000E+00	0.000E+00	0.000E+00	0.000E+00	0.000E+00
Es-253	0.000E+00	0.000E+00	0.000E+00	0.000E+00	0.000E+00	0.000E+00	0.000E+00	0.000E+00	0.000E+00	0.000E+00	0.000E+00
Total	2.959E+07	2.959E+07	2.398E+07	-1.126E+07	-3.067E+07	-4.165E+07	-1.295E+07	-2.891E+06	-1.478E+05	-7.030E+03	0.000E+00
Total/MWd	6.061E+01	6.061E+01	4.911E+01	-2.306E+01	-6.282E+01	-8.531E+01	-2.653E+01	-5.920E+00	-3.028E-01	-1.440E-02	0.000E+00

Table A.68

Converter core heavy TRU elements nuclide toxicity hazard (all regions) Basis = EQULIBRIUM FRESH FUEL CONVERTER CORE

	Initial	**1 hour**	**30 days**	**1 year**	**10 y**	**100 y**	**1000 y**	**10000 y**	**100000 y**	**1000000 y**	**Factor**
Np-236	0.000E+00	0.000E+00	0.000E+00	0.000E+00	0.000E+00	0.000E+00	0.000E+00	0.000E+00	0.000E+00	0.000E+00	0.000E+00
Np-237	4.643E+03	4.643E+03	4.644E+03	4.655E+03	4.787E+03	6.890E+03	1.834E+04	2.209E+04	2.170E+04	1.621E+04	1.972E+02
Np-238	0.000E+00	0.000E+00	0.000E+00	0.000E+00	0.000E+00	0.000E+00	0.000E+00	0.000E+00	0.000E+00	0.000E+00	0.000E+00
Np-239	0.000E+00	0.000E+00	0.000E+00	0.000E+00	0.000E+00	0.000E+00	0.000E+00	0.000E+00	0.000E+00	0.000E+00	0.000E+00
Np-240m	0.000E+00	0.000E+00	0.000E+00	0.000E+00	0.000E+00	0.000E+00	0.000E+00	0.000E+00	0.000E+00	0.000E+00	0.000E+00
Np-240	0.000E+00	0.000E+00	0.000E+00	0.000E+00	0.000E+00	0.000E+00	0.000E+00	0.000E+00	0.000E+00	0.000E+00	0.000E+00
Pu-236	0.000E+00	0.000E+00	0.000E+00	0.000E+00	0.000E+00	0.000E+00	0.000E+00	0.000E+00	0.000E+00	0.000E+00	0.000E+00
Pu-238	1.689E+08	1.689E+08	1.689E+08	1.680E+08	1.571E+08	8.043E+07	1.396E+05	2.419E-13	0.000E+00	0.000E+00	2.461E+02
Pu-239	1.299E+07	1.299E+07	1.299E+07	1.299E+07	1.298E+07	1.296E+07	1.267E+07	1.007E+07	8.065E+05	6.338E-06	2.675E+02
Pu-240	2.513E+07	2.513E+07	2.513E+07	2.517E+07	2.542E+07	2.582E+07	2.356E+07	9.365E+06	9.190E+02	0.000E+00	2.675E+02
Pu-241	0.000E+00	0.000E+00	0.000E+00	0.000E+00	0.000E+00	0.000E+00	0.000E+00	0.000E+00	0.000E+00	0.000E+00	0.000E+00
Pu-242	8.845E+04	8.845E+04	8.845E+04	8.845E+04	8.848E+04	8.868E+04	8.914E+04	8.885E+04	7.572E+04	1.460E+04	2.675E+02
Pu-243	0.000E+00	0.000E+00	0.000E+00	0.000E+00	0.000E+00	0.000E+00	0.000E+00	0.000E+00	0.000E+00	0.000E+00	0.000E+00
Pu-244	0.000E+00	0.000E+00	0.000E+00	0.000E+00	0.000E+00	0.000E+00	0.000E+00	0.000E+00	0.000E+00	0.000E+00	0.000E+00
Pu-245	0.000E+00	0.000E+00	0.000E+00	0.000E+00	0.000E+00	0.000E+00	0.000E+00	0.000E+00	0.000E+00	0.000E+00	0.000E+00
Am-241	4.771E+07	4.771E+07	4.798E+07	5.087E+07	7.321E+07	1.018E+08	2.440E+07	8.449E+04	4.698E+01	0.000E+00	2.729E+02
Am-242m	8.489E+06	8.489E+06	8.486E+06	8.450E+06	8.110E+06	5.380E+06	8.876E+04	1.325E-13	0.000E+00	0.000E+00	2.675E+02
Am-242	0.000E+00	0.000E+00	0.000E+00	0.000E+00	0.000E+00	0.000E+00	0.000E+00	0.000E+00	0.000E+00	0.000E+00	0.000E+00
Am-243	1.858E+06	1.858E+06	1.858E+06	1.858E+06	1.857E+06	1.842E+06	1.698E+06	7.512E+05	2.156E+02	0.000E+00	2.729E+02
Am-244	0.000E+00	0.000E+00	0.000E+00	0.000E+00	0.000E+00	0.000E+00	0.000E+00	0.000E+00	0.000E+00	0.000E+00	0.000E+00
Am-245	0.000E+00	0.000E+00	0.000E+00	0.000E+00	0.000E+00	0.000E+00	0.000E+00	0.000E+00	0.000E+00	0.000E+00	0.000E+00
Cm-242	2.293E+06	2.292E+06	2.039E+06	6.260E+05	1.716E+05	1.138E+05	1.877E+03	2.811E-15	0.000E+00	0.000E+00	6.900E+00
Cm-243	1.151E+06	1.151E+06	1.149E+06	1.126E+06	9.268E+05	1.320E+05	4.498E-04	0.000E+00	0.000E+00	0.000E+00	1.969E+02
Cm-244	2.230E+08	2.230E+08	2.223E+08	2.146E+08	1.521E+08	4.841E+06	5.189E-09	0.000E+00	0.000E+00	0.000E+00	1.630E+02
Cm-245	2.031E+05	2.031E+05	2.031E+05	2.029E+05	2.029E+05	2.014E+05	1.868E+05	8.777E+04	4.625E+01	0.000E+00	2.840E+02
Cm-246	1.572E+05	1.572E+05	1.572E+05	1.572E+05	1.570E+05	1.549E+05	1.357E+05	3.610E+04	6.398E-02	0.000E+00	2.845E+02
Cm-247	0.000E+00	0.000E+00	0.000E+00	0.000E+00	0.000E+00	0.000E+00	0.000E+00	0.000E+00	0.000E+00	0.000E+00	0.000E+00
Cm-248	0.000E+00	0.000E+00	0.000E+00	0.000E+00	0.000E+00	0.000E+00	0.000E+00	0.000E+00	0.000E+00	0.000E+00	0.000E+00
Cm-249	0.000E+00	0.000E+00	0.000E+00	0.000E+00	0.000E+00	0.000E+00	0.000E+00	0.000E+00	0.000E+00	0.000E+00	0.000E+00
Cm-250	0.000E+00	0.000E+00	0.000E+00	0.000E+00	0.000E+00	0.000E+00	0.000E+00	0.000E+00	0.000E+00	0.000E+00	0.000E+00
Bk-249	0.000E+00	0.000E+00	0.000E+00	0.000E+00	0.000E+00	0.000E+00	0.000E+00	0.000E+00	0.000E+00	0.000E+00	0.000E+00
Bk-250	0.000E+00	0.000E+00	0.000E+00	0.000E+00	0.000E+00	0.000E+00	0.000E+00	0.000E+00	0.000E+00	0.000E+00	0.000E+00
Cf-249	0.000E+00	0.000E+00	0.000E+00	0.000E+00	0.000E+00	0.000E+00	0.000E+00	0.000E+00	0.000E+00	0.000E+00	0.000E+00
Cf-250	0.000E+00	0.000E+00	0.000E+00	0.000E+00	0.000E+00	0.000E+00	0.000E+00	0.000E+00	0.000E+00	0.000E+00	0.000E+00
Cf-251	0.000E+00	0.000E+00	0.000E+00	0.000E+00	0.000E+00	0.000E+00	0.000E+00	0.000E+00	0.000E+00	0.000E+00	0.000E+00
Cf-252	0.000E+00	0.000E+00	0.000E+00	0.000E+00	0.000E+00	0.000E+00	0.000E+00	0.000E+00	0.000E+00	0.000E+00	0.000E+00
Cf-253	0.000E+00	0.000E+00	0.000E+00	0.000E+00	0.000E+00	0.000E+00	0.000E+00	0.000E+00	0.000E+00	0.000E+00	0.000E+00
Cf-254	0.000E+00	0.000E+00	0.000E+00	0.000E+00	0.000E+00	0.000E+00	0.000E+00	0.000E+00	0.000E+00	0.000E+00	0.000E+00
Es-253	0.000E+00	0.000E+00	0.000E+00	0.000E+00	0.000E+00	0.000E+00	0.000E+00	0.000E+00	0.000E+00	0.000E+00	0.000E+00
Total	4.920E+08	4.920E+08	4.913E+08	4.842E+08	4.323E+08	2.338E+08	6.299E+07	2.050E+07	9.052E+05	3.081E+04	0.000E+00

Table A.69

Converter core heavy TRU elements nuclide toxicity hazard (all regions) Basis = EQULIBRIUM DISCHARGE FUEL CONVERTER CORE

	Initial	1 hour	30 days	1 year	10 y	100 y	1000 y	10000 y	100000 y	1000000 y	Factor
Np-236	0.000E+00	0.000E+00	0.000E+00	0.000E+00	0.000E+00	0.000E+00	0.000E+00	0.000E+00	0.000E+00	0.000E+00	0.000E+00
Np-237	3.526E+03	3.526E+03	3.529E+03	3.539E+03	3.647E+03	5.360E+03	1.468E+04	1.780E+04	1.752E+04	1.309E+04	1.972E+02
Np-238	0.000E+00	0.000E+00	0.000E+00	0.000E+00	0.000E+00	0.000E+00	0.000E+00	0.000E+00	0.000E+00	0.000E+00	0.000E+00
Np-239	0.000E+00	0.000E+00	0.000E+00	0.000E+00	0.000E+00	0.000E+00	0.000E+00	0.000E+00	0.000E+00	0.000E+00	0.000E+00
Np-240m	0.000E+00	0.000E+00	0.000E+00	0.000E+00	0.000E+00	0.000E+00	0.000E+00	0.000E+00	0.000E+00	0.000E+00	0.000E+00
Np-240	0.000E+00	0.000E+00	0.000E+00	0.000E+00	0.000E+00	0.000E+00	0.000E+00	0.000E+00	0.000E+00	0.000E+00	0.000E+00
Pu-236	0.000E+00	0.000E+00	0.000E+00	0.000E+00	0.000E+00	0.000E+00	0.000E+00	0.000E+00	0.000E+00	0.000E+00	0.000E+00
Pu-238	1.586E+08	1.586E+08	1.596E+08	1.641E+08	1.550E+08	7.912E+07	1.319E+05	2.185E-13	0.000E+00	0.000E+00	2.461E+02
Pu-239	1.175E+07	1.175E+07	1.177E+07	1.177E+07	1.177E+07	1.174E+07	1.148E+07	9.120E+06	7.304E+05	1.066E+00	2.675E+02
Pu-240	2.337E+07	2.337E+07	2.337E+07	2.340E+07	2.366E+07	2.409E+07	2.199E+07	8.738E+06	8.576E+02	9.195E-03	2.675E+02
Pu-241	0.000E+00	0.000E+00	0.000E+00	0.000E+00	0.000E+00	0.000E+00	0.000E+00	0.000E+00	0.000E+00	0.000E+00	0.000E+00
Pu-242	8.261E+04	8.261E+04	8.261E+04	8.261E+04	8.264E+04	8.282E+04	8.325E+04	8.304E+04	7.081E+04	1.365E+04	2.675E+02
Pu-243	0.000E+00	0.000E+00	0.000E+00	0.000E+00	0.000E+00	0.000E+00	0.000E+00	0.000E+00	0.000E+00	0.000E+00	0.000E+00
Pu-244	0.000E+00	0.000E+00	0.000E+00	0.000E+00	0.000E+00	0.000E+00	0.000E+00	0.000E+00	0.000E+00	0.000E+00	0.000E+00
Pu-245	0.000E+00	0.000E+00	0.000E+00	0.000E+00	0.000E+00	0.000E+00	0.000E+00	0.000E+00	0.000E+00	0.000E+00	0.000E+00
Am-241	3.908E+07	3.908E+07	3.930E+07	4.166E+07	5.977E+07	8.295E+07	1.989E+07	8.277E+04	4.601E+01	0.000E+00	2.729E+02
Am-242m	7.667E+06	7.667E+06	7.664E+06	7.632E+06	7.325E+06	4.858E+06	8.015E+04	1.197E-13	0.000E+00	0.000E+00	2.675E+02
Am-242	0.000E+00	0.000E+00	0.000E+00	0.000E+00	0.000E+00	0.000E+00	0.000E+00	0.000E+00	0.000E+00	0.000E+00	0.000E+00
Am-243	1.665E+06	1.665E+06	1.665E+06	1.665E+06	1.664E+06	1.650E+06	1.521E+06	6.730E+05	1.944E+02	1.087E+00	2.729E+02
Am-244	0.000E+00	0.000E+00	0.000E+00	0.000E+00	0.000E+00	0.000E+00	0.000E+00	0.000E+00	0.000E+00	0.000E+00	0.000E+00
Am-245	0.000E+00	0.000E+00	0.000E+00	0.000E+00	0.000E+00	0.000E+00	0.000E+00	0.000E+00	0.000E+00	0.000E+00	0.000E+00
Cm-242	4.670E+07	4.670E+07	4.130E+07	1.005E+07	1.550E+05	1.028E+05	1.695E+03	2.539E-15	0.000E+00	0.000E+00	6.900E+00
Cm-243	1.075E+06	1.075E+06	1.073E+06	1.052E+06	8.655E+05	1.232E+05	4.200E-04	0.000E+00	0.000E+00	0.000E+00	1.969E+02
Cm-244	2.274E+08	2.274E+08	2.266E+08	2.188E+08	1.550E+08	4.936E+06	4.021E-07	5.382E-07	1.822E-06	7.283E-06	1.630E+02
Cm-245	1.989E+05	1.989E+05	1.989E+05	1.989E+05	1.987E+05	1.972E+05	1.828E+05	8.597E+04	4.529E+01	0.000E+00	2.840E+02
Cm-246	1.566E+05	1.566E+05	1.566E+05	1.566E+05	1.564E+05	1.543E+05	1.352E+05	3.595E+04	6.371E-02	0.000E+00	2.845E+02
Cm-247	0.000E+00	0.000E+00	0.000E+00	0.000E+00	0.000E+00	0.000E+00	0.000E+00	0.000E+00	0.000E+00	0.000E+00	0.000E+00
Cm-248	0.000E+00	0.000E+00	0.000E+00	0.000E+00	0.000E+00	0.000E+00	0.000E+00	0.000E+00	0.000E+00	0.000E+00	0.000E+00
Cm-249	0.000E+00	0.000E+00	0.000E+00	0.000E+00	0.000E+00	0.000E+00	0.000E+00	0.000E+00	0.000E+00	0.000E+00	0.000E+00
Cm-250	0.000E+00	0.000E+00	0.000E+00	0.000E+00	0.000E+00	0.000E+00	0.000E+00	0.000E+00	0.000E+00	0.000E+00	0.000E+00
Bk-249	0.000E+00	0.000E+00	0.000E+00	0.000E+00	0.000E+00	0.000E+00	0.000E+00	0.000E+00	0.000E+00	0.000E+00	0.000E+00
Bk-250	0.000E+00	0.000E+00	0.000E+00	0.000E+00	0.000E+00	0.000E+00	0.000E+00	0.000E+00	0.000E+00	0.000E+00	0.000E+00
Cf-249	0.000E+00	0.000E+00	0.000E+00	0.000E+00	0.000E+00	0.000E+00	0.000E+00	0.000E+00	0.000E+00	0.000E+00	0.000E+00
Cf-250	0.000E+00	0.000E+00	0.000E+00	0.000E+00	0.000E+00	0.000E+00	0.000E+00	0.000E+00	0.000E+00	0.000E+00	0.000E+00
Cf-251	0.000E+00	0.000E+00	0.000E+00	0.000E+00	0.000E+00	0.000E+00	0.000E+00	0.000E+00	0.000E+00	0.000E+00	0.000E+00
Cf-252	0.000E+00	0.000E+00	0.000E+00	0.000E+00	0.000E+00	0.000E+00	0.000E+00	0.000E+00	0.000E+00	0.000E+00	0.000E+00
Cf-253	0.000E+00	0.000E+00	0.000E+00	0.000E+00	0.000E+00	0.000E+00	0.000E+00	0.000E+00	0.000E+00	0.000E+00	0.000E+00
Cf-254	0.000E+00	0.000E+00	0.000E+00	0.000E+00	0.000E+00	0.000E+00	0.000E+00	0.000E+00	0.000E+00	0.000E+00	0.000E+00
Es-253	0.000E+00	0.000E+00	0.000E+00	0.000E+00	0.000E+00	0.000E+00	0.000E+00	0.000E+00	0.000E+00	0.000E+00	0.000E+00
Total	5.177E+08	5.177E+08	5.128E+08	4.805E+08	4.157E+08	2.100E+08	5.552E+07	1.884E+07	8.199E+05	2.674E+04	0.000E+00

Table A.70

Converter core heavy TRU elements nuclide toxicity hazard (all regions)

Basis = (EQULIBRIUM DISCHARGE FUEL - EQULIBRIUM FRESH FUEL) CONVERTER CORE

	Initial	**1 hour**	**30 days**	**1 year**	**10 y**	**100 y**	**1000 y**	**10000 y**	**100000 y**	**1000000 y**	**Factor**
Np-236	0.000E+00	0.000E+00	0.000E+00	0.000E+00	0.000E+00	0.000E+00	0.000E+00	0.000E+00	0.000E+00	0.000E+00	0.000E+00
Np-237	-1.117E+03	-1.117E+03	-1.115E+03	-1.116E+03	-1.141E+03	-1.530E+03	-3.656E+03	-4.295E+03	-4.177E+03	-3.124E+03	1.972E+02
Np-238	0.000E+00	0.000E+00	0.000E+00	0.000E+00	0.000E+00	0.000E+00	0.000E+00	0.000E+00	0.000E+00	0.000E+00	0.000E+00
Np-239	0.000E+00	0.000E+00	0.000E+00	0.000E+00	0.000E+00	0.000E+00	0.000E+00	0.000E+00	0.000E+00	0.000E+00	0.000E+00
Np-240m	0.000E+00	0.000E+00	0.000E+00	0.000E+00	0.000E+00	0.000E+00	0.000E+00	0.000E+00	0.000E+00	0.000E+00	0.000E+00
Np-240	0.000E+00	0.000E+00	0.000E+00	0.000E+00	0.000E+00	0.000E+00	0.000E+00	0.000E+00	0.000E+00	0.000E+00	0.000E+00
Pu-236	0.000E+00	0.000E+00	0.000E+00	0.000E+00	0.000E+00	0.000E+00	0.000E+00	0.000E+00	0.000E+00	0.000E+00	0.000E+00
Pu-238	-1.034E+07	-1.034E+07	-9.303E+06	-3.987E+06	-2.067E+06	-1.314E+06	-7.664E+03	-2.333E-14	0.000E+00	0.000E+00	2.461E+02
Pu-239	-1.234E+06	-1.234E+06	-1.218E+06	-1.218E+06	-1.217E+06	-1.213E+06	-1.189E+06	-9.470E+05	-7.608E+04	1.066E+00	2.675E+02
Pu-240	-1.766E+06	-1.766E+06	-1.766E+06	-1.766E+06	-1.766E+06	-1.733E+06	-1.573E+06	-6.276E+05	-6.147E+01	9.195E-03	2.675E+02
Pu-241	0.000E+00	0.000E+00	0.000E+00	0.000E+00	0.000E+00	0.000E+00	0.000E+00	0.000E+00	0.000E+00	0.000E+00	0.000E+00
Pu-242	-5.842E+03	-5.842E+03	-5.842E+03	-5.842E+03	-5.842E+03	-5.858E+03	-5.890E+03	-5.810E+03	-4.911E+03	-9.486E+02	2.675E+02
Pu-243	0.000E+00	0.000E+00	0.000E+00	0.000E+00	0.000E+00	0.000E+00	0.000E+00	0.000E+00	0.000E+00	0.000E+00	0.000E+00
Pu-244	0.000E+00	0.000E+00	0.000E+00	0.000E+00	0.000E+00	0.000E+00	0.000E+00	0.000E+00	0.000E+00	0.000E+00	0.000E+00
Pu-245	0.000E+00	0.000E+00	0.000E+00	0.000E+00	0.000E+00	0.000E+00	0.000E+00	0.000E+00	0.000E+00	0.000E+00	0.000E+00
Am-241	-8.629E+06	-8.629E+06	-8.678E+06	-9.219E+06	-1.344E+07	-1.890E+07	-4.503E+06	-1.719E+03	-9.661E-01	0.000E+00	2.729E+02
Am-242m	-8.218E+05	-8.218E+05	-8.218E+05	-8.186E+05	-7.848E+05	-5.216E+05	-8.603E+03	-1.284E-14	0.000E+00	0.000E+00	2.675E+02
Am-242	0.000E+00	0.000E+00	0.000E+00	0.000E+00	0.000E+00	0.000E+00	0.000E+00	0.000E+00	0.000E+00	0.000E+00	0.000E+00
Am-243	-1.932E+05	-1.932E+05	-1.932E+05	-1.932E+05	-1.932E+05	-1.916E+05	-1.768E+05	-7.827E+04	-2.129E+01	1.087E+00	2.729E+02
Am-244	0.000E+00	0.000E+00	0.000E+00	0.000E+00	0.000E+00	0.000E+00	0.000E+00	0.000E+00	0.000E+00	0.000E+00	0.000E+00
Am-245	0.000E+00	0.000E+00	0.000E+00	0.000E+00	0.000E+00	0.000E+00	0.000E+00	0.000E+00	0.000E+00	0.000E+00	0.000E+00
Cm-242	4.441E+07	4.441E+07	3.926E+07	9.422E+06	-1.660E+04	-1.101E+04	-1.822E+02	-2.724E-16	0.000E+00	0.000E+00	6.900E+00
Cm-243	-7.608E+04	-7.608E+04	-7.608E+04	-7.443E+04	-6.131E+04	-8.742E+03	-2.977E-05	0.000E+00	0.000E+00	0.000E+00	1.969E+02
Cm-244	4.401E+06	4.401E+06	4.303E+06	4.205E+06	2.934E+06	9.487E+04	3.969E-07	5.382E-07	1.822E-06	7.283E-06	1.630E+02
Cm-245	-4.260E+03	-4.260E+03	-4.260E+03	-4.090E+03	-4.260E+03	-4.260E+03	-3.919E+03	-1.806E+03	-9.542E-01	0.000E+00	2.840E+02
Cm-246	-6.487E+02	-6.487E+02	-6.657E+02	-6.486E+02	-6.487E+02	-6.316E+02	-5.633E+02	-1.536E+02	-2.731E-04	0.000E+00	2.845E+02
Cm-247	0.000E+00	0.000E+00	0.000E+00	0.000E+00	0.000E+00	0.000E+00	0.000E+00	0.000E+00	0.000E+00	0.000E+00	0.000E+00
Cm-248	0.000E+00	0.000E+00	0.000E+00	0.000E+00	0.000E+00	0.000E+00	0.000E+00	0.000E+00	0.000E+00	0.000E+00	0.000E+00
Cm-249	0.000E+00	0.000E+00	0.000E+00	0.000E+00	0.000E+00	0.000E+00	0.000E+00	0.000E+00	0.000E+00	0.000E+00	0.000E+00
Cm-250	0.000E+00	0.000E+00	0.000E+00	0.000E+00	0.000E+00	0.000E+00	0.000E+00	0.000E+00	0.000E+00	0.000E+00	0.000E+00
Bk-249	0.000E+00	0.000E+00	0.000E+00	0.000E+00	0.000E+00	0.000E+00	0.000E+00	0.000E+00	0.000E+00	0.000E+00	0.000E+00
Bk-250	0.000E+00	0.000E+00	0.000E+00	0.000E+00	0.000E+00	0.000E+00	0.000E+00	0.000E+00	0.000E+00	0.000E+00	0.000E+00
Cf-249	0.000E+00	0.000E+00	0.000E+00	0.000E+00	0.000E+00	0.000E+00	0.000E+00	0.000E+00	0.000E+00	0.000E+00	0.000E+00
Cf-250	0.000E+00	0.000E+00	0.000E+00	0.000E+00	0.000E+00	0.000E+00	0.000E+00	0.000E+00	0.000E+00	0.000E+00	0.000E+00
Cf-251	0.000E+00	0.000E+00	0.000E+00	0.000E+00	0.000E+00	0.000E+00	0.000E+00	0.000E+00	0.000E+00	0.000E+00	0.000E+00
Cf-252	0.000E+00	0.000E+00	0.000E+00	0.000E+00	0.000E+00	0.000E+00	0.000E+00	0.000E+00	0.000E+00	0.000E+00	0.000E+00
Cf-253	0.000E+00	0.000E+00	0.000E+00	0.000E+00	0.000E+00	0.000E+00	0.000E+00	0.000E+00	0.000E+00	0.000E+00	0.000E+00
Cf-254	0.000E+00	0.000E+00	0.000E+00	0.000E+00	0.000E+00	0.000E+00	0.000E+00	0.000E+00	0.000E+00	0.000E+00	0.000E+00
Es-253	0.000E+00	0.000E+00	0.000E+00	0.000E+00	0.000E+00	0.000E+00	0.000E+00	0.000E+00	0.000E+00	0.000E+00	0.000E+00
Total	2.574E+07	2.574E+07	2.149E+07	-3.660E+06	-1.663E+07	-2.381E+07	-7.472E+06	-1.667E+06	-8.525E+04	-4.070E+03	0.000E+00
Total/MWd	5.272E+01	5.272E+01	4.402E+01	-7.496E+00	-3.405E+01	-4.876E+01	-1.530E+01	-3.413E+00	-1.746E-01	-8.336E-03	0.000E+00

Table A.71

Breeder core heavy TRU elements nuclide toxicity hazard (all regions) Basis = EQULIBRIUM FRESH FUEL BREEDER CORE

	Initial	1 hour	30 days	1 year	10 y	100 y	1000 y	10000 y	100000 y	1000000 y	Factor
Np-236	0.000E+00	0.000E+00	0.000E+00	0.000E+00	0.000E+00	0.000E+00	0.000E+00	0.000E+00	0.000E+00	0.000E+00	0.000E+00
Np-237	1.074E+03	1.074E+03	1.074E+03	1.078E+03	1.125E+03	1.897E+03	6.110E+03	7.427E+03	7.241E+03	5.411E+03	1.972E+02
Np-238	0.000E+00	0.000E+00	0.000E+00	0.000E+00	0.000E+00	0.000E+00	0.000E+00	0.000E+00	0.000E+00	0.000E+00	0.000E+00
Np-239	0.000E+00	0.000E+00	0.000E+00	0.000E+00	0.000E+00	0.000E+00	0.000E+00	0.000E+00	0.000E+00	0.000E+00	0.000E+00
Np-240m	0.000E+00	0.000E+00	0.000E+00	0.000E+00	0.000E+00	0.000E+00	0.000E+00	0.000E+00	0.000E+00	0.000E+00	0.000E+00
Np-240	0.000E+00	0.000E+00	0.000E+00	0.000E+00	0.000E+00	0.000E+00	0.000E+00	0.000E+00	0.000E+00	0.000E+00	0.000E+00
Pu-236	0.000E+00	0.000E+00	0.000E+00	0.000E+00	0.000E+00	0.000E+00	0.000E+00	0.000E+00	0.000E+00	0.000E+00	0.000E+00
Pu-238	4.818E+07	4.818E+07	4.818E+07	4.793E+07	4.486E+07	2.312E+07	4.421E+04	8.433E-14	0.000E+00	0.000E+00	2.461E+02
Pu-239	1.510E+07	1.510E+07	1.510E+07	1.510E+07	1.510E+07	1.506E+07	1.469E+07	1.141E+07	8.877E+05	6.977E-06	2.675E+02
Pu-240	1.727E+07	1.727E+07	1.727E+07	1.727E+07	1.729E+07	1.721E+07	1.569E+07	6.234E+06	6.118E+02	0.000E+00	2.675E+02
Pu-241	0.000E+00	0.000E+00	0.000E+00	0.000E+00	0.000E+00	0.000E+00	0.000E+00	0.000E+00	0.000E+00	0.000E+00	0.000E+00
Pu-242	1.923E+04	1.923E+04	1.923E+04	1.923E+04	1.923E+04	1.929E+04	1.942E+04	1.923E+04	1.635E+04	3.152E+03	2.675E+02
Pu-243	0.000E+00	0.000E+00	0.000E+00	0.000E+00	0.000E+00	0.000E+00	0.000E+00	0.000E+00	0.000E+00	0.000E+00	0.000E+00
Pu-244	0.000E+00	0.000E+00	0.000E+00	0.000E+00	0.000E+00	0.000E+00	0.000E+00	0.000E+00	0.000E+00	0.000E+00	0.000E+00
Pu-245	0.000E+00	0.000E+00	0.000E+00	0.000E+00	0.000E+00	0.000E+00	0.000E+00	0.000E+00	0.000E+00	0.000E+00	0.000E+00
Am-241	1.647E+07	1.647E+07	1.659E+07	1.770E+07	2.630E+07	3.751E+07	8.952E+06	9.803E+03	5.448E+00	0.000E+00	2.729E+02
Am-242m	2.960E+06	2.960E+06	2.958E+06	2.945E+06	2.826E+06	1.875E+06	3.094E+04	4.619E-14	0.000E+00	0.000E+00	2.675E+02
Am-242	0.000E+00	0.000E+00	0.000E+00	0.000E+00	0.000E+00	0.000E+00	0.000E+00	0.000E+00	0.000E+00	0.000E+00	0.000E+00
Am-243	2.396E+05	2.396E+05	2.396E+05	2.396E+05	2.394E+05	2.374E+05	2.188E+05	9.680E+04	2.780E+01	0.000E+00	2.729E+02
Am-244	0.000E+00	0.000E+00	0.000E+00	0.000E+00	0.000E+00	0.000E+00	0.000E+00	0.000E+00	0.000E+00	0.000E+00	0.000E+00
Am-245	0.000E+00	0.000E+00	0.000E+00	0.000E+00	0.000E+00	0.000E+00	0.000E+00	0.000E+00	0.000E+00	0.000E+00	0.000E+00
Cm-242	8.102E+05	8.098E+05	7.204E+05	2.204E+05	5.978E+04	3.967E+04	6.545E+02	9.799E-16	0.000E+00	0.000E+00	6.900E+00
Cm-243	3.210E+05	3.210E+05	3.205E+05	3.141E+05	2.585E+05	3.679E+04	1.255E-04	0.000E+00	0.000E+00	0.000E+00	1.969E+02
Cm-244	2.637E+07	2.637E+07	2.629E+07	2.538E+07	1.798E+07	5.724E+05	6.137E-10	0.000E+00	0.000E+00	0.000E+00	1.630E+02
Cm-245	2.355E+04	2.355E+04	2.355E+04	2.355E+04	2.353E+04	2.334E+04	2.166E+04	1.018E+04	5.362E+00	0.000E+00	2.840E+02
Cm-246	1.707E+04	1.707E+04	1.707E+04	1.707E+04	1.705E+04	1.683E+04	1.474E+04	3.921E+03	6.947E-03	0.000E+00	2.845E+02
Cm-247	0.000E+00	0.000E+00	0.000E+00	0.000E+00	0.000E+00	0.000E+00	0.000E+00	0.000E+00	0.000E+00	0.000E+00	0.000E+00
Cm-248	0.000E+00	0.000E+00	0.000E+00	0.000E+00	0.000E+00	0.000E+00	0.000E+00	0.000E+00	0.000E+00	0.000E+00	0.000E+00
Cm-249	0.000E+00	0.000E+00	0.000E+00	0.000E+00	0.000E+00	0.000E+00	0.000E+00	0.000E+00	0.000E+00	0.000E+00	0.000E+00
Cm-250	0.000E+00	0.000E+00	0.000E+00	0.000E+00	0.000E+00	0.000E+00	0.000E+00	0.000E+00	0.000E+00	0.000E+00	0.000E+00
Bk-249	0.000E+00	0.000E+00	0.000E+00	0.000E+00	0.000E+00	0.000E+00	0.000E+00	0.000E+00	0.000E+00	0.000E+00	0.000E+00
Bk-250	0.000E+00	0.000E+00	0.000E+00	0.000E+00	0.000E+00	0.000E+00	0.000E+00	0.000E+00	0.000E+00	0.000E+00	0.000E+00
Cf-249	0.000E+00	0.000E+00	0.000E+00	0.000E+00	0.000E+00	0.000E+00	0.000E+00	0.000E+00	0.000E+00	0.000E+00	0.000E+00
Cf-250	0.000E+00	0.000E+00	0.000E+00	0.000E+00	0.000E+00	0.000E+00	0.000E+00	0.000E+00	0.000E+00	0.000E+00	0.000E+00
Cf-251	0.000E+00	0.000E+00	0.000E+00	0.000E+00	0.000E+00	0.000E+00	0.000E+00	0.000E+00	0.000E+00	0.000E+00	0.000E+00
Cf-252	0.000E+00	0.000E+00	0.000E+00	0.000E+00	0.000E+00	0.000E+00	0.000E+00	0.000E+00	0.000E+00	0.000E+00	0.000E+00
Cf-253	0.000E+00	0.000E+00	0.000E+00	0.000E+00	0.000E+00	0.000E+00	0.000E+00	0.000E+00	0.000E+00	0.000E+00	0.000E+00
Cf-254	0.000E+00	0.000E+00	0.000E+00	0.000E+00	0.000E+00	0.000E+00	0.000E+00	0.000E+00	0.000E+00	0.000E+00	0.000E+00
Es-253	0.000E+00	0.000E+00	0.000E+00	0.000E+00	0.000E+00	0.000E+00	0.000E+00	0.000E+00	0.000E+00	0.000E+00	0.000E+00
Total	1.278E+08	1.278E+08	1.277E+08	1.272E+08	1.250E+08	9.573E+07	3.969E+07	1.779E+07	9.120E+05	8.563E+03	0.000E+00

Table A.72

Breeder core heavy TRU elements nuclide toxicity hazard (all regions) Basis = EQULIBRIUM DISCHARGE FUEL BREEDER CORE

	Initial	1 hour	30 days	1 year	10 y	100 y	1000 y	10000 y	100000 y	1000000 y	Factor
Np-236	0.000E+00	0.000E+00	0.000E+00	0.000E+00	0.000E+00	0.000E+00	0.000E+00	0.000E+00	0.000E+00	0.000E+00	0.000E+00
Np-237	1.073E+03	1.073E+03	1.077E+03	1.080E+03	1.124E+03	1.897E+03	6.161E+03	7.493E+03	7.306E+03	5.458E+03	1.972E+02
Np-238	0.000E+00	0.000E+00	0.000E+00	0.000E+00	0.000E+00	0.000E+00	0.000E+00	0.000E+00	0.000E+00	0.000E+00	0.000E+00
Np-239	0.000E+00	0.000E+00	0.000E+00	0.000E+00	0.000E+00	0.000E+00	0.000E+00	0.000E+00	0.000E+00	0.000E+00	0.000E+00
Np-240m	0.000E+00	0.000E+00	0.000E+00	0.000E+00	0.000E+00	0.000E+00	0.000E+00	0.000E+00	0.000E+00	0.000E+00	0.000E+00
Np-240	0.000E+00	0.000E+00	0.000E+00	0.000E+00	0.000E+00	0.000E+00	0.000E+00	0.000E+00	0.000E+00	0.000E+00	0.000E+00
Pu-236	0.000E+00	0.000E+00	0.000E+00	0.000E+00	0.000E+00	0.000E+00	0.000E+00	0.000E+00	0.000E+00	0.000E+00	0.000E+00
Pu-238	4.687E+07	4.687E+07	4.722E+07	4.890E+07	4.632E+07	2.377E+07	4.284E+04	7.720E-14	0.000E+00	0.000E+00	2.461E+02
Pu-239	1.520E+07	1.520E+07	1.522E+07	1.522E+07	1.522E+07	1.518E+07	1.480E+07	1.150E+07	8.948E+05	1.164E-01	2.675E+02
Pu-240	1.741E+07	1.741E+07	1.741E+07	1.741E+07	1.743E+07	1.735E+07	1.583E+07	6.290E+06	6.173E+02	1.041E-03	2.675E+02
Pu-241	0.000E+00	0.000E+00	0.000E+00	0.000E+00	0.000E+00	0.000E+00	0.000E+00	0.000E+00	0.000E+00	0.000E+00	0.000E+00
Pu-242	1.942E+04	1.942E+04	1.942E+04	1.942E+04	1.944E+04	1.948E+04	1.961E+04	1.942E+04	1.650E+04	3.183E+03	2.675E+02
Pu-243	0.000E+00	0.000E+00	0.000E+00	0.000E+00	0.000E+00	0.000E+00	0.000E+00	0.000E+00	0.000E+00	0.000E+00	0.000E+00
Pu-244	0.000E+00	0.000E+00	0.000E+00	0.000E+00	0.000E+00	0.000E+00	0.000E+00	0.000E+00	0.000E+00	0.000E+00	0.000E+00
Pu-245	0.000E+00	0.000E+00	0.000E+00	0.000E+00	0.000E+00	0.000E+00	0.000E+00	0.000E+00	0.000E+00	0.000E+00	0.000E+00
Am-241	1.431E+07	1.431E+07	1.443E+07	1.566E+07	2.513E+07	3.794E+07	9.058E+06	9.857E+03	5.477E+00	0.000E+00	2.729E+02
Am-242m	2.709E+06	2.709E+06	2.708E+06	2.696E+06	2.587E+06	1.717E+06	2.831E+04	4.229E-14	0.000E+00	0.000E+00	2.675E+02
Am-242	0.000E+00	0.000E+00	0.000E+00	0.000E+00	0.000E+00	0.000E+00	0.000E+00	0.000E+00	0.000E+00	0.000E+00	0.000E+00
Am-243	2.415E+05	2.415E+05	2.415E+05	2.415E+05	2.414E+05	2.394E+05	2.206E+05	9.759E+04	2.815E+01	1.188E-01	2.729E+02
Am-244	0.000E+00	0.000E+00	0.000E+00	0.000E+00	0.000E+00	0.000E+00	0.000E+00	0.000E+00	0.000E+00	0.000E+00	0.000E+00
Am-245	0.000E+00	0.000E+00	0.000E+00	0.000E+00	0.000E+00	0.000E+00	0.000E+00	0.000E+00	0.000E+00	0.000E+00	0.000E+00
Cm-242	1.680E+07	1.680E+07	1.486E+07	3.614E+06	5.473E+04	3.631E+04	5.991E+02	8.971E-16	0.000E+00	0.000E+00	6.900E+00
Cm-243	3.314E+05	3.314E+05	3.308E+05	3.243E+05	2.668E+05	3.797E+04	1.295E-04	0.000E+00	0.000E+00	0.000E+00	1.969E+02
Cm-244	2.843E+07	2.843E+07	2.833E+07	2.735E+07	1.938E+07	6.172E+05	8.819E-08	1.027E-07	2.403E-07	8.250E-07	1.630E+02
Cm-245	2.369E+04	2.369E+04	2.369E+04	2.367E+04	2.365E+04	2.348E+04	2.178E+04	1.024E+04	5.393E+00	0.000E+00	2.840E+02
Cm-246	1.721E+04	1.721E+04	1.721E+04	1.721E+04	1.719E+04	1.696E+04	1.486E+04	3.952E+03	7.002E-03	0.000E+00	2.845E+02
Cm-247	0.000E+00	0.000E+00	0.000E+00	0.000E+00	0.000E+00	0.000E+00	0.000E+00	0.000E+00	0.000E+00	0.000E+00	0.000E+00
Cm-248	0.000E+00	0.000E+00	0.000E+00	0.000E+00	0.000E+00	0.000E+00	0.000E+00	0.000E+00	0.000E+00	0.000E+00	0.000E+00
Cm-249	0.000E+00	0.000E+00	0.000E+00	0.000E+00	0.000E+00	0.000E+00	0.000E+00	0.000E+00	0.000E+00	0.000E+00	0.000E+00
Cm-250	0.000E+00	0.000E+00	0.000E+00	0.000E+00	0.000E+00	0.000E+00	0.000E+00	0.000E+00	0.000E+00	0.000E+00	0.000E+00
Bk-249	0.000E+00	0.000E+00	0.000E+00	0.000E+00	0.000E+00	0.000E+00	0.000E+00	0.000E+00	0.000E+00	0.000E+00	0.000E+00
Bk-250	0.000E+00	0.000E+00	0.000E+00	0.000E+00	0.000E+00	0.000E+00	0.000E+00	0.000E+00	0.000E+00	0.000E+00	0.000E+00
Cf-249	0.000E+00	0.000E+00	0.000E+00	0.000E+00	0.000E+00	0.000E+00	0.000E+00	0.000E+00	0.000E+00	0.000E+00	0.000E+00
Cf-250	0.000E+00	0.000E+00	0.000E+00	0.000E+00	0.000E+00	0.000E+00	0.000E+00	0.000E+00	0.000E+00	0.000E+00	0.000E+00
Cf-251	0.000E+00	0.000E+00	0.000E+00	0.000E+00	0.000E+00	0.000E+00	0.000E+00	0.000E+00	0.000E+00	0.000E+00	0.000E+00
Cf-252	0.000E+00	0.000E+00	0.000E+00	0.000E+00	0.000E+00	0.000E+00	0.000E+00	0.000E+00	0.000E+00	0.000E+00	0.000E+00
Cf-253	0.000E+00	0.000E+00	0.000E+00	0.000E+00	0.000E+00	0.000E+00	0.000E+00	0.000E+00	0.000E+00	0.000E+00	0.000E+00
Cf-254	0.000E+00	0.000E+00	0.000E+00	0.000E+00	0.000E+00	0.000E+00	0.000E+00	0.000E+00	0.000E+00	0.000E+00	0.000E+00
Es-253	0.000E+00	0.000E+00	0.000E+00	0.000E+00	0.000E+00	0.000E+00	0.000E+00	0.000E+00	0.000E+00	0.000E+00	0.000E+00
Total	1.424E+08	1.424E+08	1.408E+08	1.315E+08	1.267E+08	9.695E+07	4.004E+07	1.793E+07	9.192E+05	8.641E+03	0.000E+00

Table A.73

Breeder core heavy TRU elements nuclide toxicity hazard (all regions) Basis = (EQULIBRIUM DISCHARGE FUEL - EQULIBRIUM FRESH FUEL) BREEDER CORE

	Initial	1 hour	30 days	1 year	10 y	100 y	1000 y	10000 y	100000 y	1000000 y	Factor
Np-236	0.000E+00	0.000E+00	0.000E+00	0.000E+00	0.000E+00	0.000E+00	0.000E+00	0.000E+00	0.000E+00	0.000E+00	0.000E+00
Np-237	-1.302E+00	-1.302E+00	2.485E+00	2.011E+00	-1.183E+00	0.000E+00	5.088E+01	6.626E+01	6.508E+01	4.733E+01	1.972E+02
Np-238	0.000E+00	0.000E+00	0.000E+00	0.000E+00	0.000E+00	0.000E+00	0.000E+00	0.000E+00	0.000E+00	0.000E+00	0.000E+00
Np-239	0.000E+00	0.000E+00	0.000E+00	0.000E+00	0.000E+00	0.000E+00	0.000E+00	0.000E+00	0.000E+00	0.000E+00	0.000E+00
Np-240m	0.000E+00	0.000E+00	0.000E+00	0.000E+00	0.000E+00	0.000E+00	0.000E+00	0.000E+00	0.000E+00	0.000E+00	0.000E+00
Np-240	0.000E+00	0.000E+00	0.000E+00	0.000E+00	0.000E+00	0.000E+00	0.000E+00	0.000E+00	0.000E+00	0.000E+00	0.000E+00
Pu-236	0.000E+00	0.000E+00	0.000E+00	0.000E+00	0.000E+00	0.000E+00	0.000E+00	0.000E+00	0.000E+00	0.000E+00	0.000E+00
Pu-238	-1.314E+06	-1.314E+06	-9.598E+05	9.746E+05	1.462E+06	6.497E+05	-1.373E+03	-7.132E-15	0.000E+00	0.000E+00	2.461E+02
Pu-239	9.470E+04	9.470E+04	1.188E+05	1.172E+05	1.172E+05	1.188E+05	1.140E+05	8.988E+04	7.062E+03	1.164E-01	2.675E+02
Pu-240	1.444E+05	1.444E+05	1.444E+05	1.444E+05	1.444E+05	1.444E+05	1.396E+05	5.618E+04	5.457E+00	1.041E-03	2.675E+02
Pu-241	0.000E+00	0.000E+00	0.000E+00	0.000E+00	0.000E+00	0.000E+00	0.000E+00	0.000E+00	0.000E+00	0.000E+00	0.000E+00
Pu-242	1.926E+02	1.926E+02	1.926E+02	1.926E+02	2.087E+02	1.926E+02	1.926E+02	1.926E+02	1.445E+02	3.049E+01	2.675E+02
Pu-243	0.000E+00	0.000E+00	0.000E+00	0.000E+00	0.000E+00	0.000E+00	0.000E+00	0.000E+00	0.000E+00	0.000E+00	0.000E+00
Pu-244	0.000E+00	0.000E+00	0.000E+00	0.000E+00	0.000E+00	0.000E+00	0.000E+00	0.000E+00	0.000E+00	0.000E+00	0.000E+00
Pu-245	0.000E+00	0.000E+00	0.000E+00	0.000E+00	0.000E+00	0.000E+00	0.000E+00	0.000E+00	0.000E+00	0.000E+00	0.000E+00
Am-241	-2.158E+06	-2.158E+06	-2.160E+06	-2.042E+06	-1.163E+06	4.257E+05	1.064E+05	5.403E+01	2.947E-02	0.000E+00	2.729E+02
Am-242m	-2.504E+05	-2.504E+05	-2.504E+05	-2.488E+05	-2.391E+05	-1.573E+05	-2.632E+03	-3.900E-15	0.000E+00	0.000E+00	2.675E+02
Am-242	0.000E+00	0.000E+00	0.000E+00	0.000E+00	0.000E+00	0.000E+00	0.000E+00	0.000E+00	0.000E+00	0.000E+00	0.000E+00
Am-243	1.965E+03	1.965E+03	1.965E+03	1.965E+03	1.965E+03	1.965E+03	1.801E+03	7.859E+02	3.439E-01	1.188E-01	2.729E+02
Am-244	0.000E+00	0.000E+00	0.000E+00	0.000E+00	0.000E+00	0.000E+00	0.000E+00	0.000E+00	0.000E+00	0.000E+00	0.000E+00
Am-245	0.000E+00	0.000E+00	0.000E+00	0.000E+00	0.000E+00	0.000E+00	0.000E+00	0.000E+00	0.000E+00	0.000E+00	0.000E+00
Cm-242	1.599E+07	1.599E+07	1.414E+07	3.393E+06	-5.051E+03	-3.358E+03	-5.548E+01	-8.280E-17	0.000E+00	0.000E+00	6.900E+00
Cm-243	1.040E+04	1.040E+04	1.028E+04	1.016E+04	8.270E+03	1.181E+03	4.017E-06	0.000E+00	0.000E+00	0.000E+00	1.969E+02
Cm-244	2.064E+06	2.064E+06	2.044E+06	1.976E+06	1.408E+06	4.479E+04	8.757E-08	1.027E-07	2.403E-07	8.250E-07	1.630E+02
Cm-245	1.363E+02	1.363E+02	1.363E+02	1.193E+02	1.193E+02	1.363E+02	1.193E+02	5.623E+01	3.067E-02	0.000E+00	2.840E+02
Cm-246	1.366E+02	1.366E+02	1.366E+02	1.366E+02	1.366E+02	1.314E+02	1.144E+02	3.073E+01	5.462E-05	0.000E+00	2.845E+02
Cm-247	0.000E+00	0.000E+00	0.000E+00	0.000E+00	0.000E+00	0.000E+00	0.000E+00	0.000E+00	0.000E+00	0.000E+00	0.000E+00
Cm-248	0.000E+00	0.000E+00	0.000E+00	0.000E+00	0.000E+00	0.000E+00	0.000E+00	0.000E+00	0.000E+00	0.000E+00	0.000E+00
Cm-249	0.000E+00	0.000E+00	0.000E+00	0.000E+00	0.000E+00	0.000E+00	0.000E+00	0.000E+00	0.000E+00	0.000E+00	0.000E+00
Cm-250	0.000E+00	0.000E+00	0.000E+00	0.000E+00	0.000E+00	0.000E+00	0.000E+00	0.000E+00	0.000E+00	0.000E+00	0.000E+00
Bk-249	0.000E+00	0.000E+00	0.000E+00	0.000E+00	0.000E+00	0.000E+00	0.000E+00	0.000E+00	0.000E+00	0.000E+00	0.000E+00
Bk-250	0.000E+00	0.000E+00	0.000E+00	0.000E+00	0.000E+00	0.000E+00	0.000E+00	0.000E+00	0.000E+00	0.000E+00	0.000E+00
Cf-249	0.000E+00	0.000E+00	0.000E+00	0.000E+00	0.000E+00	0.000E+00	0.000E+00	0.000E+00	0.000E+00	0.000E+00	0.000E+00
Cf-250	0.000E+00	0.000E+00	0.000E+00	0.000E+00	0.000E+00	0.000E+00	0.000E+00	0.000E+00	0.000E+00	0.000E+00	0.000E+00
Cf-251	0.000E+00	0.000E+00	0.000E+00	0.000E+00	0.000E+00	0.000E+00	0.000E+00	0.000E+00	0.000E+00	0.000E+00	0.000E+00
Cf-252	0.000E+00	0.000E+00	0.000E+00	0.000E+00	0.000E+00	0.000E+00	0.000E+00	0.000E+00	0.000E+00	0.000E+00	0.000E+00
Cf-253	0.000E+00	0.000E+00	0.000E+00	0.000E+00	0.000E+00	0.000E+00	0.000E+00	0.000E+00	0.000E+00	0.000E+00	0.000E+00
Cf-254	0.000E+00	0.000E+00	0.000E+00	0.000E+00	0.000E+00	0.000E+00	0.000E+00	0.000E+00	0.000E+00	0.000E+00	0.000E+00
Es-253	0.000E+00	0.000E+00	0.000E+00	0.000E+00	0.000E+00	0.000E+00	0.000E+00	0.000E+00	0.000E+00	0.000E+00	0.000E+00
Total	1.458E+07	1.458E+07	1.309E+07	4.327E+06	1.736E+06	1.226E+06	3.582E+05	1.472E+05	7.277E+03	7.806E+01	0.000E+00
Total/MWd	2.986E+01	2.986E+01	2.681E+01	8.862E+00	3.555E+00	2.512E+00	7.337E-01	3.016E-01	1.491E-02	1.599E-04	0.000E+00

Table A.74

Converter core heavy TRU elements nuclide toxicity hazard (core) Basis = EQULIBRIUM FRESH FUEL CONVERTER CORE

	Initial	1 hour	30 days	1 year	10 y	100 y	1000 y	10000 y	100000 y	1000000 y	Factor
Np-236	0.000E+00	0.000E+00	0.000E+00	0.000E+00	0.000E+00	0.000E+00	0.000E+00	0.000E+00	0.000E+00	0.000E+00	0.000E+00
Np-237	4.643E+03	4.643E+03	4.644E+03	4.655E+03	4.787E+03	6.890E+03	1.834E+04	2.209E+04	2.170E+04	1.621E+04	1.972E+02
Np-238	0.000E+00	0.000E+00	0.000E+00	0.000E+00	0.000E+00	0.000E+00	0.000E+00	0.000E+00	0.000E+00	0.000E+00	0.000E+00
Np-239	0.000E+00	0.000E+00	0.000E+00	0.000E+00	0.000E+00	0.000E+00	0.000E+00	0.000E+00	0.000E+00	0.000E+00	0.000E+00
Np-240m	0.000E+00	0.000E+00	0.000E+00	0.000E+00	0.000E+00	0.000E+00	0.000E+00	0.000E+00	0.000E+00	0.000E+00	0.000E+00
Np-240	0.000E+00	0.000E+00	0.000E+00	0.000E+00	0.000E+00	0.000E+00	0.000E+00	0.000E+00	0.000E+00	0.000E+00	0.000E+00
Pu-236	0.000E+00	0.000E+00	0.000E+00	0.000E+00	0.000E+00	0.000E+00	0.000E+00	0.000E+00	0.000E+00	0.000E+00	0.000E+00
Pu-238	1.689E+08	1.689E+08	1.689E+08	1.680E+08	1.571E+08	8.043E+07	1.396E+05	2.419E-13	0.000E+00	0.000E+00	2.461E+02
Pu-239	1.299E+07	1.299E+07	1.299E+07	1.299E+07	1.298E+07	1.296E+07	1.267E+07	1.007E+07	8.065E+05	6.338E-06	2.675E+02
Pu-240	2.513E+07	2.513E+07	2.513E+07	2.517E+07	2.542E+07	2.582E+07	2.356E+07	9.365E+06	9.190E+02	0.000E+00	2.675E+02
Pu-241	0.000E+00	0.000E+00	0.000E+00	0.000E+00	0.000E+00	0.000E+00	0.000E+00	0.000E+00	0.000E+00	0.000E+00	0.000E+00
Pu-242	8.845E+04	8.845E+04	8.845E+04	8.845E+04	8.848E+04	8.868E+04	8.914E+04	8.885E+04	7.572E+04	1.460E+04	2.675E+02
Pu-243	0.000E+00	0.000E+00	0.000E+00	0.000E+00	0.000E+00	0.000E+00	0.000E+00	0.000E+00	0.000E+00	0.000E+00	0.000E+00
Pu-244	0.000E+00	0.000E+00	0.000E+00	0.000E+00	0.000E+00	0.000E+00	0.000E+00	0.000E+00	0.000E+00	0.000E+00	0.000E+00
Pu-245	0.000E+00	0.000E+00	0.000E+00	0.000E+00	0.000E+00	0.000E+00	0.000E+00	0.000E+00	0.000E+00	0.000E+00	0.000E+00
Am-241	4.771E+07	4.771E+07	4.798E+07	5.087E+07	7.321E+07	1.018E+08	2.440E+07	8.449E+04	4.698E+01	0.000E+00	2.729E+02
Am-242m	8.489E+06	8.489E+06	8.486E+06	8.450E+06	8.110E+06	5.380E+06	8.876E+04	1.325E-13	0.000E+00	0.000E+00	2.675E+02
Am-242	0.000E+00	0.000E+00	0.000E+00	0.000E+00	0.000E+00	0.000E+00	0.000E+00	0.000E+00	0.000E+00	0.000E+00	0.000E+00
Am-243	1.858E+06	1.858E+06	1.858E+06	1.858E+06	1.857E+06	1.842E+06	1.698E+06	7.512E+05	2.156E+02	0.000E+00	2.729E+02
Am-244	0.000E+00	0.000E+00	0.000E+00	0.000E+00	0.000E+00	0.000E+00	0.000E+00	0.000E+00	0.000E+00	0.000E+00	0.000E+00
Am-245	0.000E+00	0.000E+00	0.000E+00	0.000E+00	0.000E+00	0.000E+00	0.000E+00	0.000E+00	0.000E+00	0.000E+00	0.000E+00
Cm-242	2.293E+06	2.292E+06	2.039E+06	6.260E+05	1.716E+05	1.138E+05	1.877E+03	2.811E-15	0.000E+00	0.000E+00	6.900E+00
Cm-243	1.151E+06	1.151E+06	1.149E+06	1.126E+06	9.268E+05	1.320E+05	4.498E-04	0.000E+00	0.000E+00	0.000E+00	1.969E+02
Cm-244	2.230E+08	2.230E+08	2.223E+08	2.146E+08	1.521E+08	4.841E+06	5.189E-09	0.000E+00	0.000E+00	0.000E+00	1.630E+02
Cm-245	2.031E+05	2.031E+05	2.031E+05	2.029E+05	2.029E+05	2.014E+05	1.868E+05	8.777E+04	4.625E+01	0.000E+00	2.840E+02
Cm-246	1.572E+05	1.572E+05	1.572E+05	1.572E+05	1.570E+05	1.549E+05	1.357E+05	3.610E+04	6.398E-02	0.000E+00	2.845E+02
Cm-247	0.000E+00	0.000E+00	0.000E+00	0.000E+00	0.000E+00	0.000E+00	0.000E+00	0.000E+00	0.000E+00	0.000E+00	0.000E+00
Cm-248	0.000E+00	0.000E+00	0.000E+00	0.000E+00	0.000E+00	0.000E+00	0.000E+00	0.000E+00	0.000E+00	0.000E+00	0.000E+00
Cm-249	0.000E+00	0.000E+00	0.000E+00	0.000E+00	0.000E+00	0.000E+00	0.000E+00	0.000E+00	0.000E+00	0.000E+00	0.000E+00
Cm-250	0.000E+00	0.000E+00	0.000E+00	0.000E+00	0.000E+00	0.000E+00	0.000E+00	0.000E+00	0.000E+00	0.000E+00	0.000E+00
Bk-249	0.000E+00	0.000E+00	0.000E+00	0.000E+00	0.000E+00	0.000E+00	0.000E+00	0.000E+00	0.000E+00	0.000E+00	0.000E+00
Bk-250	0.000E+00	0.000E+00	0.000E+00	0.000E+00	0.000E+00	0.000E+00	0.000E+00	0.000E+00	0.000E+00	0.000E+00	0.000E+00
Cf-249	0.000E+00	0.000E+00	0.000E+00	0.000E+00	0.000E+00	0.000E+00	0.000E+00	0.000E+00	0.000E+00	0.000E+00	0.000E+00
Cf-250	0.000E+00	0.000E+00	0.000E+00	0.000E+00	0.000E+00	0.000E+00	0.000E+00	0.000E+00	0.000E+00	0.000E+00	0.000E+00
Cf-251	0.000E+00	0.000E+00	0.000E+00	0.000E+00	0.000E+00	0.000E+00	0.000E+00	0.000E+00	0.000E+00	0.000E+00	0.000E+00
Cf-252	0.000E+00	0.000E+00	0.000E+00	0.000E+00	0.000E+00	0.000E+00	0.000E+00	0.000E+00	0.000E+00	0.000E+00	0.000E+00
Cf-253	0.000E+00	0.000E+00	0.000E+00	0.000E+00	0.000E+00	0.000E+00	0.000E+00	0.000E+00	0.000E+00	0.000E+00	0.000E+00
Cf-254	0.000E+00	0.000E+00	0.000E+00	0.000E+00	0.000E+00	0.000E+00	0.000E+00	0.000E+00	0.000E+00	0.000E+00	0.000E+00
Es-253	0.000E+00	0.000E+00	0.000E+00	0.000E+00	0.000E+00	0.000E+00	0.000E+00	0.000E+00	0.000E+00	0.000E+00	0.000E+00
Total	4.920E+08	4.920E+08	4.913E+08	4.842E+08	4.323E+08	2.338E+08	6.299E+07	2.050E+07	9.052E+05	3.081E+04	0.000E+00

Table A.75

Converter core heavy TRU elements nuclide toxicity hazard (core) Basis = EQULIBRIUM DISCHARGE FUEL CONVERTER CORE

	Initial	1 hour	30 days	1 year	10 y	100 y	1000 y	10000 y	100000 y	1000000 y	Factor
Np-236	0.000E+00	0.000E+00	0.000E+00	0.000E+00	0.000E+00	0.000E+00	0.000E+00	0.000E+00	0.000E+00	0.000E+00	0.000E+00
Np-237	3.488E+03	3.488E+03	3.492E+03	3.500E+03	3.609E+03	5.321E+03	1.464E+04	1.775E+04	1.748E+04	1.305E+04	1.972E+02
Np-238	0.000E+00	0.000E+00	0.000E+00	0.000E+00	0.000E+00	0.000E+00	0.000E+00	0.000E+00	0.000E+00	0.000E+00	0.000E+00
Np-239	0.000E+00	0.000E+00	0.000E+00	0.000E+00	0.000E+00	0.000E+00	0.000E+00	0.000E+00	0.000E+00	0.000E+00	0.000E+00
Np-240m	0.000E+00	0.000E+00	0.000E+00	0.000E+00	0.000E+00	0.000E+00	0.000E+00	0.000E+00	0.000E+00	0.000E+00	0.000E+00
Np-240	0.000E+00	0.000E+00	0.000E+00	0.000E+00	0.000E+00	0.000E+00	0.000E+00	0.000E+00	0.000E+00	0.000E+00	0.000E+00
Pu-236	0.000E+00	0.000E+00	0.000E+00	0.000E+00	0.000E+00	0.000E+00	0.000E+00	0.000E+00	0.000E+00	0.000E+00	0.000E+00
Pu-238	1.584E+08	1.584E+08	1.595E+08	1.639E+08	1.549E+08	7.906E+07	1.319E+05	2.185E-13	0.000E+00	0.000E+00	2.461E+02
Pu-239	1.065E+07	1.065E+07	1.066E+07	1.066E+07	1.066E+07	1.064E+07	1.041E+07	8.288E+06	6.659E+05	1.066E+00	2.675E+02
Pu-240	2.321E+07	2.321E+07	2.321E+07	2.324E+07	2.350E+07	2.395E+07	2.184E+07	8.683E+06	8.521E+02	9.195E-03	2.675E+02
Pu-241	0.000E+00	0.000E+00	0.000E+00	0.000E+00	0.000E+00	0.000E+00	0.000E+00	0.000E+00	0.000E+00	0.000E+00	0.000E+00
Pu-242	8.261E+04	8.261E+04	8.261E+04	8.261E+04	8.264E+04	8.282E+04	8.325E+04	8.304E+04	7.080E+04	1.365E+04	2.675E+02
Pu-243	0.000E+00	0.000E+00	0.000E+00	0.000E+00	0.000E+00	0.000E+00	0.000E+00	0.000E+00	0.000E+00	0.000E+00	0.000E+00
Pu-244	0.000E+00	0.000E+00	0.000E+00	0.000E+00	0.000E+00	0.000E+00	0.000E+00	0.000E+00	0.000E+00	0.000E+00	0.000E+00
Pu-245	0.000E+00	0.000E+00	0.000E+00	0.000E+00	0.000E+00	0.000E+00	0.000E+00	0.000E+00	0.000E+00	0.000E+00	0.000E+00
Am-241	3.908E+07	3.908E+07	3.930E+07	4.164E+07	5.973E+07	8.287E+07	1.988E+07	8.277E+04	4.601E+01	0.000E+00	2.729E+02
Am-242m	7.667E+06	7.667E+06	7.664E+06	7.632E+06	7.325E+06	4.858E+06	8.015E+04	1.197E-13	0.000E+00	0.000E+00	2.675E+02
Am-242	0.000E+00	0.000E+00	0.000E+00	0.000E+00	0.000E+00	0.000E+00	0.000E+00	0.000E+00	0.000E+00	0.000E+00	0.000E+00
Am-243	1.665E+06	1.665E+06	1.665E+06	1.665E+06	1.664E+06	1.650E+06	1.521E+06	6.730E+05	1.944E+02	1.087E+00	2.729E+02
Am-244	0.000E+00	0.000E+00	0.000E+00	0.000E+00	0.000E+00	0.000E+00	0.000E+00	0.000E+00	0.000E+00	0.000E+00	0.000E+00
Am-245	0.000E+00	0.000E+00	0.000E+00	0.000E+00	0.000E+00	0.000E+00	0.000E+00	0.000E+00	0.000E+00	0.000E+00	0.000E+00
Cm-242	4.670E+07	4.670E+07	4.130E+07	1.005E+07	1.550E+05	1.028E+05	1.695E+03	2.539E-15	0.000E+00	0.000E+00	6.900E+00
Cm-243	1.075E+06	1.075E+06	1.073E+06	1.052E+06	8.655E+05	1.232E+05	4.200E-04	0.000E+00	0.000E+00	0.000E+00	1.969E+02
Cm-244	2.274E+08	2.274E+08	2.266E+08	2.188E+08	1.550E+08	4.936E+06	4.021E-07	5.382E-07	1.822E-06	7.283E-06	1.630E+02
Cm-245	1.989E+05	1.989E+05	1.989E+05	1.989E+05	1.987E+05	1.972E+05	1.828E+05	8.597E+04	4.529E+01	0.000E+00	2.840E+02
Cm-246	1.566E+05	1.566E+05	1.566E+05	1.566E+05	1.564E+05	1.543E+05	1.352E+05	3.595E+04	6.371E-02	0.000E+00	2.845E+02
Cm-247	0.000E+00	0.000E+00	0.000E+00	0.000E+00	0.000E+00	0.000E+00	0.000E+00	0.000E+00	0.000E+00	0.000E+00	0.000E+00
Cm-248	0.000E+00	0.000E+00	0.000E+00	0.000E+00	0.000E+00	0.000E+00	0.000E+00	0.000E+00	0.000E+00	0.000E+00	0.000E+00
Cm-249	0.000E+00	0.000E+00	0.000E+00	0.000E+00	0.000E+00	0.000E+00	0.000E+00	0.000E+00	0.000E+00	0.000E+00	0.000E+00
Cm-250	0.000E+00	0.000E+00	0.000E+00	0.000E+00	0.000E+00	0.000E+00	0.000E+00	0.000E+00	0.000E+00	0.000E+00	0.000E+00
Bk-249	0.000E+00	0.000E+00	0.000E+00	0.000E+00	0.000E+00	0.000E+00	0.000E+00	0.000E+00	0.000E+00	0.000E+00	0.000E+00
Bk-250	0.000E+00	0.000E+00	0.000E+00	0.000E+00	0.000E+00	0.000E+00	0.000E+00	0.000E+00	0.000E+00	0.000E+00	0.000E+00
Cf-249	0.000E+00	0.000E+00	0.000E+00	0.000E+00	0.000E+00	0.000E+00	0.000E+00	0.000E+00	0.000E+00	0.000E+00	0.000E+00
Cf-250	0.000E+00	0.000E+00	0.000E+00	0.000E+00	0.000E+00	0.000E+00	0.000E+00	0.000E+00	0.000E+00	0.000E+00	0.000E+00
Cf-251	0.000E+00	0.000E+00	0.000E+00	0.000E+00	0.000E+00	0.000E+00	0.000E+00	0.000E+00	0.000E+00	0.000E+00	0.000E+00
Cf-252	0.000E+00	0.000E+00	0.000E+00	0.000E+00	0.000E+00	0.000E+00	0.000E+00	0.000E+00	0.000E+00	0.000E+00	0.000E+00
Cf-253	0.000E+00	0.000E+00	0.000E+00	0.000E+00	0.000E+00	0.000E+00	0.000E+00	0.000E+00	0.000E+00	0.000E+00	0.000E+00
Cf-254	0.000E+00	0.000E+00	0.000E+00	0.000E+00	0.000E+00	0.000E+00	0.000E+00	0.000E+00	0.000E+00	0.000E+00	0.000E+00
Es-253	0.000E+00	0.000E+00	0.000E+00	0.000E+00	0.000E+00	0.000E+00	0.000E+00	0.000E+00	0.000E+00	0.000E+00	0.000E+00
Total	5.163E+08	5.163E+08	5.114E+08	4.791E+08	4.143E+08	2.086E+08	5.428E+07	1.795E+07	7.553E+05	2.670E+04	0.000E+00

Table A.76

Converter core heavy TRU elements nuclide toxicity hazard (core) Basis = (EQULIBRIUM DISCHARGE FUEL - EQULIBRIUM FRESH FUEL) CONVERTER CORE

	Initial	1 hour	30 days	1 year	10 y	100 y	1000 y	10000 y	100000 y	1000000 y	Factor
Np-236	0.000E+00	0.000E+00	0.000E+00	0.000E+00	0.000E+00	0.000E+00	0.000E+00	0.000E+00	0.000E+00	0.000E+00	0.000E+00
Np-237	-1.155E+03	-1.155E+03	-1.152E+03	-1.155E+03	-1.178E+03	-1.569E+03	-3.703E+03	-4.342E+03	-4.224E+03	-3.159E+03	1.972E+02
Np-238	0.000E+00	0.000E+00	0.000E+00	0.000E+00	0.000E+00	0.000E+00	0.000E+00	0.000E+00	0.000E+00	0.000E+00	0.000E+00
Np-239	0.000E+00	0.000E+00	0.000E+00	0.000E+00	0.000E+00	0.000E+00	0.000E+00	0.000E+00	0.000E+00	0.000E+00	0.000E+00
Np-240m	0.000E+00	0.000E+00	0.000E+00	0.000E+00	0.000E+00	0.000E+00	0.000E+00	0.000E+00	0.000E+00	0.000E+00	0.000E+00
Np-240	0.000E+00	0.000E+00	0.000E+00	0.000E+00	0.000E+00	0.000E+00	0.000E+00	0.000E+00	0.000E+00	0.000E+00	0.000E+00
Pu-236	0.000E+00	0.000E+00	0.000E+00	0.000E+00	0.000E+00	0.000E+00	0.000E+00	0.000E+00	0.000E+00	0.000E+00	0.000E+00
Pu-238	-1.048E+07	-1.048E+07	-9.450E+06	-4.134E+06	-2.215E+06	-1.373E+06	-7.723E+03	-2.333E-14	0.000E+00	0.000E+00	2.461E+02
Pu-239	-2.334E+06	-2.334E+06	-2.322E+06	-2.322E+06	-2.321E+06	-2.316E+06	-2.263E+06	-1.778E+06	-1.406E+05	1.066E+00	2.675E+02
Pu-240	-1.926E+06	-1.926E+06	-1.926E+06	-1.926E+06	-1.926E+06	-1.878E+06	-1.717E+06	-6.821E+05	-6.693E+01	9.195E-03	2.675E+02
Pu-241	0.000E+00	0.000E+00	0.000E+00	0.000E+00	0.000E+00	0.000E+00	0.000E+00	0.000E+00	0.000E+00	0.000E+00	0.000E+00
Pu-242	-5.842E+03	-5.842E+03	-5.842E+03	-5.842E+03	-5.842E+03	-5.858E+03	-5.890E+03	-5.810E+03	-4.927E+03	-9.486E+02	2.675E+02
Pu-243	0.000E+00	0.000E+00	0.000E+00	0.000E+00	0.000E+00	0.000E+00	0.000E+00	0.000E+00	0.000E+00	0.000E+00	0.000E+00
Pu-244	0.000E+00	0.000E+00	0.000E+00	0.000E+00	0.000E+00	0.000E+00	0.000E+00	0.000E+00	0.000E+00	0.000E+00	0.000E+00
Pu-245	0.000E+00	0.000E+00	0.000E+00	0.000E+00	0.000E+00	0.000E+00	0.000E+00	0.000E+00	0.000E+00	0.000E+00	0.000E+00
Am-241	-8.629E+06	-8.629E+06	-8.678E+06	-9.235E+06	-1.348E+07	-1.898E+07	-4.519E+06	-1.719E+03	-9.661E-01	0.000E+00	2.729E+02
Am-242m	-8.218E+05	-8.218E+05	-8.218E+05	-8.186E+05	-7.848E+05	-5.216E+05	-8.603E+03	-1.284E-14	0.000E+00	0.000E+00	2.675E+02
Am-242	0.000E+00	0.000E+00	0.000E+00	0.000E+00	0.000E+00	0.000E+00	0.000E+00	0.000E+00	0.000E+00	0.000E+00	0.000E+00
Am-243	-1.932E+05	-1.932E+05	-1.932E+05	-1.932E+05	-1.932E+05	-1.916E+05	-1.768E+05	-7.827E+04	-2.129E+01	1.087E+00	2.729E+02
Am-244	0.000E+00	0.000E+00	0.000E+00	0.000E+00	0.000E+00	0.000E+00	0.000E+00	0.000E+00	0.000E+00	0.000E+00	0.000E+00
Am-245	0.000E+00	0.000E+00	0.000E+00	0.000E+00	0.000E+00	0.000E+00	0.000E+00	0.000E+00	0.000E+00	0.000E+00	0.000E+00
Cm-242	4.441E+07	4.441E+07	3.926E+07	9.422E+06	-1.660E+04	-1.101E+04	-1.822E+02	-2.724E-16	0.000E+00	0.000E+00	6.900E+00
Cm-243	-7.620E+04	-7.620E+04	-7.608E+04	-7.443E+04	-6.131E+04	-8.742E+03	-2.977E-05	0.000E+00	0.000E+00	0.000E+00	1.969E+02
Cm-244	4.401E+06	4.401E+06	4.303E+06	4.205E+06	2.934E+06	9.487E+04	3.969E-07	5.382E-07	1.822E-06	7.283E-06	1.630E+02
Cm-245	-4.260E+03	-4.260E+03	-4.260E+03	-4.090E+03	-4.260E+03	-4.260E+03	-3.919E+03	-1.806E+03	-9.542E-01	0.000E+00	2.840E+02
Cm-246	-6.487E+02	-6.487E+02	-6.657E+02	-6.486E+02	-6.487E+02	-6.316E+02	-5.633E+02	-1.536E+02	-2.731E-04	0.000E+00	2.845E+02
Cm-247	0.000E+00	0.000E+00	0.000E+00	0.000E+00	0.000E+00	0.000E+00	0.000E+00	0.000E+00	0.000E+00	0.000E+00	0.000E+00
Cm-248	0.000E+00	0.000E+00	0.000E+00	0.000E+00	0.000E+00	0.000E+00	0.000E+00	0.000E+00	0.000E+00	0.000E+00	0.000E+00
Cm-249	0.000E+00	0.000E+00	0.000E+00	0.000E+00	0.000E+00	0.000E+00	0.000E+00	0.000E+00	0.000E+00	0.000E+00	0.000E+00
Cm-250	0.000E+00	0.000E+00	0.000E+00	0.000E+00	0.000E+00	0.000E+00	0.000E+00	0.000E+00	0.000E+00	0.000E+00	0.000E+00
Bk-249	0.000E+00	0.000E+00	0.000E+00	0.000E+00	0.000E+00	0.000E+00	0.000E+00	0.000E+00	0.000E+00	0.000E+00	0.000E+00
Bk-250	0.000E+00	0.000E+00	0.000E+00	0.000E+00	0.000E+00	0.000E+00	0.000E+00	0.000E+00	0.000E+00	0.000E+00	0.000E+00
Cf-249	0.000E+00	0.000E+00	0.000E+00	0.000E+00	0.000E+00	0.000E+00	0.000E+00	0.000E+00	0.000E+00	0.000E+00	0.000E+00
Cf-250	0.000E+00	0.000E+00	0.000E+00	0.000E+00	0.000E+00	0.000E+00	0.000E+00	0.000E+00	0.000E+00	0.000E+00	0.000E+00
Cf-251	0.000E+00	0.000E+00	0.000E+00	0.000E+00	0.000E+00	0.000E+00	0.000E+00	0.000E+00	0.000E+00	0.000E+00	0.000E+00
Cf-252	0.000E+00	0.000E+00	0.000E+00	0.000E+00	0.000E+00	0.000E+00	0.000E+00	0.000E+00	0.000E+00	0.000E+00	0.000E+00
Cf-253	0.000E+00	0.000E+00	0.000E+00	0.000E+00	0.000E+00	0.000E+00	0.000E+00	0.000E+00	0.000E+00	0.000E+00	0.000E+00
Cf-254	0.000E+00	0.000E+00	0.000E+00	0.000E+00	0.000E+00	0.000E+00	0.000E+00	0.000E+00	0.000E+00	0.000E+00	0.000E+00
Es-253	0.000E+00	0.000E+00	0.000E+00	0.000E+00	0.000E+00	0.000E+00	0.000E+00	0.000E+00	0.000E+00	0.000E+00	0.000E+00
Total	2.433E+07	2.433E+07	2.008E+07	-5.089E+06	-1.807E+07	-2.519E+07	-8.707E+06	-2.553E+06	-1.498E+05	-4.106E+03	0.000E+00
Total/MWd	4.983E+01	4.984E+01	4.113E+01	-1.042E+01	-3.701E+01	-5.160E+01	-1.783E+01	-5.228E+00	-3.069E-01	-8.409E-03	0.000E+00

Table A.77

Converter core heavy TRU elements nuclide toxicity hazard (axial blanket) Basis = EQULIBRIUM FRESH FUEL CONVERTER CORE

	Initial	1 hour	30 days	1 year	10 y	100 y	1000 y	10000 y	100000 y	1000000 y	Factor
Np-236	0.000E+00	0.000E+00	0.000E+00	0.000E+00	0.000E+00	0.000E+00	0.000E+00	0.000E+00	0.000E+00	0.000E+00	0.000E+00
Np-237	0.000E+00	0.000E+00	0.000E+00	0.000E+00	0.000E+00	0.000E+00	0.000E+00	0.000E+00	0.000E+00	0.000E+00	1.972E+02
Np-238	0.000E+00	0.000E+00	0.000E+00	0.000E+00	0.000E+00	0.000E+00	0.000E+00	0.000E+00	0.000E+00	0.000E+00	0.000E+00
Np-239	0.000E+00	0.000E+00	0.000E+00	0.000E+00	0.000E+00	0.000E+00	0.000E+00	0.000E+00	0.000E+00	0.000E+00	0.000E+00
Np-240m	0.000E+00	0.000E+00	0.000E+00	0.000E+00	0.000E+00	0.000E+00	0.000E+00	0.000E+00	0.000E+00	0.000E+00	0.000E+00
Np-240	0.000E+00	0.000E+00	0.000E+00	0.000E+00	0.000E+00	0.000E+00	0.000E+00	0.000E+00	0.000E+00	0.000E+00	0.000E+00
Pu-236	0.000E+00	0.000E+00	0.000E+00	0.000E+00	0.000E+00	0.000E+00	0.000E+00	0.000E+00	0.000E+00	0.000E+00	0.000E+00
Pu-238	0.000E+00	0.000E+00	0.000E+00	0.000E+00	0.000E+00	0.000E+00	0.000E+00	0.000E+00	0.000E+00	0.000E+00	2.461E+02
Pu-239	0.000E+00	0.000E+00	0.000E+00	0.000E+00	0.000E+00	0.000E+00	0.000E+00	0.000E+00	0.000E+00	0.000E+00	2.675E+02
Pu-240	0.000E+00	0.000E+00	0.000E+00	0.000E+00	0.000E+00	0.000E+00	0.000E+00	0.000E+00	0.000E+00	0.000E+00	2.675E+02
Pu-241	0.000E+00	0.000E+00	0.000E+00	0.000E+00	0.000E+00	0.000E+00	0.000E+00	0.000E+00	0.000E+00	0.000E+00	0.000E+00
Pu-242	0.000E+00	0.000E+00	0.000E+00	0.000E+00	0.000E+00	0.000E+00	0.000E+00	0.000E+00	0.000E+00	0.000E+00	2.675E+02
Pu-243	0.000E+00	0.000E+00	0.000E+00	0.000E+00	0.000E+00	0.000E+00	0.000E+00	0.000E+00	0.000E+00	0.000E+00	0.000E+00
Pu-244	0.000E+00	0.000E+00	0.000E+00	0.000E+00	0.000E+00	0.000E+00	0.000E+00	0.000E+00	0.000E+00	0.000E+00	0.000E+00
Pu-245	0.000E+00	0.000E+00	0.000E+00	0.000E+00	0.000E+00	0.000E+00	0.000E+00	0.000E+00	0.000E+00	0.000E+00	0.000E+00
Am-241	0.000E+00	0.000E+00	0.000E+00	0.000E+00	0.000E+00	0.000E+00	0.000E+00	0.000E+00	0.000E+00	0.000E+00	2.729E+02
Am-242m	0.000E+00	0.000E+00	0.000E+00	0.000E+00	0.000E+00	0.000E+00	0.000E+00	0.000E+00	0.000E+00	0.000E+00	2.675E+02
Am-242	0.000E+00	0.000E+00	0.000E+00	0.000E+00	0.000E+00	0.000E+00	0.000E+00	0.000E+00	0.000E+00	0.000E+00	0.000E+00
Am-243	0.000E+00	0.000E+00	0.000E+00	0.000E+00	0.000E+00	0.000E+00	0.000E+00	0.000E+00	0.000E+00	0.000E+00	2.729E+02
Am-244	0.000E+00	0.000E+00	0.000E+00	0.000E+00	0.000E+00	0.000E+00	0.000E+00	0.000E+00	0.000E+00	0.000E+00	0.000E+00
Am-245	0.000E+00	0.000E+00	0.000E+00	0.000E+00	0.000E+00	0.000E+00	0.000E+00	0.000E+00	0.000E+00	0.000E+00	0.000E+00
Cm-242	0.000E+00	0.000E+00	0.000E+00	0.000E+00	0.000E+00	0.000E+00	0.000E+00	0.000E+00	0.000E+00	0.000E+00	6.900E+00
Cm-243	0.000E+00	0.000E+00	0.000E+00	0.000E+00	0.000E+00	0.000E+00	0.000E+00	0.000E+00	0.000E+00	0.000E+00	1.969E+02
Cm-244	0.000E+00	0.000E+00	0.000E+00	0.000E+00	0.000E+00	0.000E+00	0.000E+00	0.000E+00	0.000E+00	0.000E+00	1.630E+02
Cm-245	0.000E+00	0.000E+00	0.000E+00	0.000E+00	0.000E+00	0.000E+00	0.000E+00	0.000E+00	0.000E+00	0.000E+00	2.840E+02
Cm-246	0.000E+00	0.000E+00	0.000E+00	0.000E+00	0.000E+00	0.000E+00	0.000E+00	0.000E+00	0.000E+00	0.000E+00	2.845E+02
Cm-247	0.000E+00	0.000E+00	0.000E+00	0.000E+00	0.000E+00	0.000E+00	0.000E+00	0.000E+00	0.000E+00	0.000E+00	0.000E+00
Cm-248	0.000E+00	0.000E+00	0.000E+00	0.000E+00	0.000E+00	0.000E+00	0.000E+00	0.000E+00	0.000E+00	0.000E+00	0.000E+00
Cm-249	0.000E+00	0.000E+00	0.000E+00	0.000E+00	0.000E+00	0.000E+00	0.000E+00	0.000E+00	0.000E+00	0.000E+00	0.000E+00
Cm-250	0.000E+00	0.000E+00	0.000E+00	0.000E+00	0.000E+00	0.000E+00	0.000E+00	0.000E+00	0.000E+00	0.000E+00	0.000E+00
Bk-249	0.000E+00	0.000E+00	0.000E+00	0.000E+00	0.000E+00	0.000E+00	0.000E+00	0.000E+00	0.000E+00	0.000E+00	0.000E+00
Bk-250	0.000E+00	0.000E+00	0.000E+00	0.000E+00	0.000E+00	0.000E+00	0.000E+00	0.000E+00	0.000E+00	0.000E+00	0.000E+00
Cf-249	0.000E+00	0.000E+00	0.000E+00	0.000E+00	0.000E+00	0.000E+00	0.000E+00	0.000E+00	0.000E+00	0.000E+00	0.000E+00
Cf-250	0.000E+00	0.000E+00	0.000E+00	0.000E+00	0.000E+00	0.000E+00	0.000E+00	0.000E+00	0.000E+00	0.000E+00	0.000E+00
Cf-251	0.000E+00	0.000E+00	0.000E+00	0.000E+00	0.000E+00	0.000E+00	0.000E+00	0.000E+00	0.000E+00	0.000E+00	0.000E+00
Cf-252	0.000E+00	0.000E+00	0.000E+00	0.000E+00	0.000E+00	0.000E+00	0.000E+00	0.000E+00	0.000E+00	0.000E+00	0.000E+00
Cf-253	0.000E+00	0.000E+00	0.000E+00	0.000E+00	0.000E+00	0.000E+00	0.000E+00	0.000E+00	0.000E+00	0.000E+00	0.000E+00
Cf-254	0.000E+00	0.000E+00	0.000E+00	0.000E+00	0.000E+00	0.000E+00	0.000E+00	0.000E+00	0.000E+00	0.000E+00	0.000E+00
Es-253	0.000E+00	0.000E+00	0.000E+00	0.000E+00	0.000E+00	0.000E+00	0.000E+00	0.000E+00	0.000E+00	0.000E+00	0.000E+00
Total	0.000E+00	0.000E+00	0.000E+00	0.000E+00	0.000E+00	0.000E+00	0.000E+00	0.000E+00	0.000E+00	0.000E+00	0.000E+00

Table A.78

Converter core heavy TRU elements nuclide toxicity hazard (axial blanket) Basis = EQULIBRIUM DISCHARGE FUEL CONVERTER CORE

	Initial	1 hour	30 days	1 year	10 y	100 y	1000 y	10000 y	100000 y	1000000 y	Factor
Np-236	0.000E+00	0.000E+00	0.000E+00	0.000E+00	0.000E+00	0.000E+00	0.000E+00	0.000E+00	0.000E+00	0.000E+00	0.000E+00
Np-237	3.792E+01	3.792E+01	3.837E+01	3.839E+01	3.843E+01	3.983E+01	4.788E+01	5.024E+01	4.880E+01	3.645E+01	1.972E+02
Np-238	0.000E+00	0.000E+00	0.000E+00	0.000E+00	0.000E+00	0.000E+00	0.000E+00	0.000E+00	0.000E+00	0.000E+00	0.000E+00
Np-239	0.000E+00	0.000E+00	0.000E+00	0.000E+00	0.000E+00	0.000E+00	0.000E+00	0.000E+00	0.000E+00	0.000E+00	0.000E+00
Np-240m	0.000E+00	0.000E+00	0.000E+00	0.000E+00	0.000E+00	0.000E+00	0.000E+00	0.000E+00	0.000E+00	0.000E+00	0.000E+00
Np-240	0.000E+00	0.000E+00	0.000E+00	0.000E+00	0.000E+00	0.000E+00	0.000E+00	0.000E+00	0.000E+00	0.000E+00	0.000E+00
Pu-236	0.000E+00	0.000E+00	0.000E+00	0.000E+00	0.000E+00	0.000E+00	0.000E+00	0.000E+00	0.000E+00	0.000E+00	0.000E+00
Pu-238	1.389E+05	1.389E+05	1.397E+05	1.389E+05	1.296E+05	6.431E+04	5.859E+01	1.943E-18	0.000E+00	0.000E+00	2.461E+02
Pu-239	1.100E+06	1.100E+06	1.105E+06	1.105E+06	1.104E+06	1.102E+06	1.074E+06	8.314E+05	6.447E+04	5.067E-07	2.675E+02
Pu-240	1.548E+05	1.548E+05	1.548E+05	1.548E+05	1.546E+05	1.532E+05	1.397E+05	5.552E+04	5.447E+00	2.602E-09	2.675E+02
Pu-241	0.000E+00	0.000E+00	0.000E+00	0.000E+00	0.000E+00	0.000E+00	0.000E+00	0.000E+00	0.000E+00	0.000E+00	0.000E+00
Pu-242	1.770E+00	1.770E+00	1.770E+00	1.770E+00	1.770E+00	1.772E+00	1.772E+00	1.743E+00	1.479E+00	2.850E-01	2.675E+02
Pu-243	0.000E+00	0.000E+00	0.000E+00	0.000E+00	0.000E+00	0.000E+00	0.000E+00	0.000E+00	0.000E+00	0.000E+00	0.000E+00
Pu-244	0.000E+00	0.000E+00	0.000E+00	0.000E+00	0.000E+00	0.000E+00	0.000E+00	0.000E+00	0.000E+00	0.000E+00	0.000E+00
Pu-245	0.000E+00	0.000E+00	0.000E+00	0.000E+00	0.000E+00	0.000E+00	0.000E+00	0.000E+00	0.000E+00	0.000E+00	0.000E+00
Am-241	2.585E+03	2.585E+03	2.895E+03	6.268E+03	3.236E+04	7.165E+04	1.713E+04	9.926E-03	2.558E-07	0.000E+00	2.729E+02
Am-242m	6.818E+01	6.818E+01	6.815E+01	6.788E+01	6.513E+01	4.321E+01	7.128E-01	1.064E-18	0.000E+00	0.000E+00	2.675E+02
Am-242	0.000E+00	0.000E+00	0.000E+00	0.000E+00	0.000E+00	0.000E+00	0.000E+00	0.000E+00	0.000E+00	0.000E+00	0.000E+00
Am-243	1.398E+00	1.398E+00	1.400E+00	1.400E+00	1.399E+00	1.388E+00	1.279E+00	5.659E-01	1.625E-04	1.646E-11	2.729E+02
Am-244	0.000E+00	0.000E+00	0.000E+00	0.000E+00	0.000E+00	0.000E+00	0.000E+00	0.000E+00	0.000E+00	0.000E+00	0.000E+00
Am-245	0.000E+00	0.000E+00	0.000E+00	0.000E+00	0.000E+00	0.000E+00	0.000E+00	0.000E+00	0.000E+00	0.000E+00	0.000E+00
Cm-242	1.598E+03	1.599E+03	1.419E+03	3.421E+02	1.378E+00	9.141E-01	1.508E-02	0.000E+00	0.000E+00	0.000E+00	6.900E+00
Cm-243	5.905E+00	5.905E+00	5.894E+00	5.778E+00	4.755E+00	6.767E-01	2.307E-09	0.000E+00	0.000E+00	0.000E+00	1.969E+02
Cm-244	1.356E+01	1.356E+01	1.352E+01	1.305E+01	9.244E+00	2.944E-01	2.078E-12	2.078E-12	2.076E-12	2.061E-12	1.630E+02
Cm-245	1.106E-03	1.106E-03	1.106E-03	1.106E-03	1.105E-03	1.097E-03	1.017E-03	4.780E-04	2.519E-07	0.000E+00	2.840E+02
Cm-246	1.370E-05	1.370E-05	1.370E-05	1.370E-05	1.368E-05	1.350E-05	1.182E-05	3.144E-06	5.573E-12	0.000E+00	2.845E+02
Cm-247	0.000E+00	0.000E+00	0.000E+00	0.000E+00	0.000E+00	0.000E+00	0.000E+00	0.000E+00	0.000E+00	0.000E+00	0.000E+00
Cm-248	0.000E+00	0.000E+00	0.000E+00	0.000E+00	0.000E+00	0.000E+00	0.000E+00	0.000E+00	0.000E+00	0.000E+00	0.000E+00
Cm-249	0.000E+00	0.000E+00	0.000E+00	0.000E+00	0.000E+00	0.000E+00	0.000E+00	0.000E+00	0.000E+00	0.000E+00	0.000E+00
Cm-250	0.000E+00	0.000E+00	0.000E+00	0.000E+00	0.000E+00	0.000E+00	0.000E+00	0.000E+00	0.000E+00	0.000E+00	0.000E+00
Bk-249	0.000E+00	0.000E+00	0.000E+00	0.000E+00	0.000E+00	0.000E+00	0.000E+00	0.000E+00	0.000E+00	0.000E+00	0.000E+00
Bk-250	0.000E+00	0.000E+00	0.000E+00	0.000E+00	0.000E+00	0.000E+00	0.000E+00	0.000E+00	0.000E+00	0.000E+00	0.000E+00
Cf-249	0.000E+00	0.000E+00	0.000E+00	0.000E+00	0.000E+00	0.000E+00	0.000E+00	0.000E+00	0.000E+00	0.000E+00	0.000E+00
Cf-250	0.000E+00	0.000E+00	0.000E+00	0.000E+00	0.000E+00	0.000E+00	0.000E+00	0.000E+00	0.000E+00	0.000E+00	0.000E+00
Cf-251	0.000E+00	0.000E+00	0.000E+00	0.000E+00	0.000E+00	0.000E+00	0.000E+00	0.000E+00	0.000E+00	0.000E+00	0.000E+00
Cf-252	0.000E+00	0.000E+00	0.000E+00	0.000E+00	0.000E+00	0.000E+00	0.000E+00	0.000E+00	0.000E+00	0.000E+00	0.000E+00
Cf-253	0.000E+00	0.000E+00	0.000E+00	0.000E+00	0.000E+00	0.000E+00	0.000E+00	0.000E+00	0.000E+00	0.000E+00	0.000E+00
Cf-254	0.000E+00	0.000E+00	0.000E+00	0.000E+00	0.000E+00	0.000E+00	0.000E+00	0.000E+00	0.000E+00	0.000E+00	0.000E+00
Es-253	0.000E+00	0.000E+00	0.000E+00	0.000E+00	0.000E+00	0.000E+00	0.000E+00	0.000E+00	0.000E+00	0.000E+00	0.000E+00
Total	1.398E+06	1.398E+06	1.404E+06	1.405E+06	1.421E+06	1.391E+06	1.231E+06	8.870E+05	6.453E+04	3.674E+01	0.000E+00

Table A.79

Converter core heavy TRU elements nuclide toxicity hazard (axial blanket)

Basis = (EQULIBRIUM DISCHARGE FUEL - EQULIBRIUM FRESH FUEL) CONVERTER CORE

	Initial	1 hour	30 days	1 year	10 y	100 y	1000 y	10000 y	100000 y	1000000 y	Factor
Np-236	0.000E+00	0.000E+00	0.000E+00	0.000E+00	0.000E+00	0.000E+00	0.000E+00	0.000E+00	0.000E+00	0.000E+00	0.000E+00
Np-237	3.792E+01	3.792E+01	3.837E+01	3.839E+01	3.843E+01	3.983E+01	4.788E+01	5.024E+01	4.880E+01	3.645E+01	1.972E+02
Np-238	0.000E+00	0.000E+00	0.000E+00	0.000E+00	0.000E+00	0.000E+00	0.000E+00	0.000E+00	0.000E+00	0.000E+00	0.000E+00
Np-239	0.000E+00	0.000E+00	0.000E+00	0.000E+00	0.000E+00	0.000E+00	0.000E+00	0.000E+00	0.000E+00	0.000E+00	0.000E+00
Np-240m	0.000E+00	0.000E+00	0.000E+00	0.000E+00	0.000E+00	0.000E+00	0.000E+00	0.000E+00	0.000E+00	0.000E+00	0.000E+00
Np-240	0.000E+00	0.000E+00	0.000E+00	0.000E+00	0.000E+00	0.000E+00	0.000E+00	0.000E+00	0.000E+00	0.000E+00	0.000E+00
Pu-236	0.000E+00	0.000E+00	0.000E+00	0.000E+00	0.000E+00	0.000E+00	0.000E+00	0.000E+00	0.000E+00	0.000E+00	0.000E+00
Pu-238	1.389E+05	1.389E+05	1.397E+05	1.389E+05	1.296E+05	6.431E+04	5.859E+01	1.943E-18	0.000E+00	0.000E+00	2.461E+02
Pu-239	1.100E+06	1.100E+06	1.105E+06	1.105E+06	1.104E+06	1.102E+06	1.074E+06	8.314E+05	6.447E+04	5.067E-07	2.675E+02
Pu-240	1.548E+05	1.548E+05	1.548E+05	1.548E+05	1.546E+05	1.532E+05	1.397E+05	5.552E+04	5.447E+00	2.602E-09	2.675E+02
Pu-241	0.000E+00	0.000E+00	0.000E+00	0.000E+00	0.000E+00	0.000E+00	0.000E+00	0.000E+00	0.000E+00	0.000E+00	0.000E+00
Pu-242	1.770E+00	1.770E+00	1.770E+00	1.770E+00	1.770E+00	1.772E+00	1.772E+00	1.743E+00	1.479E+00	2.850E-01	2.675E+02
Pu-243	0.000E+00	0.000E+00	0.000E+00	0.000E+00	0.000E+00	0.000E+00	0.000E+00	0.000E+00	0.000E+00	0.000E+00	0.000E+00
Pu-244	0.000E+00	0.000E+00	0.000E+00	0.000E+00	0.000E+00	0.000E+00	0.000E+00	0.000E+00	0.000E+00	0.000E+00	0.000E+00
Pu-245	0.000E+00	0.000E+00	0.000E+00	0.000E+00	0.000E+00	0.000E+00	0.000E+00	0.000E+00	0.000E+00	0.000E+00	0.000E+00
Am-241	2.585E+03	2.585E+03	2.895E+03	6.268E+03	3.236E+04	7.165E+04	1.713E+04	9.926E-03	2.558E-07	0.000E+00	2.729E+02
Am-242m	6.818E+01	6.818E+01	6.815E+01	6.788E+01	6.513E+01	4.321E+01	7.128E-01	1.064E-18	0.000E+00	0.000E+00	2.675E+02
Am-242	0.000E+00	0.000E+00	0.000E+00	0.000E+00	0.000E+00	0.000E+00	0.000E+00	0.000E+00	0.000E+00	0.000E+00	0.000E+00
Am-243	1.398E+00	1.398E+00	1.400E+00	1.400E+00	1.399E+00	1.388E+00	1.279E+00	5.659E-01	1.625E-04	1.646E-11	2.729E+02
Am-244	0.000E+00	0.000E+00	0.000E+00	0.000E+00	0.000E+00	0.000E+00	0.000E+00	0.000E+00	0.000E+00	0.000E+00	0.000E+00
Am-245	0.000E+00	0.000E+00	0.000E+00	0.000E+00	0.000E+00	0.000E+00	0.000E+00	0.000E+00	0.000E+00	0.000E+00	0.000E+00
Cm-242	1.598E+03	1.599E+03	1.419E+03	3.421E+02	1.378E+00	9.141E-01	1.508E-02	0.000E+00	0.000E+00	0.000E+00	6.900E+00
Cm-243	5.905E+00	5.905E+00	5.894E+00	5.778E+00	4.755E+00	6.767E-01	2.307E-09	0.000E+00	0.000E+00	0.000E+00	1.969E+02
Cm-244	1.356E+01	1.356E+01	1.352E+01	1.305E+01	9.244E+00	2.944E-01	2.078E-12	2.078E-12	2.076E-12	2.061E-12	1.630E+02
Cm-245	1.106E-03	1.106E-03	1.106E-03	1.106E-03	1.105E-03	1.097E-03	1.017E-03	4.780E-04	2.519E-07	0.000E+00	2.840E+02
Cm-246	1.370E-05	1.370E-05	1.370E-05	1.370E-05	1.368E-05	1.350E-05	1.182E-05	3.144E-06	5.573E-12	0.000E+00	2.845E+02
Cm-247	0.000E+00	0.000E+00	0.000E+00	0.000E+00	0.000E+00	0.000E+00	0.000E+00	0.000E+00	0.000E+00	0.000E+00	0.000E+00
Cm-248	0.000E+00	0.000E+00	0.000E+00	0.000E+00	0.000E+00	0.000E+00	0.000E+00	0.000E+00	0.000E+00	0.000E+00	0.000E+00
Cm-249	0.000E+00	0.000E+00	0.000E+00	0.000E+00	0.000E+00	0.000E+00	0.000E+00	0.000E+00	0.000E+00	0.000E+00	0.000E+00
Cm-250	0.000E+00	0.000E+00	0.000E+00	0.000E+00	0.000E+00	0.000E+00	0.000E+00	0.000E+00	0.000E+00	0.000E+00	0.000E+00
Bk-249	0.000E+00	0.000E+00	0.000E+00	0.000E+00	0.000E+00	0.000E+00	0.000E+00	0.000E+00	0.000E+00	0.000E+00	0.000E+00
Bk-250	0.000E+00	0.000E+00	0.000E+00	0.000E+00	0.000E+00	0.000E+00	0.000E+00	0.000E+00	0.000E+00	0.000E+00	0.000E+00
Cf-249	0.000E+00	0.000E+00	0.000E+00	0.000E+00	0.000E+00	0.000E+00	0.000E+00	0.000E+00	0.000E+00	0.000E+00	0.000E+00
Cf-250	0.000E+00	0.000E+00	0.000E+00	0.000E+00	0.000E+00	0.000E+00	0.000E+00	0.000E+00	0.000E+00	0.000E+00	0.000E+00
Cf-251	0.000E+00	0.000E+00	0.000E+00	0.000E+00	0.000E+00	0.000E+00	0.000E+00	0.000E+00	0.000E+00	0.000E+00	0.000E+00
Cf-252	0.000E+00	0.000E+00	0.000E+00	0.000E+00	0.000E+00	0.000E+00	0.000E+00	0.000E+00	0.000E+00	0.000E+00	0.000E+00
Cf-253	0.000E+00	0.000E+00	0.000E+00	0.000E+00	0.000E+00	0.000E+00	0.000E+00	0.000E+00	0.000E+00	0.000E+00	0.000E+00
Cf-254	0.000E+00	0.000E+00	0.000E+00	0.000E+00	0.000E+00	0.000E+00	0.000E+00	0.000E+00	0.000E+00	0.000E+00	0.000E+00
Es-253	0.000E+00	0.000E+00	0.000E+00	0.000E+00	0.000E+00	0.000E+00	0.000E+00	0.000E+00	0.000E+00	0.000E+00	0.000E+00
Total	1.398E+06	1.398E+06	1.404E+06	1.405E+06	1.421E+06	1.391E+06	1.231E+06	8.870E+05	6.453E+04	3.674E+01	0.000E+00
Total/MWd	2.863E+00	2.863E+00	2.875E+00	2.878E+00	2.910E+00	2.848E+00	2.520E+00	1.817E+00	1.322E-01	7.525E-05	0.000E+00

Table A.80

Breeder core heavy TRU elements nuclide toxicity hazard (core) Basis = EQULIBRIUM FRESH FUEL BREEDER CORE

	Initial	1 hour	30 days	1 year	10 y	100 y	1000 y	10000 y	100000 y	1000000 y	Factor
Np-236	0.000E+00	0.000E+00	0.000E+00	0.000E+00	0.000E+00	0.000E+00	0.000E+00	0.000E+00	0.000E+00	0.000E+00	0.000E+00
Np-237	1.074E+03	1.074E+03	1.074E+03	1.078E+03	1.125E+03	1.897E+03	6.110E+03	7.427E+03	7.241E+03	5.411E+03	1.972E+02
Np-238	0.000E+00	0.000E+00	0.000E+00	0.000E+00	0.000E+00	0.000E+00	0.000E+00	0.000E+00	0.000E+00	0.000E+00	0.000E+00
Np-239	0.000E+00	0.000E+00	0.000E+00	0.000E+00	0.000E+00	0.000E+00	0.000E+00	0.000E+00	0.000E+00	0.000E+00	0.000E+00
Np-240m	0.000E+00	0.000E+00	0.000E+00	0.000E+00	0.000E+00	0.000E+00	0.000E+00	0.000E+00	0.000E+00	0.000E+00	0.000E+00
Np-240	0.000E+00	0.000E+00	0.000E+00	0.000E+00	0.000E+00	0.000E+00	0.000E+00	0.000E+00	0.000E+00	0.000E+00	0.000E+00
Pu-236	0.000E+00	0.000E+00	0.000E+00	0.000E+00	0.000E+00	0.000E+00	0.000E+00	0.000E+00	0.000E+00	0.000E+00	0.000E+00
Pu-238	4.818E+07	4.818E+07	4.818E+07	4.793E+07	4.486E+07	2.312E+07	4.421E+04	8.433E-14	0.000E+00	0.000E+00	2.461E+02
Pu-239	1.510E+07	1.510E+07	1.510E+07	1.510E+07	1.510E+07	1.506E+07	1.469E+07	1.141E+07	8.877E+05	6.977E-06	2.675E+02
Pu-240	1.727E+07	1.727E+07	1.727E+07	1.727E+07	1.729E+07	1.721E+07	1.569E+07	6.234E+06	6.118E+02	0.000E+00	2.675E+02
Pu-241	0.000E+00	0.000E+00	0.000E+00	0.000E+00	0.000E+00	0.000E+00	0.000E+00	0.000E+00	0.000E+00	0.000E+00	0.000E+00
Pu-242	1.923E+04	1.923E+04	1.923E+04	1.923E+04	1.923E+04	1.929E+04	1.942E+04	1.923E+04	1.635E+04	3.152E+03	2.675E+02
Pu-243	0.000E+00	0.000E+00	0.000E+00	0.000E+00	0.000E+00	0.000E+00	0.000E+00	0.000E+00	0.000E+00	0.000E+00	0.000E+00
Pu-244	0.000E+00	0.000E+00	0.000E+00	0.000E+00	0.000E+00	0.000E+00	0.000E+00	0.000E+00	0.000E+00	0.000E+00	0.000E+00
Pu-245	0.000E+00	0.000E+00	0.000E+00	0.000E+00	0.000E+00	0.000E+00	0.000E+00	0.000E+00	0.000E+00	0.000E+00	0.000E+00
Am-241	1.647E+07	1.647E+07	1.659E+07	1.770E+07	2.630E+07	3.751E+07	8.952E+06	9.803E+03	5.448E+00	0.000E+00	2.729E+02
Am-242m	2.960E+06	2.960E+06	2.958E+06	2.945E+06	2.826E+06	1.875E+06	3.094E+04	4.619E-14	0.000E+00	0.000E+00	2.675E+02
Am-242	0.000E+00	0.000E+00	0.000E+00	0.000E+00	0.000E+00	0.000E+00	0.000E+00	0.000E+00	0.000E+00	0.000E+00	0.000E+00
Am-243	2.396E+05	2.396E+05	2.396E+05	2.396E+05	2.394E+05	2.374E+05	2.188E+05	9.680E+04	2.780E+01	0.000E+00	2.729E+02
Am-244	0.000E+00	0.000E+00	0.000E+00	0.000E+00	0.000E+00	0.000E+00	0.000E+00	0.000E+00	0.000E+00	0.000E+00	0.000E+00
Am-245	0.000E+00	0.000E+00	0.000E+00	0.000E+00	0.000E+00	0.000E+00	0.000E+00	0.000E+00	0.000E+00	0.000E+00	0.000E+00
Cm-242	8.102E+05	8.098E+05	7.204E+05	2.204E+05	5.978E+04	3.967E+04	6.545E+02	9.799E-16	0.000E+00	0.000E+00	6.900E+00
Cm-243	3.210E+05	3.210E+05	3.205E+05	3.141E+05	2.585E+05	3.679E+04	1.255E-04	0.000E+00	0.000E+00	0.000E+00	1.969E+02
Cm-244	2.637E+07	2.637E+07	2.629E+07	2.538E+07	1.798E+07	5.724E+05	6.137E-10	0.000E+00	0.000E+00	0.000E+00	1.630E+02
Cm-245	2.355E+04	2.355E+04	2.355E+04	2.355E+04	2.353E+04	2.334E+04	2.166E+04	1.018E+04	5.362E+00	0.000E+00	2.840E+02
Cm-246	1.707E+04	1.707E+04	1.707E+04	1.707E+04	1.705E+04	1.683E+04	1.474E+04	3.921E+03	6.947E-03	0.000E+00	2.845E+02
Cm-247	0.000E+00	0.000E+00	0.000E+00	0.000E+00	0.000E+00	0.000E+00	0.000E+00	0.000E+00	0.000E+00	0.000E+00	0.000E+00
Cm-248	0.000E+00	0.000E+00	0.000E+00	0.000E+00	0.000E+00	0.000E+00	0.000E+00	0.000E+00	0.000E+00	0.000E+00	0.000E+00
Cm-249	0.000E+00	0.000E+00	0.000E+00	0.000E+00	0.000E+00	0.000E+00	0.000E+00	0.000E+00	0.000E+00	0.000E+00	0.000E+00
Cm-250	0.000E+00	0.000E+00	0.000E+00	0.000E+00	0.000E+00	0.000E+00	0.000E+00	0.000E+00	0.000E+00	0.000E+00	0.000E+00
Bk-249	0.000E+00	0.000E+00	0.000E+00	0.000E+00	0.000E+00	0.000E+00	0.000E+00	0.000E+00	0.000E+00	0.000E+00	0.000E+00
Bk-250	0.000E+00	0.000E+00	0.000E+00	0.000E+00	0.000E+00	0.000E+00	0.000E+00	0.000E+00	0.000E+00	0.000E+00	0.000E+00
Cf-249	0.000E+00	0.000E+00	0.000E+00	0.000E+00	0.000E+00	0.000E+00	0.000E+00	0.000E+00	0.000E+00	0.000E+00	0.000E+00
Cf-250	0.000E+00	0.000E+00	0.000E+00	0.000E+00	0.000E+00	0.000E+00	0.000E+00	0.000E+00	0.000E+00	0.000E+00	0.000E+00
Cf-251	0.000E+00	0.000E+00	0.000E+00	0.000E+00	0.000E+00	0.000E+00	0.000E+00	0.000E+00	0.000E+00	0.000E+00	0.000E+00
Cf-252	0.000E+00	0.000E+00	0.000E+00	0.000E+00	0.000E+00	0.000E+00	0.000E+00	0.000E+00	0.000E+00	0.000E+00	0.000E+00
Cf-253	0.000E+00	0.000E+00	0.000E+00	0.000E+00	0.000E+00	0.000E+00	0.000E+00	0.000E+00	0.000E+00	0.000E+00	0.000E+00
Cf-254	0.000E+00	0.000E+00	0.000E+00	0.000E+00	0.000E+00	0.000E+00	0.000E+00	0.000E+00	0.000E+00	0.000E+00	0.000E+00
Es-253	0.000E+00	0.000E+00	0.000E+00	0.000E+00	0.000E+00	0.000E+00	0.000E+00	0.000E+00	0.000E+00	0.000E+00	0.000E+00
Total	1.278E+08	1.278E+08	1.277E+08	1.272E+08	1.250E+08	9.573E+07	3.969E+07	1.779E+07	9.120E+05	8.563E+03	0.000E+00

Table A.81

Breeder core heavy TRU elements nuclide toxicity hazard (core) Basis = EQULIBRIUM DISCHARGE FUEL BREEDER CORE

	Initial	1 hour	30 days	1 year	10 y	100 y	1000 y	10000 y	100000 y	1000000 y	Factor
Np-236	0.000E+00	0.000E+00	0.000E+00	0.000E+00	0.000E+00	0.000E+00	0.000E+00	0.000E+00	0.000E+00	0.000E+00	0.000E+00
Np-237	9.998E+02	9.998E+02	1.003E+03	1.006E+03	1.050E+03	1.821E+03	6.075E+03	7.403E+03	7.219E+03	5.393E+03	1.972E+02
Np-238	0.000E+00	0.000E+00	0.000E+00	0.000E+00	0.000E+00	0.000E+00	0.000E+00	0.000E+00	0.000E+00	0.000E+00	0.000E+00
Np-239	0.000E+00	0.000E+00	0.000E+00	0.000E+00	0.000E+00	0.000E+00	0.000E+00	0.000E+00	0.000E+00	0.000E+00	0.000E+00
Np-240m	0.000E+00	0.000E+00	0.000E+00	0.000E+00	0.000E+00	0.000E+00	0.000E+00	0.000E+00	0.000E+00	0.000E+00	0.000E+00
Np-240	0.000E+00	0.000E+00	0.000E+00	0.000E+00	0.000E+00	0.000E+00	0.000E+00	0.000E+00	0.000E+00	0.000E+00	0.000E+00
Pu-236	0.000E+00	0.000E+00	0.000E+00	0.000E+00	0.000E+00	0.000E+00	0.000E+00	0.000E+00	0.000E+00	0.000E+00	0.000E+00
Pu-238	4.666E+07	4.666E+07	4.701E+07	4.870E+07	4.613E+07	2.367E+07	4.275E+04	7.720E-14	0.000E+00	0.000E+00	2.461E+02
Pu-239	1.212E+07	1.212E+07	1.213E+07	1.213E+07	1.213E+07	1.210E+07	1.180E+07	9.171E+06	7.145E+05	1.164E-01	2.675E+02
Pu-240	1.713E+07	1.713E+07	1.713E+07	1.713E+07	1.714E+07	1.708E+07	1.557E+07	6.187E+06	6.072E+02	1.041E-03	2.675E+02
Pu-241	0.000E+00	0.000E+00	0.000E+00	0.000E+00	0.000E+00	0.000E+00	0.000E+00	0.000E+00	0.000E+00	0.000E+00	0.000E+00
Pu-242	1.942E+04	1.942E+04	1.942E+04	1.942E+04	1.942E+04	1.948E+04	1.960E+04	1.942E+04	1.650E+04	3.183E+03	2.675E+02
Pu-243	0.000E+00	0.000E+00	0.000E+00	0.000E+00	0.000E+00	0.000E+00	0.000E+00	0.000E+00	0.000E+00	0.000E+00	0.000E+00
Pu-244	0.000E+00	0.000E+00	0.000E+00	0.000E+00	0.000E+00	0.000E+00	0.000E+00	0.000E+00	0.000E+00	0.000E+00	0.000E+00
Pu-245	0.000E+00	0.000E+00	0.000E+00	0.000E+00	0.000E+00	0.000E+00	0.000E+00	0.000E+00	0.000E+00	0.000E+00	0.000E+00
Am-241	1.431E+07	1.431E+07	1.442E+07	1.565E+07	2.508E+07	3.784E+07	9.035E+06	9.857E+03	5.477E+00	0.000E+00	2.729E+02
Am-242m	2.709E+06	2.709E+06	2.708E+06	2.696E+06	2.587E+06	1.716E+06	2.831E+04	4.228E-14	0.000E+00	0.000E+00	2.675E+02
Am-242	0.000E+00	0.000E+00	0.000E+00	0.000E+00	0.000E+00	0.000E+00	0.000E+00	0.000E+00	0.000E+00	0.000E+00	0.000E+00
Am-243	2.415E+05	2.415E+05	2.415E+05	2.415E+05	2.414E+05	2.394E+05	2.206E+05	9.759E+04	2.815E+01	1.188E-01	2.729E+02
Am-244	0.000E+00	0.000E+00	0.000E+00	0.000E+00	0.000E+00	0.000E+00	0.000E+00	0.000E+00	0.000E+00	0.000E+00	0.000E+00
Am-245	0.000E+00	0.000E+00	0.000E+00	0.000E+00	0.000E+00	0.000E+00	0.000E+00	0.000E+00	0.000E+00	0.000E+00	0.000E+00
Cm-242	1.680E+07	1.680E+07	1.485E+07	3.613E+06	5.473E+04	3.631E+04	5.991E+02	8.971E-16	0.000E+00	0.000E+00	6.900E+00
Cm-243	3.314E+05	3.314E+05	3.308E+05	3.243E+05	2.668E+05	3.797E+04	1.295E-04	0.000E+00	0.000E+00	0.000E+00	1.969E+02
Cm-244	2.843E+07	2.843E+07	2.833E+07	2.735E+07	1.938E+07	6.172E+05	8.819E-08	1.027E-07	2.403E-07	8.250E-07	1.630E+02
Cm-245	2.369E+04	2.369E+04	2.369E+04	2.367E+04	2.365E+04	2.348E+04	2.178E+04	1.024E+04	5.393E+00	0.000E+00	2.840E+02
Cm-246	1.721E+04	1.721E+04	1.721E+04	1.721E+04	1.719E+04	1.696E+04	1.486E+04	3.952E+03	7.002E-03	0.000E+00	2.845E+02
Cm-247	0.000E+00	0.000E+00	0.000E+00	0.000E+00	0.000E+00	0.000E+00	0.000E+00	0.000E+00	0.000E+00	0.000E+00	0.000E+00
Cm-248	0.000E+00	0.000E+00	0.000E+00	0.000E+00	0.000E+00	0.000E+00	0.000E+00	0.000E+00	0.000E+00	0.000E+00	0.000E+00
Cm-249	0.000E+00	0.000E+00	0.000E+00	0.000E+00	0.000E+00	0.000E+00	0.000E+00	0.000E+00	0.000E+00	0.000E+00	0.000E+00
Cm-250	0.000E+00	0.000E+00	0.000E+00	0.000E+00	0.000E+00	0.000E+00	0.000E+00	0.000E+00	0.000E+00	0.000E+00	0.000E+00
Bk-249	0.000E+00	0.000E+00	0.000E+00	0.000E+00	0.000E+00	0.000E+00	0.000E+00	0.000E+00	0.000E+00	0.000E+00	0.000E+00
Bk-250	0.000E+00	0.000E+00	0.000E+00	0.000E+00	0.000E+00	0.000E+00	0.000E+00	0.000E+00	0.000E+00	0.000E+00	0.000E+00
Cf-249	0.000E+00	0.000E+00	0.000E+00	0.000E+00	0.000E+00	0.000E+00	0.000E+00	0.000E+00	0.000E+00	0.000E+00	0.000E+00
Cf-250	0.000E+00	0.000E+00	0.000E+00	0.000E+00	0.000E+00	0.000E+00	0.000E+00	0.000E+00	0.000E+00	0.000E+00	0.000E+00
Cf-251	0.000E+00	0.000E+00	0.000E+00	0.000E+00	0.000E+00	0.000E+00	0.000E+00	0.000E+00	0.000E+00	0.000E+00	0.000E+00
Cf-252	0.000E+00	0.000E+00	0.000E+00	0.000E+00	0.000E+00	0.000E+00	0.000E+00	0.000E+00	0.000E+00	0.000E+00	0.000E+00
Cf-253	0.000E+00	0.000E+00	0.000E+00	0.000E+00	0.000E+00	0.000E+00	0.000E+00	0.000E+00	0.000E+00	0.000E+00	0.000E+00
Cf-254	0.000E+00	0.000E+00	0.000E+00	0.000E+00	0.000E+00	0.000E+00	0.000E+00	0.000E+00	0.000E+00	0.000E+00	0.000E+00
Es-253	0.000E+00	0.000E+00	0.000E+00	0.000E+00	0.000E+00	0.000E+00	0.000E+00	0.000E+00	0.000E+00	0.000E+00	0.000E+00
Total	1.388E+08	1.388E+08	1.372E+08	1.279E+08	1.231E+08	9.340E+07	3.676E+07	1.551E+07	7.389E+05	8.576E+03	0.000E+00

Table A.82

Breeder core heavy TRU elements nuclide toxicity hazard (core) Basis = (EQULIBRIUM DISCHARGE FUEL - EQULIBRIUM FRESH FUEL) BREEDER CORE

	Initial	**1 hour**	**30 days**	**1 year**	**10 y**	**100 y**	**1000 y**	**10000 y**	**100000 y**	**1000000 y**	**Factor**
Np-236	0.000E+00	0.000E+00	0.000E+00	0.000E+00	0.000E+00	0.000E+00	0.000E+00	0.000E+00	0.000E+00	0.000E+00	0.000E+00
Np-237	-7.419E+01	-7.419E+01	-7.135E+01	-7.170E+01	-7.501E+01	-7.572E+01	-3.550E+01	-2.366E+01	-2.248E+01	-1.775E+01	1.972E+02
Np-238	0.000E+00	0.000E+00	0.000E+00	0.000E+00	0.000E+00	0.000E+00	0.000E+00	0.000E+00	0.000E+00	0.000E+00	0.000E+00
Np-239	0.000E+00	0.000E+00	0.000E+00	0.000E+00	0.000E+00	0.000E+00	0.000E+00	0.000E+00	0.000E+00	0.000E+00	0.000E+00
Np-240m	0.000E+00	0.000E+00	0.000E+00	0.000E+00	0.000E+00	0.000E+00	0.000E+00	0.000E+00	0.000E+00	0.000E+00	0.000E+00
Np-240	0.000E+00	0.000E+00	0.000E+00	0.000E+00	0.000E+00	0.000E+00	0.000E+00	0.000E+00	0.000E+00	0.000E+00	0.000E+00
Pu-236	0.000E+00	0.000E+00	0.000E+00	0.000E+00	0.000E+00	0.000E+00	0.000E+00	0.000E+00	0.000E+00	0.000E+00	0.000E+00
Pu-238	-1.521E+06	-1.521E+06	-1.167E+06	7.678E+05	1.270E+06	5.463E+05	-1.462E+03	-7.132E-15	0.000E+00	0.000E+00	2.461E+02
Pu-239	-2.980E+06	-2.980E+06	-2.969E+06	-2.969E+06	-2.969E+06	-2.960E+06	-2.886E+06	-2.234E+06	-1.732E+05	1.164E-01	2.675E+02
Pu-240	-1.444E+05	-1.444E+05	-1.444E+05	-1.444E+05	-1.444E+05	-1.284E+05	-1.204E+05	-4.654E+04	-4.654E+00	1.041E-03	2.675E+02
Pu-241	0.000E+00	0.000E+00	0.000E+00	0.000E+00	0.000E+00	0.000E+00	0.000E+00	0.000E+00	0.000E+00	0.000E+00	0.000E+00
Pu-242	1.926E+02	1.926E+02	1.926E+02	1.926E+02	1.926E+02	1.926E+02	1.765E+02	1.926E+02	1.445E+02	3.049E+01	2.675E+02
Pu-243	0.000E+00	0.000E+00	0.000E+00	0.000E+00	0.000E+00	0.000E+00	0.000E+00	0.000E+00	0.000E+00	0.000E+00	0.000E+00
Pu-244	0.000E+00	0.000E+00	0.000E+00	0.000E+00	0.000E+00	0.000E+00	0.000E+00	0.000E+00	0.000E+00	0.000E+00	0.000E+00
Pu-245	0.000E+00	0.000E+00	0.000E+00	0.000E+00	0.000E+00	0.000E+00	0.000E+00	0.000E+00	0.000E+00	0.000E+00	0.000E+00
Am-241	-2.161E+06	-2.161E+06	-2.163E+06	-2.052E+06	-1.212E+06	3.275E+05	8.351E+04	5.403E+01	2.947E-02	0.000E+00	2.729E+02
Am-242m	-2.504E+05	-2.504E+05	-2.504E+05	-2.488E+05	-2.391E+05	-1.589E+05	-2.632E+03	-3.916E-15	0.000E+00	0.000E+00	2.675E+02
Am-242	0.000E+00	0.000E+00	0.000E+00	0.000E+00	0.000E+00	0.000E+00	0.000E+00	0.000E+00	0.000E+00	0.000E+00	0.000E+00
Am-243	1.965E+03	1.965E+03	1.965E+03	1.965E+03	1.965E+03	1.965E+03	1.801E+03	7.859E+02	3.439E-01	1.188E-01	2.729E+02
Am-244	0.000E+00	0.000E+00	0.000E+00	0.000E+00	0.000E+00	0.000E+00	0.000E+00	0.000E+00	0.000E+00	0.000E+00	0.000E+00
Am-245	0.000E+00	0.000E+00	0.000E+00	0.000E+00	0.000E+00	0.000E+00	0.000E+00	0.000E+00	0.000E+00	0.000E+00	0.000E+00
Cm-242	1.599E+07	1.599E+07	1.413E+07	3.393E+06	-5.051E+03	-3.358E+03	-5.548E+01	-8.280E-17	0.000E+00	0.000E+00	6.900E+00
Cm-243	1.040E+04	1.040E+04	1.028E+04	1.016E+04	8.270E+03	1.181E+03	4.017E-06	0.000E+00	0.000E+00	0.000E+00	1.969E+02
Cm-244	2.064E+06	2.064E+06	2.044E+06	1.976E+06	1.408E+06	4.479E+04	8.757E-08	1.027E-07	2.403E-07	8.250E-07	1.630E+02
Cm-245	1.363E+02	1.363E+02	1.363E+02	1.193E+02	1.193E+02	1.363E+02	1.193E+02	5.623E+01	3.067E-02	0.000E+00	2.840E+02
Cm-246	1.366E+02	1.366E+02	1.366E+02	1.366E+02	1.366E+02	1.314E+02	1.144E+02	3.073E+01	5.462E-05	0.000E+00	2.845E+02
Cm-247	0.000E+00	0.000E+00	0.000E+00	0.000E+00	0.000E+00	0.000E+00	0.000E+00	0.000E+00	0.000E+00	0.000E+00	0.000E+00
Cm-248	0.000E+00	0.000E+00	0.000E+00	0.000E+00	0.000E+00	0.000E+00	0.000E+00	0.000E+00	0.000E+00	0.000E+00	0.000E+00
Cm-249	0.000E+00	0.000E+00	0.000E+00	0.000E+00	0.000E+00	0.000E+00	0.000E+00	0.000E+00	0.000E+00	0.000E+00	0.000E+00
Cm-250	0.000E+00	0.000E+00	0.000E+00	0.000E+00	0.000E+00	0.000E+00	0.000E+00	0.000E+00	0.000E+00	0.000E+00	0.000E+00
Bk-249	0.000E+00	0.000E+00	0.000E+00	0.000E+00	0.000E+00	0.000E+00	0.000E+00	0.000E+00	0.000E+00	0.000E+00	0.000E+00
Bk-250	0.000E+00	0.000E+00	0.000E+00	0.000E+00	0.000E+00	0.000E+00	0.000E+00	0.000E+00	0.000E+00	0.000E+00	0.000E+00
Cf-249	0.000E+00	0.000E+00	0.000E+00	0.000E+00	0.000E+00	0.000E+00	0.000E+00	0.000E+00	0.000E+00	0.000E+00	0.000E+00
Cf-250	0.000E+00	0.000E+00	0.000E+00	0.000E+00	0.000E+00	0.000E+00	0.000E+00	0.000E+00	0.000E+00	0.000E+00	0.000E+00
Cf-251	0.000E+00	0.000E+00	0.000E+00	0.000E+00	0.000E+00	0.000E+00	0.000E+00	0.000E+00	0.000E+00	0.000E+00	0.000E+00
Cf-252	0.000E+00	0.000E+00	0.000E+00	0.000E+00	0.000E+00	0.000E+00	0.000E+00	0.000E+00	0.000E+00	0.000E+00	0.000E+00
Cf-253	0.000E+00	0.000E+00	0.000E+00	0.000E+00	0.000E+00	0.000E+00	0.000E+00	0.000E+00	0.000E+00	0.000E+00	0.000E+00
Cf-254	0.000E+00	0.000E+00	0.000E+00	0.000E+00	0.000E+00	0.000E+00	0.000E+00	0.000E+00	0.000E+00	0.000E+00	0.000E+00
Es-253	0.000E+00	0.000E+00	0.000E+00	0.000E+00	0.000E+00	0.000E+00	0.000E+00	0.000E+00	0.000E+00	0.000E+00	0.000E+00
Total	1.100E+07	1.100E+07	9.497E+06	7.347E+05	-1.881E+06	-2.328E+06	-2.925E+06	-2.280E+06	-1.731E+05	1.298E+01	0.000E+00
Total/MWd	2.254E+01	2.254E+01	1.945E+01	1.505E+00	-3.852E+00	-4.768E+00	-5.990E+00	-4.669E+00	-3.545E-01	2.659E-05	0.000E+00

Table A.83

Breeder core heavy TRU elements nuclide toxicity hazard (axial blanket) Basis = EQULIBRIUM FRESH FUEL BREEDER CORE

	Initial	1 hour	30 days	1 year	10 y	100 y	1000 y	10000 y	100000 y	1000000 y	Factor
Np-236	0.000E+00	0.000E+00	0.000E+00	0.000E+00	0.000E+00	0.000E+00	0.000E+00	0.000E+00	0.000E+00	0.000E+00	0.000E+00
Np-237	0.000E+00	0.000E+00	0.000E+00	0.000E+00	0.000E+00	0.000E+00	0.000E+00	0.000E+00	0.000E+00	0.000E+00	1.972E+02
Np-238	0.000E+00	0.000E+00	0.000E+00	0.000E+00	0.000E+00	0.000E+00	0.000E+00	0.000E+00	0.000E+00	0.000E+00	0.000E+00
Np-239	0.000E+00	0.000E+00	0.000E+00	0.000E+00	0.000E+00	0.000E+00	0.000E+00	0.000E+00	0.000E+00	0.000E+00	0.000E+00
Np-240m	0.000E+00	0.000E+00	0.000E+00	0.000E+00	0.000E+00	0.000E+00	0.000E+00	0.000E+00	0.000E+00	0.000E+00	0.000E+00
Np-240	0.000E+00	0.000E+00	0.000E+00	0.000E+00	0.000E+00	0.000E+00	0.000E+00	0.000E+00	0.000E+00	0.000E+00	0.000E+00
Pu-236	0.000E+00	0.000E+00	0.000E+00	0.000E+00	0.000E+00	0.000E+00	0.000E+00	0.000E+00	0.000E+00	0.000E+00	0.000E+00
Pu-238	0.000E+00	0.000E+00	0.000E+00	0.000E+00	0.000E+00	0.000E+00	0.000E+00	0.000E+00	0.000E+00	0.000E+00	2.461E+02
Pu-239	0.000E+00	0.000E+00	0.000E+00	0.000E+00	0.000E+00	0.000E+00	0.000E+00	0.000E+00	0.000E+00	0.000E+00	2.675E+02
Pu-240	0.000E+00	0.000E+00	0.000E+00	0.000E+00	0.000E+00	0.000E+00	0.000E+00	0.000E+00	0.000E+00	0.000E+00	2.675E+02
Pu-241	0.000E+00	0.000E+00	0.000E+00	0.000E+00	0.000E+00	0.000E+00	0.000E+00	0.000E+00	0.000E+00	0.000E+00	0.000E+00
Pu-242	0.000E+00	0.000E+00	0.000E+00	0.000E+00	0.000E+00	0.000E+00	0.000E+00	0.000E+00	0.000E+00	0.000E+00	2.675E+02
Pu-243	0.000E+00	0.000E+00	0.000E+00	0.000E+00	0.000E+00	0.000E+00	0.000E+00	0.000E+00	0.000E+00	0.000E+00	0.000E+00
Pu-244	0.000E+00	0.000E+00	0.000E+00	0.000E+00	0.000E+00	0.000E+00	0.000E+00	0.000E+00	0.000E+00	0.000E+00	0.000E+00
Pu-245	0.000E+00	0.000E+00	0.000E+00	0.000E+00	0.000E+00	0.000E+00	0.000E+00	0.000E+00	0.000E+00	0.000E+00	0.000E+00
Am-241	0.000E+00	0.000E+00	0.000E+00	0.000E+00	0.000E+00	0.000E+00	0.000E+00	0.000E+00	0.000E+00	0.000E+00	2.729E+02
Am-242m	0.000E+00	0.000E+00	0.000E+00	0.000E+00	0.000E+00	0.000E+00	0.000E+00	0.000E+00	0.000E+00	0.000E+00	2.675E+02
Am-242	0.000E+00	0.000E+00	0.000E+00	0.000E+00	0.000E+00	0.000E+00	0.000E+00	0.000E+00	0.000E+00	0.000E+00	0.000E+00
Am-243	0.000E+00	0.000E+00	0.000E+00	0.000E+00	0.000E+00	0.000E+00	0.000E+00	0.000E+00	0.000E+00	0.000E+00	2.729E+02
Am-244	0.000E+00	0.000E+00	0.000E+00	0.000E+00	0.000E+00	0.000E+00	0.000E+00	0.000E+00	0.000E+00	0.000E+00	0.000E+00
Am-245	0.000E+00	0.000E+00	0.000E+00	0.000E+00	0.000E+00	0.000E+00	0.000E+00	0.000E+00	0.000E+00	0.000E+00	0.000E+00
Cm-242	0.000E+00	0.000E+00	0.000E+00	0.000E+00	0.000E+00	0.000E+00	0.000E+00	0.000E+00	0.000E+00	0.000E+00	6.900E+00
Cm-243	0.000E+00	0.000E+00	0.000E+00	0.000E+00	0.000E+00	0.000E+00	0.000E+00	0.000E+00	0.000E+00	0.000E+00	1.969E+02
Cm-244	0.000E+00	0.000E+00	0.000E+00	0.000E+00	0.000E+00	0.000E+00	0.000E+00	0.000E+00	0.000E+00	0.000E+00	1.630E+02
Cm-245	0.000E+00	0.000E+00	0.000E+00	0.000E+00	0.000E+00	0.000E+00	0.000E+00	0.000E+00	0.000E+00	0.000E+00	2.840E+02
Cm-246	0.000E+00	0.000E+00	0.000E+00	0.000E+00	0.000E+00	0.000E+00	0.000E+00	0.000E+00	0.000E+00	0.000E+00	2.845E+02
Cm-247	0.000E+00	0.000E+00	0.000E+00	0.000E+00	0.000E+00	0.000E+00	0.000E+00	0.000E+00	0.000E+00	0.000E+00	0.000E+00
Cm-248	0.000E+00	0.000E+00	0.000E+00	0.000E+00	0.000E+00	0.000E+00	0.000E+00	0.000E+00	0.000E+00	0.000E+00	0.000E+00
Cm-249	0.000E+00	0.000E+00	0.000E+00	0.000E+00	0.000E+00	0.000E+00	0.000E+00	0.000E+00	0.000E+00	0.000E+00	0.000E+00
Cm-250	0.000E+00	0.000E+00	0.000E+00	0.000E+00	0.000E+00	0.000E+00	0.000E+00	0.000E+00	0.000E+00	0.000E+00	0.000E+00
Bk-249	0.000E+00	0.000E+00	0.000E+00	0.000E+00	0.000E+00	0.000E+00	0.000E+00	0.000E+00	0.000E+00	0.000E+00	0.000E+00
Bk-250	0.000E+00	0.000E+00	0.000E+00	0.000E+00	0.000E+00	0.000E+00	0.000E+00	0.000E+00	0.000E+00	0.000E+00	0.000E+00
Cf-249	0.000E+00	0.000E+00	0.000E+00	0.000E+00	0.000E+00	0.000E+00	0.000E+00	0.000E+00	0.000E+00	0.000E+00	0.000E+00
Cf-250	0.000E+00	0.000E+00	0.000E+00	0.000E+00	0.000E+00	0.000E+00	0.000E+00	0.000E+00	0.000E+00	0.000E+00	0.000E+00
Cf-251	0.000E+00	0.000E+00	0.000E+00	0.000E+00	0.000E+00	0.000E+00	0.000E+00	0.000E+00	0.000E+00	0.000E+00	0.000E+00
Cf-252	0.000E+00	0.000E+00	0.000E+00	0.000E+00	0.000E+00	0.000E+00	0.000E+00	0.000E+00	0.000E+00	0.000E+00	0.000E+00
Cf-253	0.000E+00	0.000E+00	0.000E+00	0.000E+00	0.000E+00	0.000E+00	0.000E+00	0.000E+00	0.000E+00	0.000E+00	0.000E+00
Cf-254	0.000E+00	0.000E+00	0.000E+00	0.000E+00	0.000E+00	0.000E+00	0.000E+00	0.000E+00	0.000E+00	0.000E+00	0.000E+00
Es-253	0.000E+00	0.000E+00	0.000E+00	0.000E+00	0.000E+00	0.000E+00	0.000E+00	0.000E+00	0.000E+00	0.000E+00	0.000E+00
Total	0.000E+00	0.000E+00	0.000E+00	0.000E+00	0.000E+00	0.000E+00	0.000E+00	0.000E+00	0.000E+00	0.000E+00	0.000E+00

Table A.84

Breeder core heavy TRU elements nuclide toxicity hazard (axial blanket) Basis = EQULIBRIUM DISCHARGE FUEL BREEDER CORE

	Initial	1 hour	30 days	1 year	10 y	100 y	1000 y	10000 y	100000 y	1000000 y	Factor
Np-236	0.000E+00	0.000E+00	0.000E+00	0.000E+00	0.000E+00	0.000E+00	0.000E+00	0.000E+00	0.000E+00	0.000E+00	0.000E+00
Np-237	5.240E+01	5.240E+01	5.301E+01	5.304E+01	5.309E+01	5.459E+01	6.330E+01	6.581E+01	6.392E+01	4.775E+01	1.972E+02
Np-238	0.000E+00	0.000E+00	0.000E+00	0.000E+00	0.000E+00	0.000E+00	0.000E+00	0.000E+00	0.000E+00	0.000E+00	0.000E+00
Np-239	0.000E+00	0.000E+00	0.000E+00	0.000E+00	0.000E+00	0.000E+00	0.000E+00	0.000E+00	0.000E+00	0.000E+00	0.000E+00
Np-240m	0.000E+00	0.000E+00	0.000E+00	0.000E+00	0.000E+00	0.000E+00	0.000E+00	0.000E+00	0.000E+00	0.000E+00	0.000E+00
Np-240	0.000E+00	0.000E+00	0.000E+00	0.000E+00	0.000E+00	0.000E+00	0.000E+00	0.000E+00	0.000E+00	0.000E+00	0.000E+00
Pu-236	0.000E+00	0.000E+00	0.000E+00	0.000E+00	0.000E+00	0.000E+00	0.000E+00	0.000E+00	0.000E+00	0.000E+00	0.000E+00
Pu-238	1.562E+05	1.562E+05	1.571E+05	1.562E+05	1.457E+05	7.229E+04	6.574E+01	1.726E-18	0.000E+00	0.000E+00	2.461E+02
Pu-239	2.287E+06	2.287E+06	2.297E+06	2.297E+06	2.297E+06	2.290E+06	2.233E+06	1.729E+06	1.340E+05	1.053E-06	2.675E+02
Pu-240	2.250E+05	2.250E+05	2.250E+05	2.250E+05	2.249E+05	2.228E+05	2.030E+05	8.070E+04	7.919E+00	1.532E-09	2.675E+02
Pu-241	0.000E+00	0.000E+00	0.000E+00	0.000E+00	0.000E+00	0.000E+00	0.000E+00	0.000E+00	0.000E+00	0.000E+00	0.000E+00
Pu-242	1.526E+00	1.526E+00	1.526E+00	1.526E+00	1.527E+00	1.528E+00	1.528E+00	1.503E+00	1.275E+00	2.457E-01	2.675E+02
Pu-243	0.000E+00	0.000E+00	0.000E+00	0.000E+00	0.000E+00	0.000E+00	0.000E+00	0.000E+00	0.000E+00	0.000E+00	0.000E+00
Pu-244	0.000E+00	0.000E+00	0.000E+00	0.000E+00	0.000E+00	0.000E+00	0.000E+00	0.000E+00	0.000E+00	0.000E+00	0.000E+00
Pu-245	0.000E+00	0.000E+00	0.000E+00	0.000E+00	0.000E+00	0.000E+00	0.000E+00	0.000E+00	0.000E+00	0.000E+00	0.000E+00
Am-241	2.802E+03	2.802E+03	3.136E+03	6.780E+03	3.497E+04	7.743E+04	1.852E+04	1.046E-02	1.301E-07	0.000E+00	2.729E+02
Am-242m	6.059E+01	6.059E+01	6.056E+01	6.032E+01	5.789E+01	3.839E+01	6.335E-01	9.458E-19	0.000E+00	0.000E+00	2.675E+02
Am-242	0.000E+00	0.000E+00	0.000E+00	0.000E+00	0.000E+00	0.000E+00	0.000E+00	0.000E+00	0.000E+00	0.000E+00	0.000E+00
Am-243	9.885E-01	9.887E-01	9.901E-01	9.900E-01	9.892E-01	9.811E-01	9.043E-01	4.000E-01	1.149E-04	6.165E-12	2.729E+02
Am-244	0.000E+00	0.000E+00	0.000E+00	0.000E+00	0.000E+00	0.000E+00	0.000E+00	0.000E+00	0.000E+00	0.000E+00	0.000E+00
Am-245	0.000E+00	0.000E+00	0.000E+00	0.000E+00	0.000E+00	0.000E+00	0.000E+00	0.000E+00	0.000E+00	0.000E+00	0.000E+00
Cm-242	1.397E+03	1.397E+03	1.240E+03	2.990E+02	1.225E+00	8.123E-01	1.340E-02	0.000E+00	0.000E+00	0.000E+00	6.900E+00
Cm-243	4.004E+00	4.004E+00	3.997E+00	3.918E+00	3.224E+00	4.587E-01	1.564E-09	0.000E+00	0.000E+00	0.000E+00	1.969E+02
Cm-244	8.049E+00	8.050E+00	8.025E+00	7.748E+00	5.490E+00	1.748E-01	1.224E-12	1.223E-12	1.223E-12	1.214E-12	1.630E+02
Cm-245	5.623E-04	5.623E-04	5.623E-04	5.623E-04	5.620E-04	5.577E-04	5.172E-04	2.432E-04	1.281E-07	0.000E+00	2.840E+02
Cm-246	6.155E-06	6.155E-06	6.155E-06	6.155E-06	6.147E-06	6.067E-06	5.314E-06	1.413E-06	2.504E-12	0.000E+00	2.845E+02
Cm-247	0.000E+00	0.000E+00	0.000E+00	0.000E+00	0.000E+00	0.000E+00	0.000E+00	0.000E+00	0.000E+00	0.000E+00	0.000E+00
Cm-248	0.000E+00	0.000E+00	0.000E+00	0.000E+00	0.000E+00	0.000E+00	0.000E+00	0.000E+00	0.000E+00	0.000E+00	0.000E+00
Cm-249	0.000E+00	0.000E+00	0.000E+00	0.000E+00	0.000E+00	0.000E+00	0.000E+00	0.000E+00	0.000E+00	0.000E+00	0.000E+00
Cm-250	0.000E+00	0.000E+00	0.000E+00	0.000E+00	0.000E+00	0.000E+00	0.000E+00	0.000E+00	0.000E+00	0.000E+00	0.000E+00
Bk-249	0.000E+00	0.000E+00	0.000E+00	0.000E+00	0.000E+00	0.000E+00	0.000E+00	0.000E+00	0.000E+00	0.000E+00	0.000E+00
Bk-250	0.000E+00	0.000E+00	0.000E+00	0.000E+00	0.000E+00	0.000E+00	0.000E+00	0.000E+00	0.000E+00	0.000E+00	0.000E+00
Cf-249	0.000E+00	0.000E+00	0.000E+00	0.000E+00	0.000E+00	0.000E+00	0.000E+00	0.000E+00	0.000E+00	0.000E+00	0.000E+00
Cf-250	0.000E+00	0.000E+00	0.000E+00	0.000E+00	0.000E+00	0.000E+00	0.000E+00	0.000E+00	0.000E+00	0.000E+00	0.000E+00
Cf-251	0.000E+00	0.000E+00	0.000E+00	0.000E+00	0.000E+00	0.000E+00	0.000E+00	0.000E+00	0.000E+00	0.000E+00	0.000E+00
Cf-252	0.000E+00	0.000E+00	0.000E+00	0.000E+00	0.000E+00	0.000E+00	0.000E+00	0.000E+00	0.000E+00	0.000E+00	0.000E+00
Cf-253	0.000E+00	0.000E+00	0.000E+00	0.000E+00	0.000E+00	0.000E+00	0.000E+00	0.000E+00	0.000E+00	0.000E+00	0.000E+00
Cf-254	0.000E+00	0.000E+00	0.000E+00	0.000E+00	0.000E+00	0.000E+00	0.000E+00	0.000E+00	0.000E+00	0.000E+00	0.000E+00
Es-253	0.000E+00	0.000E+00	0.000E+00	0.000E+00	0.000E+00	0.000E+00	0.000E+00	0.000E+00	0.000E+00	0.000E+00	0.000E+00
Total	2.673E+06	2.673E+06	2.683E+06	2.685E+06	2.702E+06	2.663E+06	2.454E+06	1.809E+06	1.341E+05	4.800E+01	0.000E+00

Table A.85

Breeder core heavy TRU elements nuclide toxicity hazard (axial blanket) Basis = (EQULIBRIUM DISCHARGE FUEL - EQULIBRIUM FRESH FUEL) BREEDER CORE

	Initial	1 hour	30 days	1 year	10 y	100 y	1000 y	10000 y	100000 y	1000000 y	Factor
Np-236	0.000E+00	0.000E+00	0.000E+00	0.000E+00	0.000E+00	0.000E+00	0.000E+00	0.000E+00	0.000E+00	0.000E+00	0.000E+00
Np-237	5.240E+01	5.240E+01	5.301E+01	5.304E+01	5.309E+01	5.459E+01	6.330E+01	6.581E+01	6.392E+01	4.775E+01	1.972E+02
Np-238	0.000E+00	0.000E+00	0.000E+00	0.000E+00	0.000E+00	0.000E+00	0.000E+00	0.000E+00	0.000E+00	0.000E+00	0.000E+00
Np-239	0.000E+00	0.000E+00	0.000E+00	0.000E+00	0.000E+00	0.000E+00	0.000E+00	0.000E+00	0.000E+00	0.000E+00	0.000E+00
Np-240m	0.000E+00	0.000E+00	0.000E+00	0.000E+00	0.000E+00	0.000E+00	0.000E+00	0.000E+00	0.000E+00	0.000E+00	0.000E+00
Np-240	0.000E+00	0.000E+00	0.000E+00	0.000E+00	0.000E+00	0.000E+00	0.000E+00	0.000E+00	0.000E+00	0.000E+00	0.000E+00
Pu-236	0.000E+00	0.000E+00	0.000E+00	0.000E+00	0.000E+00	0.000E+00	0.000E+00	0.000E+00	0.000E+00	0.000E+00	0.000E+00
Pu-238	1.562E+05	1.562E+05	1.571E+05	1.562E+05	1.457E+05	7.229E+04	6.574E+01	1.726E-18	0.000E+00	0.000E+00	2.461E+02
Pu-239	2.287E+06	2.287E+06	2.297E+06	2.297E+06	2.297E+06	2.290E+06	2.233E+06	1.729E+06	1.340E+05	1.053E-06	2.675E+02
Pu-240	2.250E+05	2.250E+05	2.250E+05	2.250E+05	2.249E+05	2.228E+05	2.030E+05	8.070E+04	7.919E+00	1.532E-09	2.675E+02
Pu-241	0.000E+00	0.000E+00	0.000E+00	0.000E+00	0.000E+00	0.000E+00	0.000E+00	0.000E+00	0.000E+00	0.000E+00	0.000E+00
Pu-242	1.526E+00	1.526E+00	1.526E+00	1.526E+00	1.527E+00	1.528E+00	1.528E+00	1.503E+00	1.275E+00	2.457E-01	2.675E+02
Pu-243	0.000E+00	0.000E+00	0.000E+00	0.000E+00	0.000E+00	0.000E+00	0.000E+00	0.000E+00	0.000E+00	0.000E+00	0.000E+00
Pu-244	0.000E+00	0.000E+00	0.000E+00	0.000E+00	0.000E+00	0.000E+00	0.000E+00	0.000E+00	0.000E+00	0.000E+00	0.000E+00
Pu-245	0.000E+00	0.000E+00	0.000E+00	0.000E+00	0.000E+00	0.000E+00	0.000E+00	0.000E+00	0.000E+00	0.000E+00	0.000E+00
Am-241	2.802E+03	2.802E+03	3.136E+03	6.780E+03	3.497E+04	7.743E+04	1.852E+04	1.046E-02	1.301E-07	0.000E+00	2.729E+02
Am-242m	6.059E+01	6.059E+01	6.056E+01	6.032E+01	5.789E+01	3.839E+01	6.335E-01	9.458E-19	0.000E+00	0.000E+00	2.675E+02
Am-242	0.000E+00	0.000E+00	0.000E+00	0.000E+00	0.000E+00	0.000E+00	0.000E+00	0.000E+00	0.000E+00	0.000E+00	0.000E+00
Am-243	9.885E-01	9.887E-01	9.901E-01	9.900E-01	9.892E-01	9.811E-01	9.043E-01	4.000E-01	1.149E-04	6.165E-12	2.729E+02
Am-244	0.000E+00	0.000E+00	0.000E+00	0.000E+00	0.000E+00	0.000E+00	0.000E+00	0.000E+00	0.000E+00	0.000E+00	0.000E+00
Am-245	0.000E+00	0.000E+00	0.000E+00	0.000E+00	0.000E+00	0.000E+00	0.000E+00	0.000E+00	0.000E+00	0.000E+00	0.000E+00
Cm-242	1.397E+03	1.397E+03	1.240E+03	2.990E+02	1.225E+00	8.123E-01	1.340E-02	0.000E+00	0.000E+00	0.000E+00	6.900E+00
Cm-243	4.004E+00	4.004E+00	3.997E+00	3.918E+00	3.224E+00	4.587E-01	1.564E-09	0.000E+00	0.000E+00	0.000E+00	1.969E+02
Cm-244	8.049E+00	8.050E+00	8.025E+00	7.748E+00	5.490E+00	1.748E-01	1.224E-12	1.223E-12	1.223E-12	1.214E-12	1.630E+02
Cm-245	5.623E-04	5.623E-04	5.623E-04	5.623E-04	5.620E-04	5.577E-04	5.172E-04	2.432E-04	1.281E-07	0.000E+00	2.840E+02
Cm-246	6.155E-06	6.155E-06	6.155E-06	6.155E-06	6.147E-06	6.067E-06	5.314E-06	1.413E-06	2.504E-12	0.000E+00	2.845E+02
Cm-247	0.000E+00	0.000E+00	0.000E+00	0.000E+00	0.000E+00	0.000E+00	0.000E+00	0.000E+00	0.000E+00	0.000E+00	0.000E+00
Cm-248	0.000E+00	0.000E+00	0.000E+00	0.000E+00	0.000E+00	0.000E+00	0.000E+00	0.000E+00	0.000E+00	0.000E+00	0.000E+00
Cm-249	0.000E+00	0.000E+00	0.000E+00	0.000E+00	0.000E+00	0.000E+00	0.000E+00	0.000E+00	0.000E+00	0.000E+00	0.000E+00
Cm-250	0.000E+00	0.000E+00	0.000E+00	0.000E+00	0.000E+00	0.000E+00	0.000E+00	0.000E+00	0.000E+00	0.000E+00	0.000E+00
Bk-249	0.000E+00	0.000E+00	0.000E+00	0.000E+00	0.000E+00	0.000E+00	0.000E+00	0.000E+00	0.000E+00	0.000E+00	0.000E+00
Bk-250	0.000E+00	0.000E+00	0.000E+00	0.000E+00	0.000E+00	0.000E+00	0.000E+00	0.000E+00	0.000E+00	0.000E+00	0.000E+00
Cf-249	0.000E+00	0.000E+00	0.000E+00	0.000E+00	0.000E+00	0.000E+00	0.000E+00	0.000E+00	0.000E+00	0.000E+00	0.000E+00
Cf-250	0.000E+00	0.000E+00	0.000E+00	0.000E+00	0.000E+00	0.000E+00	0.000E+00	0.000E+00	0.000E+00	0.000E+00	0.000E+00
Cf-251	0.000E+00	0.000E+00	0.000E+00	0.000E+00	0.000E+00	0.000E+00	0.000E+00	0.000E+00	0.000E+00	0.000E+00	0.000E+00
Cf-252	0.000E+00	0.000E+00	0.000E+00	0.000E+00	0.000E+00	0.000E+00	0.000E+00	0.000E+00	0.000E+00	0.000E+00	0.000E+00
Cf-253	0.000E+00	0.000E+00	0.000E+00	0.000E+00	0.000E+00	0.000E+00	0.000E+00	0.000E+00	0.000E+00	0.000E+00	0.000E+00
Cf-254	0.000E+00	0.000E+00	0.000E+00	0.000E+00	0.000E+00	0.000E+00	0.000E+00	0.000E+00	0.000E+00	0.000E+00	0.000E+00
Es-253	0.000E+00	0.000E+00	0.000E+00	0.000E+00	0.000E+00	0.000E+00	0.000E+00	0.000E+00	0.000E+00	0.000E+00	0.000E+00
Total	2.673E+06	2.673E+06	2.683E+06	2.685E+06	2.702E+06	2.663E+06	2.454E+06	1.809E+06	1.341E+05	4.800E+01	0.000E+00
Total/MWd	5.474E+00	5.474E+00	5.496E+00	5.500E+00	5.535E+00	5.454E+00	5.027E+00	3.706E+00	2.747E-01	9.831E-05	0.000E+00

Table A.86

Breeder core heavy TRU elements nuclide toxicity hazard (inner blanket) Basis = EQULIBRIUM FRESH FUEL BREEDER CORE

	Initial	1 hour	30 days	1 year	10 y	100 y	1000 y	10000 y	100000 y	1000000 y	Factor
Np-236	0.000E+00	0.000E+00	0.000E+00	0.000E+00	0.000E+00	0.000E+00	0.000E+00	0.000E+00	0.000E+00	0.000E+00	0.000E+00
Np-237	0.000E+00	0.000E+00	0.000E+00	0.000E+00	0.000E+00	0.000E+00	0.000E+00	0.000E+00	0.000E+00	0.000E+00	1.972E+02
Np-238	0.000E+00	0.000E+00	0.000E+00	0.000E+00	0.000E+00	0.000E+00	0.000E+00	0.000E+00	0.000E+00	0.000E+00	0.000E+00
Np-239	0.000E+00	0.000E+00	0.000E+00	0.000E+00	0.000E+00	0.000E+00	0.000E+00	0.000E+00	0.000E+00	0.000E+00	0.000E+00
Np-240m	0.000E+00	0.000E+00	0.000E+00	0.000E+00	0.000E+00	0.000E+00	0.000E+00	0.000E+00	0.000E+00	0.000E+00	0.000E+00
Np-240	0.000E+00	0.000E+00	0.000E+00	0.000E+00	0.000E+00	0.000E+00	0.000E+00	0.000E+00	0.000E+00	0.000E+00	0.000E+00
Pu-236	0.000E+00	0.000E+00	0.000E+00	0.000E+00	0.000E+00	0.000E+00	0.000E+00	0.000E+00	0.000E+00	0.000E+00	0.000E+00
Pu-238	0.000E+00	0.000E+00	0.000E+00	0.000E+00	0.000E+00	0.000E+00	0.000E+00	0.000E+00	0.000E+00	0.000E+00	2.461E+02
Pu-239	0.000E+00	0.000E+00	0.000E+00	0.000E+00	0.000E+00	0.000E+00	0.000E+00	0.000E+00	0.000E+00	0.000E+00	2.675E+02
Pu-240	0.000E+00	0.000E+00	0.000E+00	0.000E+00	0.000E+00	0.000E+00	0.000E+00	0.000E+00	0.000E+00	0.000E+00	2.675E+02
Pu-241	0.000E+00	0.000E+00	0.000E+00	0.000E+00	0.000E+00	0.000E+00	0.000E+00	0.000E+00	0.000E+00	0.000E+00	0.000E+00
Pu-242	0.000E+00	0.000E+00	0.000E+00	0.000E+00	0.000E+00	0.000E+00	0.000E+00	0.000E+00	0.000E+00	0.000E+00	2.675E+02
Pu-243	0.000E+00	0.000E+00	0.000E+00	0.000E+00	0.000E+00	0.000E+00	0.000E+00	0.000E+00	0.000E+00	0.000E+00	0.000E+00
Pu-244	0.000E+00	0.000E+00	0.000E+00	0.000E+00	0.000E+00	0.000E+00	0.000E+00	0.000E+00	0.000E+00	0.000E+00	0.000E+00
Pu-245	0.000E+00	0.000E+00	0.000E+00	0.000E+00	0.000E+00	0.000E+00	0.000E+00	0.000E+00	0.000E+00	0.000E+00	0.000E+00
Am-241	0.000E+00	0.000E+00	0.000E+00	0.000E+00	0.000E+00	0.000E+00	0.000E+00	0.000E+00	0.000E+00	0.000E+00	2.729E+02
Am-242m	0.000E+00	0.000E+00	0.000E+00	0.000E+00	0.000E+00	0.000E+00	0.000E+00	0.000E+00	0.000E+00	0.000E+00	2.675E+02
Am-242	0.000E+00	0.000E+00	0.000E+00	0.000E+00	0.000E+00	0.000E+00	0.000E+00	0.000E+00	0.000E+00	0.000E+00	0.000E+00
Am-243	0.000E+00	0.000E+00	0.000E+00	0.000E+00	0.000E+00	0.000E+00	0.000E+00	0.000E+00	0.000E+00	0.000E+00	2.729E+02
Am-244	0.000E+00	0.000E+00	0.000E+00	0.000E+00	0.000E+00	0.000E+00	0.000E+00	0.000E+00	0.000E+00	0.000E+00	0.000E+00
Am-245	0.000E+00	0.000E+00	0.000E+00	0.000E+00	0.000E+00	0.000E+00	0.000E+00	0.000E+00	0.000E+00	0.000E+00	0.000E+00
Cm-242	0.000E+00	0.000E+00	0.000E+00	0.000E+00	0.000E+00	0.000E+00	0.000E+00	0.000E+00	0.000E+00	0.000E+00	6.900E+00
Cm-243	0.000E+00	0.000E+00	0.000E+00	0.000E+00	0.000E+00	0.000E+00	0.000E+00	0.000E+00	0.000E+00	0.000E+00	1.969E+02
Cm-244	0.000E+00	0.000E+00	0.000E+00	0.000E+00	0.000E+00	0.000E+00	0.000E+00	0.000E+00	0.000E+00	0.000E+00	1.630E+02
Cm-245	0.000E+00	0.000E+00	0.000E+00	0.000E+00	0.000E+00	0.000E+00	0.000E+00	0.000E+00	0.000E+00	0.000E+00	2.840E+02
Cm-246	0.000E+00	0.000E+00	0.000E+00	0.000E+00	0.000E+00	0.000E+00	0.000E+00	0.000E+00	0.000E+00	0.000E+00	2.845E+02
Cm-247	0.000E+00	0.000E+00	0.000E+00	0.000E+00	0.000E+00	0.000E+00	0.000E+00	0.000E+00	0.000E+00	0.000E+00	0.000E+00
Cm-248	0.000E+00	0.000E+00	0.000E+00	0.000E+00	0.000E+00	0.000E+00	0.000E+00	0.000E+00	0.000E+00	0.000E+00	0.000E+00
Cm-249	0.000E+00	0.000E+00	0.000E+00	0.000E+00	0.000E+00	0.000E+00	0.000E+00	0.000E+00	0.000E+00	0.000E+00	0.000E+00
Cm-250	0.000E+00	0.000E+00	0.000E+00	0.000E+00	0.000E+00	0.000E+00	0.000E+00	0.000E+00	0.000E+00	0.000E+00	0.000E+00
Bk-249	0.000E+00	0.000E+00	0.000E+00	0.000E+00	0.000E+00	0.000E+00	0.000E+00	0.000E+00	0.000E+00	0.000E+00	0.000E+00
Bk-250	0.000E+00	0.000E+00	0.000E+00	0.000E+00	0.000E+00	0.000E+00	0.000E+00	0.000E+00	0.000E+00	0.000E+00	0.000E+00
Cf-249	0.000E+00	0.000E+00	0.000E+00	0.000E+00	0.000E+00	0.000E+00	0.000E+00	0.000E+00	0.000E+00	0.000E+00	0.000E+00
Cf-250	0.000E+00	0.000E+00	0.000E+00	0.000E+00	0.000E+00	0.000E+00	0.000E+00	0.000E+00	0.000E+00	0.000E+00	0.000E+00
Cf-251	0.000E+00	0.000E+00	0.000E+00	0.000E+00	0.000E+00	0.000E+00	0.000E+00	0.000E+00	0.000E+00	0.000E+00	0.000E+00
Cf-252	0.000E+00	0.000E+00	0.000E+00	0.000E+00	0.000E+00	0.000E+00	0.000E+00	0.000E+00	0.000E+00	0.000E+00	0.000E+00
Cf-253	0.000E+00	0.000E+00	0.000E+00	0.000E+00	0.000E+00	0.000E+00	0.000E+00	0.000E+00	0.000E+00	0.000E+00	0.000E+00
Cf-254	0.000E+00	0.000E+00	0.000E+00	0.000E+00	0.000E+00	0.000E+00	0.000E+00	0.000E+00	0.000E+00	0.000E+00	0.000E+00
Es-253	0.000E+00	0.000E+00	0.000E+00	0.000E+00	0.000E+00	0.000E+00	0.000E+00	0.000E+00	0.000E+00	0.000E+00	0.000E+00
Total	0.000E+00	0.000E+00	0.000E+00	0.000E+00	0.000E+00	0.000E+00	0.000E+00	0.000E+00	0.000E+00	0.000E+00	0.000E+00

Table A.87

Breeder core heavy TRU elements nuclide toxicity hazard (inner blanket) Basis = EQULIBRIUM DISCHARGE FUEL BREEDER CORE

	Initial	1 hour	30 days	1 year	10 y	100 y	1000 y	10000 y	100000 y	1000000 y	Factor
Np-236	0.000E+00	0.000E+00	0.000E+00	0.000E+00	0.000E+00	0.000E+00	0.000E+00	0.000E+00	0.000E+00	0.000E+00	0.000E+00
Np-237	7.516E-01	7.516E-01	7.575E-01	7.577E-01	7.578E-01	7.594E-01	7.680E-01	7.685E-01	7.464E-01	5.576E-01	1.972E+02
Np-238	0.000E+00	0.000E+00	0.000E+00	0.000E+00	0.000E+00	0.000E+00	0.000E+00	0.000E+00	0.000E+00	0.000E+00	0.000E+00
Np-239	0.000E+00	0.000E+00	0.000E+00	0.000E+00	0.000E+00	0.000E+00	0.000E+00	0.000E+00	0.000E+00	0.000E+00	0.000E+00
Np-240m	0.000E+00	0.000E+00	0.000E+00	0.000E+00	0.000E+00	0.000E+00	0.000E+00	0.000E+00	0.000E+00	0.000E+00	0.000E+00
Np-240	0.000E+00	0.000E+00	0.000E+00	0.000E+00	0.000E+00	0.000E+00	0.000E+00	0.000E+00	0.000E+00	0.000E+00	0.000E+00
Pu-236	0.000E+00	0.000E+00	0.000E+00	0.000E+00	0.000E+00	0.000E+00	0.000E+00	0.000E+00	0.000E+00	0.000E+00	0.000E+00
Pu-238	5.193E+02	5.193E+02	5.215E+02	5.178E+02	4.828E+02	2.395E+02	2.163E-01	0.000E+00	0.000E+00	0.000E+00	2.461E+02
Pu-239	5.242E+04	5.242E+04	5.256E+04	5.256E+04	5.255E+04	5.242E+04	5.110E+04	3.956E+04	3.069E+03	2.411E-08	2.675E+02
Pu-240	1.146E+03	1.146E+03	1.146E+03	1.146E+03	1.145E+03	1.135E+03	1.034E+03	4.110E+02	4.033E-02	1.077E-14	2.675E+02
Pu-241	0.000E+00	0.000E+00	0.000E+00	0.000E+00	0.000E+00	0.000E+00	0.000E+00	0.000E+00	0.000E+00	0.000E+00	0.000E+00
Pu-242	3.316E-04	3.316E-04	3.316E-04	3.316E-04	3.318E-04	3.321E-04	3.324E-04	3.269E-04	2.773E-04	5.348E-05	2.675E+02
Pu-243	0.000E+00	0.000E+00	0.000E+00	0.000E+00	0.000E+00	0.000E+00	0.000E+00	0.000E+00	0.000E+00	0.000E+00	0.000E+00
Pu-244	0.000E+00	0.000E+00	0.000E+00	0.000E+00	0.000E+00	0.000E+00	0.000E+00	0.000E+00	0.000E+00	0.000E+00	0.000E+00
Pu-245	0.000E+00	0.000E+00	0.000E+00	0.000E+00	0.000E+00	0.000E+00	0.000E+00	0.000E+00	0.000E+00	0.000E+00	0.000E+00
Am-241	3.858E+00	3.859E+00	4.193E+00	7.853E+00	3.615E+01	7.864E+01	1.880E+01	1.039E-05	2.495E-13	0.000E+00	2.729E+02
Am-242m	1.918E-02	1.918E-02	1.918E-02	1.910E-02	1.833E-02	1.216E-02	2.006E-04	0.000E+00	0.000E+00	0.000E+00	2.675E+02
Am-242	0.000E+00	0.000E+00	0.000E+00	0.000E+00	0.000E+00	0.000E+00	0.000E+00	0.000E+00	0.000E+00	0.000E+00	0.000E+00
Am-243	4.442E-05	4.442E-05	4.447E-05	4.446E-05	4.442E-05	4.406E-05	4.061E-05	1.796E-05	5.159E-09	4.901E-19	2.729E+02
Am-244	0.000E+00	0.000E+00	0.000E+00	0.000E+00	0.000E+00	0.000E+00	0.000E+00	0.000E+00	0.000E+00	0.000E+00	0.000E+00
Am-245	0.000E+00	0.000E+00	0.000E+00	0.000E+00	0.000E+00	0.000E+00	0.000E+00	0.000E+00	0.000E+00	0.000E+00	0.000E+00
Cm-242	3.616E-01	3.617E-01	3.207E-01	7.738E-02	3.877E-04	2.571E-04	4.244E-06	0.000E+00	0.000E+00	0.000E+00	6.900E+00
Cm-243	2.145E-04	2.145E-04	2.142E-04	2.099E-04	1.728E-04	2.460E-05	8.384E-14	0.000E+00	0.000E+00	0.000E+00	1.969E+02
Cm-244	7.548E-05	7.548E-05	7.525E-05	7.266E-05	5.147E-05	1.639E-06	8.602E-18	8.600E-18	8.593E-18	8.528E-18	1.630E+02
Cm-245	1.079E-09	1.079E-09	1.079E-09	1.079E-09	1.078E-09	1.070E-09	9.922E-10	4.664E-10	2.457E-13	0.000E+00	2.840E+02
Cm-246	2.446E-12	2.446E-12	2.446E-12	2.444E-12	2.443E-12	2.410E-12	2.112E-12	5.614E-13	9.950E-19	0.000E+00	2.845E+02
Cm-247	0.000E+00	0.000E+00	0.000E+00	0.000E+00	0.000E+00	0.000E+00	0.000E+00	0.000E+00	0.000E+00	0.000E+00	0.000E+00
Cm-248	0.000E+00	0.000E+00	0.000E+00	0.000E+00	0.000E+00	0.000E+00	0.000E+00	0.000E+00	0.000E+00	0.000E+00	0.000E+00
Cm-249	0.000E+00	0.000E+00	0.000E+00	0.000E+00	0.000E+00	0.000E+00	0.000E+00	0.000E+00	0.000E+00	0.000E+00	0.000E+00
Cm-250	0.000E+00	0.000E+00	0.000E+00	0.000E+00	0.000E+00	0.000E+00	0.000E+00	0.000E+00	0.000E+00	0.000E+00	0.000E+00
Bk-249	0.000E+00	0.000E+00	0.000E+00	0.000E+00	0.000E+00	0.000E+00	0.000E+00	0.000E+00	0.000E+00	0.000E+00	0.000E+00
Bk-250	0.000E+00	0.000E+00	0.000E+00	0.000E+00	0.000E+00	0.000E+00	0.000E+00	0.000E+00	0.000E+00	0.000E+00	0.000E+00
Cf-249	0.000E+00	0.000E+00	0.000E+00	0.000E+00	0.000E+00	0.000E+00	0.000E+00	0.000E+00	0.000E+00	0.000E+00	0.000E+00
Cf-250	0.000E+00	0.000E+00	0.000E+00	0.000E+00	0.000E+00	0.000E+00	0.000E+00	0.000E+00	0.000E+00	0.000E+00	0.000E+00
Cf-251	0.000E+00	0.000E+00	0.000E+00	0.000E+00	0.000E+00	0.000E+00	0.000E+00	0.000E+00	0.000E+00	0.000E+00	0.000E+00
Cf-252	0.000E+00	0.000E+00	0.000E+00	0.000E+00	0.000E+00	0.000E+00	0.000E+00	0.000E+00	0.000E+00	0.000E+00	0.000E+00
Cf-253	0.000E+00	0.000E+00	0.000E+00	0.000E+00	0.000E+00	0.000E+00	0.000E+00	0.000E+00	0.000E+00	0.000E+00	0.000E+00
Cf-254	0.000E+00	0.000E+00	0.000E+00	0.000E+00	0.000E+00	0.000E+00	0.000E+00	0.000E+00	0.000E+00	0.000E+00	0.000E+00
Es-253	0.000E+00	0.000E+00	0.000E+00	0.000E+00	0.000E+00	0.000E+00	0.000E+00	0.000E+00	0.000E+00	0.000E+00	0.000E+00
Total	5.409E+04	5.409E+04	5.424E+04	5.424E+04	5.421E+04	5.387E+04	5.216E+04	3.998E+04	3.070E+03	5.577E-01	0.000E+00

Table A.88

Breeder core heavy TRU elements nuclide toxicity hazard (inner blanket) Basis = (EQULIBRIUM DISCHARGE FUEL - EQULIBRIUM FRESH FUEL) BREEDER CORE

	Initial	1 hour	30 days	1 year	10 y	100 y	1000 y	10000 y	100000 y	1000000 y	Factor
Np-236	0.000E+00	0.000E+00	0.000E+00	0.000E+00	0.000E+00	0.000E+00	0.000E+00	0.000E+00	0.000E+00	0.000E+00	0.000E+00
Np-237	7.516E-01	7.516E-01	7.575E-01	7.577E-01	7.578E-01	7.594E-01	7.680E-01	7.685E-01	7.464E-01	5.576E-01	1.972E+02
Np-238	0.000E+00	0.000E+00	0.000E+00	0.000E+00	0.000E+00	0.000E+00	0.000E+00	0.000E+00	0.000E+00	0.000E+00	0.000E+00
Np-239	0.000E+00	0.000E+00	0.000E+00	0.000E+00	0.000E+00	0.000E+00	0.000E+00	0.000E+00	0.000E+00	0.000E+00	0.000E+00
Np-240m	0.000E+00	0.000E+00	0.000E+00	0.000E+00	0.000E+00	0.000E+00	0.000E+00	0.000E+00	0.000E+00	0.000E+00	0.000E+00
Np-240	0.000E+00	0.000E+00	0.000E+00	0.000E+00	0.000E+00	0.000E+00	0.000E+00	0.000E+00	0.000E+00	0.000E+00	0.000E+00
Pu-236	0.000E+00	0.000E+00	0.000E+00	0.000E+00	0.000E+00	0.000E+00	0.000E+00	0.000E+00	0.000E+00	0.000E+00	0.000E+00
Pu-238	5.193E+02	5.193E+02	5.215E+02	5.178E+02	4.828E+02	2.395E+02	2.163E-01	0.000E+00	0.000E+00	0.000E+00	2.461E+02
Pu-239	5.242E+04	5.242E+04	5.256E+04	5.256E+04	5.255E+04	5.242E+04	5.110E+04	3.956E+04	3.069E+03	2.411E-08	2.675E+02
Pu-240	1.146E+03	1.146E+03	1.146E+03	1.146E+03	1.145E+03	1.135E+03	1.034E+03	4.110E+02	4.033E-02	1.077E-14	2.675E+02
Pu-241	0.000E+00	0.000E+00	0.000E+00	0.000E+00	0.000E+00	0.000E+00	0.000E+00	0.000E+00	0.000E+00	0.000E+00	0.000E+00
Pu-242	3.316E-04	3.316E-04	3.316E-04	3.316E-04	3.318E-04	3.321E-04	3.324E-04	3.269E-04	2.773E-04	5.348E-05	2.675E+02
Pu-243	0.000E+00	0.000E+00	0.000E+00	0.000E+00	0.000E+00	0.000E+00	0.000E+00	0.000E+00	0.000E+00	0.000E+00	0.000E+00
Pu-244	0.000E+00	0.000E+00	0.000E+00	0.000E+00	0.000E+00	0.000E+00	0.000E+00	0.000E+00	0.000E+00	0.000E+00	0.000E+00
Pu-245	0.000E+00	0.000E+00	0.000E+00	0.000E+00	0.000E+00	0.000E+00	0.000E+00	0.000E+00	0.000E+00	0.000E+00	0.000E+00
Am-241	3.858E+00	3.859E+00	4.193E+00	7.853E+00	3.615E+01	7.864E+01	1.880E+01	1.039E-05	2.495E-13	0.000E+00	2.729E+02
Am-242m	1.918E-02	1.918E-02	1.918E-02	1.910E-02	1.833E-02	1.216E-02	2.006E-04	0.000E+00	0.000E+00	0.000E+00	2.675E+02
Am-242	0.000E+00	0.000E+00	0.000E+00	0.000E+00	0.000E+00	0.000E+00	0.000E+00	0.000E+00	0.000E+00	0.000E+00	0.000E+00
Am-243	4.442E-05	4.442E-05	4.447E-05	4.446E-05	4.442E-05	4.406E-05	4.061E-05	1.796E-05	5.159E-09	4.901E-19	2.729E+02
Am-244	0.000E+00	0.000E+00	0.000E+00	0.000E+00	0.000E+00	0.000E+00	0.000E+00	0.000E+00	0.000E+00	0.000E+00	0.000E+00
Am-245	0.000E+00	0.000E+00	0.000E+00	0.000E+00	0.000E+00	0.000E+00	0.000E+00	0.000E+00	0.000E+00	0.000E+00	0.000E+00
Cm-242	3.616E-01	3.617E-01	3.207E-01	7.738E-02	3.877E-04	2.571E-04	4.244E-06	0.000E+00	0.000E+00	0.000E+00	6.900E+00
Cm-243	2.145E-04	2.145E-04	2.142E-04	2.099E-04	1.728E-04	2.460E-05	8.384E-14	0.000E+00	0.000E+00	0.000E+00	1.969E+02
Cm-244	7.548E-05	7.548E-05	7.525E-05	7.266E-05	5.147E-05	1.639E-06	8.602E-18	8.600E-18	8.593E-18	8.528E-18	1.630E+02
Cm-245	1.079E-09	1.079E-09	1.079E-09	1.079E-09	1.078E-09	1.070E-09	9.922E-10	4.664E-10	2.457E-13	0.000E+00	2.840E+02
Cm-246	2.446E-12	2.446E-12	2.446E-12	2.444E-12	2.443E-12	2.410E-12	2.112E-12	5.614E-13	9.950E-19	0.000E+00	2.845E+02
Cm-247	0.000E+00	0.000E+00	0.000E+00	0.000E+00	0.000E+00	0.000E+00	0.000E+00	0.000E+00	0.000E+00	0.000E+00	0.000E+00
Cm-248	0.000E+00	0.000E+00	0.000E+00	0.000E+00	0.000E+00	0.000E+00	0.000E+00	0.000E+00	0.000E+00	0.000E+00	0.000E+00
Cm-249	0.000E+00	0.000E+00	0.000E+00	0.000E+00	0.000E+00	0.000E+00	0.000E+00	0.000E+00	0.000E+00	0.000E+00	0.000E+00
Cm-250	0.000E+00	0.000E+00	0.000E+00	0.000E+00	0.000E+00	0.000E+00	0.000E+00	0.000E+00	0.000E+00	0.000E+00	0.000E+00
Bk-249	0.000E+00	0.000E+00	0.000E+00	0.000E+00	0.000E+00	0.000E+00	0.000E+00	0.000E+00	0.000E+00	0.000E+00	0.000E+00
Bk-250	0.000E+00	0.000E+00	0.000E+00	0.000E+00	0.000E+00	0.000E+00	0.000E+00	0.000E+00	0.000E+00	0.000E+00	0.000E+00
Cf-249	0.000E+00	0.000E+00	0.000E+00	0.000E+00	0.000E+00	0.000E+00	0.000E+00	0.000E+00	0.000E+00	0.000E+00	0.000E+00
Cf-250	0.000E+00	0.000E+00	0.000E+00	0.000E+00	0.000E+00	0.000E+00	0.000E+00	0.000E+00	0.000E+00	0.000E+00	0.000E+00
Cf-251	0.000E+00	0.000E+00	0.000E+00	0.000E+00	0.000E+00	0.000E+00	0.000E+00	0.000E+00	0.000E+00	0.000E+00	0.000E+00
Cf-252	0.000E+00	0.000E+00	0.000E+00	0.000E+00	0.000E+00	0.000E+00	0.000E+00	0.000E+00	0.000E+00	0.000E+00	0.000E+00
Cf-253	0.000E+00	0.000E+00	0.000E+00	0.000E+00	0.000E+00	0.000E+00	0.000E+00	0.000E+00	0.000E+00	0.000E+00	0.000E+00
Cf-254	0.000E+00	0.000E+00	0.000E+00	0.000E+00	0.000E+00	0.000E+00	0.000E+00	0.000E+00	0.000E+00	0.000E+00	0.000E+00
Es-253	0.000E+00	0.000E+00	0.000E+00	0.000E+00	0.000E+00	0.000E+00	0.000E+00	0.000E+00	0.000E+00	0.000E+00	0.000E+00
Total	5.409E+04	5.409E+04	5.424E+04	5.424E+04	5.421E+04	5.387E+04	5.216E+04	3.998E+04	3.070E+03	5.577E-01	0.000E+00
Total/MWd	1.108E-01	1.108E-01	1.111E-01	1.111E-01	1.110E-01	1.103E-01	1.068E-01	8.187E-02	6.287E-03	1.142E-06	0.000E+00

Table A.89

Breeder core heavy TRU elements nuclide toxicity hazard (outer blanket) Basis = EQULIBRIUM FRESH FUEL BREEDER CORE

	Initial	1 hour	30 days	1 year	10 y	100 y	1000 y	10000 y	100000 y	1000000 y	Factor
Np-236	0.000E+00	0.000E+00	0.000E+00	0.000E+00	0.000E+00	0.000E+00	0.000E+00	0.000E+00	0.000E+00	0.000E+00	0.000E+00
Np-237	0.000E+00	0.000E+00	0.000E+00	0.000E+00	0.000E+00	0.000E+00	0.000E+00	0.000E+00	0.000E+00	0.000E+00	1.972E+02
Np-238	0.000E+00	0.000E+00	0.000E+00	0.000E+00	0.000E+00	0.000E+00	0.000E+00	0.000E+00	0.000E+00	0.000E+00	0.000E+00
Np-239	0.000E+00	0.000E+00	0.000E+00	0.000E+00	0.000E+00	0.000E+00	0.000E+00	0.000E+00	0.000E+00	0.000E+00	0.000E+00
Np-240m	0.000E+00	0.000E+00	0.000E+00	0.000E+00	0.000E+00	0.000E+00	0.000E+00	0.000E+00	0.000E+00	0.000E+00	0.000E+00
Np-240	0.000E+00	0.000E+00	0.000E+00	0.000E+00	0.000E+00	0.000E+00	0.000E+00	0.000E+00	0.000E+00	0.000E+00	0.000E+00
Pu-236	0.000E+00	0.000E+00	0.000E+00	0.000E+00	0.000E+00	0.000E+00	0.000E+00	0.000E+00	0.000E+00	0.000E+00	0.000E+00
Pu-238	0.000E+00	0.000E+00	0.000E+00	0.000E+00	0.000E+00	0.000E+00	0.000E+00	0.000E+00	0.000E+00	0.000E+00	2.461E+02
Pu-239	0.000E+00	0.000E+00	0.000E+00	0.000E+00	0.000E+00	0.000E+00	0.000E+00	0.000E+00	0.000E+00	0.000E+00	2.675E+02
Pu-240	0.000E+00	0.000E+00	0.000E+00	0.000E+00	0.000E+00	0.000E+00	0.000E+00	0.000E+00	0.000E+00	0.000E+00	2.675E+02
Pu-241	0.000E+00	0.000E+00	0.000E+00	0.000E+00	0.000E+00	0.000E+00	0.000E+00	0.000E+00	0.000E+00	0.000E+00	0.000E+00
Pu-242	0.000E+00	0.000E+00	0.000E+00	0.000E+00	0.000E+00	0.000E+00	0.000E+00	0.000E+00	0.000E+00	0.000E+00	2.675E+02
Pu-243	0.000E+00	0.000E+00	0.000E+00	0.000E+00	0.000E+00	0.000E+00	0.000E+00	0.000E+00	0.000E+00	0.000E+00	0.000E+00
Pu-244	0.000E+00	0.000E+00	0.000E+00	0.000E+00	0.000E+00	0.000E+00	0.000E+00	0.000E+00	0.000E+00	0.000E+00	0.000E+00
Pu-245	0.000E+00	0.000E+00	0.000E+00	0.000E+00	0.000E+00	0.000E+00	0.000E+00	0.000E+00	0.000E+00	0.000E+00	0.000E+00
Am-241	0.000E+00	0.000E+00	0.000E+00	0.000E+00	0.000E+00	0.000E+00	0.000E+00	0.000E+00	0.000E+00	0.000E+00	2.729E+02
Am-242m	0.000E+00	0.000E+00	0.000E+00	0.000E+00	0.000E+00	0.000E+00	0.000E+00	0.000E+00	0.000E+00	0.000E+00	2.675E+02
Am-242	0.000E+00	0.000E+00	0.000E+00	0.000E+00	0.000E+00	0.000E+00	0.000E+00	0.000E+00	0.000E+00	0.000E+00	0.000E+00
Am-243	0.000E+00	0.000E+00	0.000E+00	0.000E+00	0.000E+00	0.000E+00	0.000E+00	0.000E+00	0.000E+00	0.000E+00	2.729E+02
Am-244	0.000E+00	0.000E+00	0.000E+00	0.000E+00	0.000E+00	0.000E+00	0.000E+00	0.000E+00	0.000E+00	0.000E+00	0.000E+00
Am-245	0.000E+00	0.000E+00	0.000E+00	0.000E+00	0.000E+00	0.000E+00	0.000E+00	0.000E+00	0.000E+00	0.000E+00	0.000E+00
Cm-242	0.000E+00	0.000E+00	0.000E+00	0.000E+00	0.000E+00	0.000E+00	0.000E+00	0.000E+00	0.000E+00	0.000E+00	6.900E+00
Cm-243	0.000E+00	0.000E+00	0.000E+00	0.000E+00	0.000E+00	0.000E+00	0.000E+00	0.000E+00	0.000E+00	0.000E+00	1.969E+02
Cm-244	0.000E+00	0.000E+00	0.000E+00	0.000E+00	0.000E+00	0.000E+00	0.000E+00	0.000E+00	0.000E+00	0.000E+00	1.630E+02
Cm-245	0.000E+00	0.000E+00	0.000E+00	0.000E+00	0.000E+00	0.000E+00	0.000E+00	0.000E+00	0.000E+00	0.000E+00	2.840E+02
Cm-246	0.000E+00	0.000E+00	0.000E+00	0.000E+00	0.000E+00	0.000E+00	0.000E+00	0.000E+00	0.000E+00	0.000E+00	2.845E+02
Cm-247	0.000E+00	0.000E+00	0.000E+00	0.000E+00	0.000E+00	0.000E+00	0.000E+00	0.000E+00	0.000E+00	0.000E+00	0.000E+00
Cm-248	0.000E+00	0.000E+00	0.000E+00	0.000E+00	0.000E+00	0.000E+00	0.000E+00	0.000E+00	0.000E+00	0.000E+00	0.000E+00
Cm-249	0.000E+00	0.000E+00	0.000E+00	0.000E+00	0.000E+00	0.000E+00	0.000E+00	0.000E+00	0.000E+00	0.000E+00	0.000E+00
Cm-250	0.000E+00	0.000E+00	0.000E+00	0.000E+00	0.000E+00	0.000E+00	0.000E+00	0.000E+00	0.000E+00	0.000E+00	0.000E+00
Bk-249	0.000E+00	0.000E+00	0.000E+00	0.000E+00	0.000E+00	0.000E+00	0.000E+00	0.000E+00	0.000E+00	0.000E+00	0.000E+00
Bk-250	0.000E+00	0.000E+00	0.000E+00	0.000E+00	0.000E+00	0.000E+00	0.000E+00	0.000E+00	0.000E+00	0.000E+00	0.000E+00
Cf-249	0.000E+00	0.000E+00	0.000E+00	0.000E+00	0.000E+00	0.000E+00	0.000E+00	0.000E+00	0.000E+00	0.000E+00	0.000E+00
Cf-250	0.000E+00	0.000E+00	0.000E+00	0.000E+00	0.000E+00	0.000E+00	0.000E+00	0.000E+00	0.000E+00	0.000E+00	0.000E+00
Cf-251	0.000E+00	0.000E+00	0.000E+00	0.000E+00	0.000E+00	0.000E+00	0.000E+00	0.000E+00	0.000E+00	0.000E+00	0.000E+00
Cf-252	0.000E+00	0.000E+00	0.000E+00	0.000E+00	0.000E+00	0.000E+00	0.000E+00	0.000E+00	0.000E+00	0.000E+00	0.000E+00
Cf-253	0.000E+00	0.000E+00	0.000E+00	0.000E+00	0.000E+00	0.000E+00	0.000E+00	0.000E+00	0.000E+00	0.000E+00	0.000E+00
Cf-254	0.000E+00	0.000E+00	0.000E+00	0.000E+00	0.000E+00	0.000E+00	0.000E+00	0.000E+00	0.000E+00	0.000E+00	0.000E+00
Es-253	0.000E+00	0.000E+00	0.000E+00	0.000E+00	0.000E+00	0.000E+00	0.000E+00	0.000E+00	0.000E+00	0.000E+00	0.000E+00
Total	0.000E+00	0.000E+00	0.000E+00	0.000E+00	0.000E+00	0.000E+00	0.000E+00	0.000E+00	0.000E+00	0.000E+00	0.000E+00

Table A.90

Breeder core heavy TRU elements nuclide toxicity hazard (outer blanket) Basis = EQULIBRIUM DISCHARGE FUEL BREEDER CORE

	Initial	1 hour	30 days	1 year	10 y	100 y	1000 y	10000 y	100000 y	1000000 y	Factor
Np-236	0.000E+00	0.000E+00	0.000E+00	0.000E+00	0.000E+00	0.000E+00	0.000E+00	0.000E+00	0.000E+00	0.000E+00	0.000E+00
Np-237	1.979E+01	1.979E+01	1.997E+01	1.997E+01	1.998E+01	2.036E+01	2.255E+01	2.317E+01	2.249E+01	1.681E+01	1.972E+02
Np-238	0.000E+00	0.000E+00	0.000E+00	0.000E+00	0.000E+00	0.000E+00	0.000E+00	0.000E+00	0.000E+00	0.000E+00	0.000E+00
Np-239	0.000E+00	0.000E+00	0.000E+00	0.000E+00	0.000E+00	0.000E+00	0.000E+00	0.000E+00	0.000E+00	0.000E+00	0.000E+00
Np-240m	0.000E+00	0.000E+00	0.000E+00	0.000E+00	0.000E+00	0.000E+00	0.000E+00	0.000E+00	0.000E+00	0.000E+00	0.000E+00
Np-240	0.000E+00	0.000E+00	0.000E+00	0.000E+00	0.000E+00	0.000E+00	0.000E+00	0.000E+00	0.000E+00	0.000E+00	0.000E+00
Pu-236	0.000E+00	0.000E+00	0.000E+00	0.000E+00	0.000E+00	0.000E+00	0.000E+00	0.000E+00	0.000E+00	0.000E+00	0.000E+00
Pu-238	4.828E+04	4.828E+04	4.851E+04	4.820E+04	4.495E+04	2.230E+04	2.026E+01	0.000E+00	0.000E+00	0.000E+00	2.461E+02
Pu-239	7.353E+05	7.354E+05	7.375E+05	7.375E+05	7.373E+05	7.354E+05	7.170E+05	5.552E+05	4.305E+04	3.383E-07	2.675E+02
Pu-240	6.221E+04	6.221E+04	6.221E+04	6.219E+04	6.215E+04	6.157E+04	5.614E+04	2.231E+04	2.189E+00	1.674E-10	2.675E+02
Pu-241	0.000E+00	0.000E+00	0.000E+00	0.000E+00	0.000E+00	0.000E+00	0.000E+00	0.000E+00	0.000E+00	0.000E+00	0.000E+00
Pu-242	3.152E-01	3.152E-01	3.152E-01	3.152E-01	3.152E-01	3.155E-01	3.159E-01	3.107E-01	2.635E-01	5.080E-02	2.675E+02
Pu-243	0.000E+00	0.000E+00	0.000E+00	0.000E+00	0.000E+00	0.000E+00	0.000E+00	0.000E+00	0.000E+00	0.000E+00	0.000E+00
Pu-244	0.000E+00	0.000E+00	0.000E+00	0.000E+00	0.000E+00	0.000E+00	0.000E+00	0.000E+00	0.000E+00	0.000E+00	0.000E+00
Pu-245	0.000E+00	0.000E+00	0.000E+00	0.000E+00	0.000E+00	0.000E+00	0.000E+00	0.000E+00	0.000E+00	0.000E+00	0.000E+00
Am-241	9.482E+02	9.484E+02	1.031E+03	1.935E+03	8.934E+03	1.944E+04	4.649E+03	2.595E-03	1.520E-08	0.000E+00	2.729E+02
Am-242m	1.679E+01	1.679E+01	1.679E+01	1.672E+01	1.605E+01	1.064E+01	1.756E-01	0.000E+00	0.000E+00	0.000E+00	2.675E+02
Am-242	0.000E+00	0.000E+00	0.000E+00	0.000E+00	0.000E+00	0.000E+00	0.000E+00	0.000E+00	0.000E+00	0.000E+00	0.000E+00
Am-243	1.726E-01	1.726E-01	1.727E-01	1.727E-01	1.727E-01	1.713E-01	1.578E-01	6.984E-02	2.006E-05	4.670E-13	2.729E+02
Am-244	0.000E+00	0.000E+00	0.000E+00	0.000E+00	0.000E+00	0.000E+00	0.000E+00	0.000E+00	0.000E+00	0.000E+00	0.000E+00
Am-245	0.000E+00	0.000E+00	0.000E+00	0.000E+00	0.000E+00	0.000E+00	0.000E+00	0.000E+00	0.000E+00	0.000E+00	0.000E+00
Cm-242	3.316E+02	3.317E+02	2.941E+02	7.096E+01	3.395E-01	2.252E-01	3.714E-03	0.000E+00	0.000E+00	0.000E+00	6.900E+00
Cm-243	8.989E-01	8.988E-01	8.973E-01	8.796E-01	7.238E-01	1.030E-01	3.512E-10	0.000E+00	0.000E+00	0.000E+00	1.969E+02
Cm-244	1.155E+00	1.155E+00	1.151E+00	1.111E+00	7.873E-01	2.507E-02	1.338E-13	1.337E-13	1.336E-13	1.326E-13	1.630E+02
Cm-245	6.572E-05	6.572E-05	6.572E-05	6.572E-05	6.567E-05	6.518E-05	6.044E-05	2.841E-05	1.497E-08	0.000E+00	2.840E+02
Cm-246	5.281E-07	5.281E-07	5.281E-07	5.280E-07	5.273E-07	5.203E-07	4.558E-07	1.212E-07	2.149E-13	0.000E+00	2.845E+02
Cm-247	0.000E+00	0.000E+00	0.000E+00	0.000E+00	0.000E+00	0.000E+00	0.000E+00	0.000E+00	0.000E+00	0.000E+00	0.000E+00
Cm-248	0.000E+00	0.000E+00	0.000E+00	0.000E+00	0.000E+00	0.000E+00	0.000E+00	0.000E+00	0.000E+00	0.000E+00	0.000E+00
Cm-249	0.000E+00	0.000E+00	0.000E+00	0.000E+00	0.000E+00	0.000E+00	0.000E+00	0.000E+00	0.000E+00	0.000E+00	0.000E+00
Cm-250	0.000E+00	0.000E+00	0.000E+00	0.000E+00	0.000E+00	0.000E+00	0.000E+00	0.000E+00	0.000E+00	0.000E+00	0.000E+00
Bk-249	0.000E+00	0.000E+00	0.000E+00	0.000E+00	0.000E+00	0.000E+00	0.000E+00	0.000E+00	0.000E+00	0.000E+00	0.000E+00
Bk-250	0.000E+00	0.000E+00	0.000E+00	0.000E+00	0.000E+00	0.000E+00	0.000E+00	0.000E+00	0.000E+00	0.000E+00	0.000E+00
Cf-249	0.000E+00	0.000E+00	0.000E+00	0.000E+00	0.000E+00	0.000E+00	0.000E+00	0.000E+00	0.000E+00	0.000E+00	0.000E+00
Cf-250	0.000E+00	0.000E+00	0.000E+00	0.000E+00	0.000E+00	0.000E+00	0.000E+00	0.000E+00	0.000E+00	0.000E+00	0.000E+00
Cf-251	0.000E+00	0.000E+00	0.000E+00	0.000E+00	0.000E+00	0.000E+00	0.000E+00	0.000E+00	0.000E+00	0.000E+00	0.000E+00
Cf-252	0.000E+00	0.000E+00	0.000E+00	0.000E+00	0.000E+00	0.000E+00	0.000E+00	0.000E+00	0.000E+00	0.000E+00	0.000E+00
Cf-253	0.000E+00	0.000E+00	0.000E+00	0.000E+00	0.000E+00	0.000E+00	0.000E+00	0.000E+00	0.000E+00	0.000E+00	0.000E+00
Cf-254	0.000E+00	0.000E+00	0.000E+00	0.000E+00	0.000E+00	0.000E+00	0.000E+00	0.000E+00	0.000E+00	0.000E+00	0.000E+00
Es-253	0.000E+00	0.000E+00	0.000E+00	0.000E+00	0.000E+00	0.000E+00	0.000E+00	0.000E+00	0.000E+00	0.000E+00	0.000E+00
Total	8.471E+05	8.472E+05	8.496E+05	8.499E+05	8.534E+05	8.387E+05	7.778E+05	5.775E+05	4.307E+04	1.686E+01	0.000E+00

Table A.91

Breeder core heavy TRU elements nuclide toxicity hazard (outer blanket) Basis = (EQULIBRIUM DISCHARGE FUEL - EQULIBRIUM FRESH FUEL) BREEDER CORE

	Initial	1 hour	30 days	1 year	10 y	100 y	1000 y	10000 y	100000 y	1000000 y	Factor
Np-236	0.000E+00	0.000E+00	0.000E+00	0.000E+00	0.000E+00	0.000E+00	0.000E+00	0.000E+00	0.000E+00	0.000E+00	0.000E+00
Np-237	1.979E+01	1.979E+01	1.997E+01	1.997E+01	1.998E+01	2.036E+01	2.255E+01	2.317E+01	2.249E+01	1.681E+01	1.972E+02
Np-238	0.000E+00	0.000E+00	0.000E+00	0.000E+00	0.000E+00	0.000E+00	0.000E+00	0.000E+00	0.000E+00	0.000E+00	0.000E+00
Np-239	0.000E+00	0.000E+00	0.000E+00	0.000E+00	0.000E+00	0.000E+00	0.000E+00	0.000E+00	0.000E+00	0.000E+00	0.000E+00
Np-240m	0.000E+00	0.000E+00	0.000E+00	0.000E+00	0.000E+00	0.000E+00	0.000E+00	0.000E+00	0.000E+00	0.000E+00	0.000E+00
Np-240	0.000E+00	0.000E+00	0.000E+00	0.000E+00	0.000E+00	0.000E+00	0.000E+00	0.000E+00	0.000E+00	0.000E+00	0.000E+00
Pu-236	0.000E+00	0.000E+00	0.000E+00	0.000E+00	0.000E+00	0.000E+00	0.000E+00	0.000E+00	0.000E+00	0.000E+00	0.000E+00
Pu-238	4.828E+04	4.828E+04	4.851E+04	4.820E+04	4.495E+04	2.230E+04	2.026E+01	0.000E+00	0.000E+00	0.000E+00	2.461E+02
Pu-239	7.353E+05	7.354E+05	7.375E+05	7.375E+05	7.373E+05	7.354E+05	7.170E+05	5.552E+05	4.305E+04	3.383E-07	2.675E+02
Pu-240	6.221E+04	6.221E+04	6.221E+04	6.219E+04	6.215E+04	6.157E+04	5.614E+04	2.231E+04	2.189E+00	1.674E-10	2.675E+02
Pu-241	0.000E+00	0.000E+00	0.000E+00	0.000E+00	0.000E+00	0.000E+00	0.000E+00	0.000E+00	0.000E+00	0.000E+00	0.000E+00
Pu-242	3.152E-01	3.152E-01	3.152E-01	3.152E-01	3.152E-01	3.155E-01	3.159E-01	3.107E-01	2.635E-01	5.080E-02	2.675E+02
Pu-243	0.000E+00	0.000E+00	0.000E+00	0.000E+00	0.000E+00	0.000E+00	0.000E+00	0.000E+00	0.000E+00	0.000E+00	0.000E+00
Pu-244	0.000E+00	0.000E+00	0.000E+00	0.000E+00	0.000E+00	0.000E+00	0.000E+00	0.000E+00	0.000E+00	0.000E+00	0.000E+00
Pu-245	0.000E+00	0.000E+00	0.000E+00	0.000E+00	0.000E+00	0.000E+00	0.000E+00	0.000E+00	0.000E+00	0.000E+00	0.000E+00
Am-241	9.482E+02	9.484E+02	1.031E+03	1.935E+03	8.934E+03	1.944E+04	4.649E+03	2.595E-03	1.520E-08	0.000E+00	2.729E+02
Am-242m	1.679E+01	1.679E+01	1.679E+01	1.672E+01	1.605E+01	1.064E+01	1.756E-01	0.000E+00	0.000E+00	0.000E+00	2.675E+02
Am-242	0.000E+00	0.000E+00	0.000E+00	0.000E+00	0.000E+00	0.000E+00	0.000E+00	0.000E+00	0.000E+00	0.000E+00	0.000E+00
Am-243	1.726E-01	1.726E-01	1.727E-01	1.727E-01	1.727E-01	1.713E-01	1.578E-01	6.984E-02	2.006E-05	4.670E-13	2.729E+02
Am-244	0.000E+00	0.000E+00	0.000E+00	0.000E+00	0.000E+00	0.000E+00	0.000E+00	0.000E+00	0.000E+00	0.000E+00	0.000E+00
Am-245	0.000E+00	0.000E+00	0.000E+00	0.000E+00	0.000E+00	0.000E+00	0.000E+00	0.000E+00	0.000E+00	0.000E+00	0.000E+00
Cm-242	3.316E+02	3.317E+02	2.941E+02	7.096E+01	3.395E-01	2.252E-01	3.714E-03	0.000E+00	0.000E+00	0.000E+00	6.900E+00
Cm-243	8.989E-01	8.988E-01	8.973E-01	8.796E-01	7.238E-01	1.030E-01	3.512E-10	0.000E+00	0.000E+00	0.000E+00	1.969E+02
Cm-244	1.155E+00	1.155E+00	1.151E+00	1.111E+00	7.873E-01	2.507E-02	1.338E-13	1.337E-13	1.336E-13	1.326E-13	1.630E+02
Cm-245	6.572E-05	6.572E-05	6.572E-05	6.572E-05	6.567E-05	6.518E-05	6.044E-05	2.841E-05	1.497E-08	0.000E+00	2.840E+02
Cm-246	5.281E-07	5.281E-07	5.281E-07	5.280E-07	5.273E-07	5.203E-07	4.558E-07	1.212E-07	2.149E-13	0.000E+00	2.845E+02
Cm-247	0.000E+00	0.000E+00	0.000E+00	0.000E+00	0.000E+00	0.000E+00	0.000E+00	0.000E+00	0.000E+00	0.000E+00	0.000E+00
Cm-248	0.000E+00	0.000E+00	0.000E+00	0.000E+00	0.000E+00	0.000E+00	0.000E+00	0.000E+00	0.000E+00	0.000E+00	0.000E+00
Cm-249	0.000E+00	0.000E+00	0.000E+00	0.000E+00	0.000E+00	0.000E+00	0.000E+00	0.000E+00	0.000E+00	0.000E+00	0.000E+00
Cm-250	0.000E+00	0.000E+00	0.000E+00	0.000E+00	0.000E+00	0.000E+00	0.000E+00	0.000E+00	0.000E+00	0.000E+00	0.000E+00
Bk-249	0.000E+00	0.000E+00	0.000E+00	0.000E+00	0.000E+00	0.000E+00	0.000E+00	0.000E+00	0.000E+00	0.000E+00	0.000E+00
Bk-250	0.000E+00	0.000E+00	0.000E+00	0.000E+00	0.000E+00	0.000E+00	0.000E+00	0.000E+00	0.000E+00	0.000E+00	0.000E+00
Cf-249	0.000E+00	0.000E+00	0.000E+00	0.000E+00	0.000E+00	0.000E+00	0.000E+00	0.000E+00	0.000E+00	0.000E+00	0.000E+00
Cf-250	0.000E+00	0.000E+00	0.000E+00	0.000E+00	0.000E+00	0.000E+00	0.000E+00	0.000E+00	0.000E+00	0.000E+00	0.000E+00
Cf-251	0.000E+00	0.000E+00	0.000E+00	0.000E+00	0.000E+00	0.000E+00	0.000E+00	0.000E+00	0.000E+00	0.000E+00	0.000E+00
Cf-252	0.000E+00	0.000E+00	0.000E+00	0.000E+00	0.000E+00	0.000E+00	0.000E+00	0.000E+00	0.000E+00	0.000E+00	0.000E+00
Cf-253	0.000E+00	0.000E+00	0.000E+00	0.000E+00	0.000E+00	0.000E+00	0.000E+00	0.000E+00	0.000E+00	0.000E+00	0.000E+00
Cf-254	0.000E+00	0.000E+00	0.000E+00	0.000E+00	0.000E+00	0.000E+00	0.000E+00	0.000E+00	0.000E+00	0.000E+00	0.000E+00
Es-253	0.000E+00	0.000E+00	0.000E+00	0.000E+00	0.000E+00	0.000E+00	0.000E+00	0.000E+00	0.000E+00	0.000E+00	0.000E+00
Total	8.471E+05	8.472E+05	8.496E+05	8.499E+05	8.534E+05	8.387E+05	7.778E+05	5.775E+05	4.307E+04	1.686E+01	0.000E+00
Total/MWd	1.735E+00	1.735E+00	1.740E+00	1.741E+00	1.748E+00	1.718E+00	1.593E+00	1.183E+00	8.822E-02	3.454E-05	0.000E+00

Appendix B

Specification of a metal-fuelled multiple recycle burner core benchmark

Introduction and goals

While the once-through burner core benchmark [1] addresses the rate of reduction of the transuranics (TRU) from LWR spent fuel and the buildup rate of TRU from such a reactor, for a fixed breeding ratio – which are key issues to plutonium recycle – and also provides a first comparison of the sensitivity of predicted safety parameters to neutronic data and modelling, it is still unrealistically idealised in that it uses a once-through cycle. Any fast reactor deployment for burnup of TRU will, of necessity, employ recycle of the fast reactor spent fuel, because the structural endurance of the fuel pins is reached well in advance of the substantial burnout of all the TRU material they contain. **Thus, recycle for recovery of the remaining TRU for blending with the makeup feedstream, and refabrication of fresh assemblies is an essential element of any fast-reactor based system for TRU burning.** The study of that realistic case is the purpose of the third benchmark which is discussed here (see Figure B.1).

In this benchmark, the geometry is specified and a $^1/_3$ core refuelling scheme and the burn cycle time duration and energy extraction are also specified. The dwell time durations in the cooling/recycle/refabrication out-of-core segments of the closed fuel cycle and the chemical partitioning fractions between recycle and waste streams in the recycle/refabrication steps are specified as well. Finally, the isotopic mass fraction composition of the makeup TRU feedstream is specified.

Then, the blending ratio between the recycle and makeup feedstreams is to be determined by each participant such that the EOEC eigenvalue is 1.0 when all control rods are fully withdrawn – and after an infinite number of recycle steps have been taken such that the recycle isotopic mass fractions have relaxed to their asymptotic values.

This benchmark is done parametrically in three specified geometries which yield breeding ratios near 0.5, 0.75 and 1.0.

The edits of interest include:

- The Blending Ratio between LMR recycle and LWR makeup TRU for the fresh fuel fabrication, and,
- The isotopic mass fraction of the infinitely-recycle feedstream from the PYRØ recycle step to fabrication.

[1] See Volume IV, Appendix B.

- The rate of reduction of the external feedstream expressed in:
  - *isotopic mass / MWe year of fast-reactor energy*
  - *Ci / MWe year of fast-reactor energy*
  - *Toxicity / MWe year of fast-reactor energy*
  - *Watts / MWe year of fast-reactor energy*

- The buildup rate of the waste stream from the fast reactor cycle – expressed in the same units.

- The TRU breeding ratio, and

- The BOEC safety parameters.

All edits are given for each of the three breeding ratio cases.

The goal of this benchmark is:

- To ascertain, for a realistic scenario, the systematics relative to core breeding ratio of the relevant mass and toxicity flows and safety parameters,

- To determine the degree of sensitivity in prediction of the relevant mass and toxicity etc. flows and in safety parameters which derive from the diversity of basic data and modelling choices among participants.

Specification of the model

The planar geometry is given in Figure B.2.I (BR ≈ 0.5 and ≈ 0.75) and in Figure B.2.II (BR ≈ 1.0). Three cases – BR ≈ 0.5, 0.75 and 1.0 – are to be considered, and the axial geometries are specified in Figure B.2.I (BR ≈ 0.5), Figure B.2.II (BR ≈ 1.0) and B.2.III (BR ≈ 0.75). All three cores use the same assembly and pin dimensions and the same grid plate and control rod layout pattern; they only differ in the heights of the fuel and axial blankets and in the type of assemblies positioned in the radial blanket locations.

The compositions of the non-fuelled – non blanket regions are given in Table B.1. The fresh blanket compositions are given in Table B.2.

The isotopic fractions of the external feedstock to be used in blending the driver assemblies are given in Table B.3 [2].

The burn cycle duration and energy extraction are given in Table B.4 [2]. For benchmark purposes, the control rods are to (unphysically) remain fixed at the fully withdrawn position for the burn step.

The recycle/refabrication dwell time durations and the chemical partition fractions between recycle and waste streams are given in Table B.4 [2].

---

2 These values are unchanged from the once-through burner core benchmark, Volume IV.

The blending ratio between the (100% utilised – except for the 0.1% recycle losses) recycle actinide stream and the external makeup TRU which is to be used in fabricating fresh fuel assemblies is to be determined by each participant such that at EOEC the eigenvalue of the core (comprised of one-cycle, two-cycle, and three-cycle burnt assemblies) is 1.0 when all control rods are fully withdrawn to the top of the fuelled zone.

BOEC neutron balance reporting – to be completed for each of the three models

1. Narrative synopsis of the fuel management representation **in the core**, e.g., discrete vs. spatially smeared representation of composition of fresh, once-burnt, and twice burnt assemblies; fission product representation in partially burnt assemblies;

2. Narrative synopsis of the fuel management representation **in the recycle/refabrication segments of the cycle**, e.g.,

- How the radioactive decay and branching is modelled during cooling, subsequent to reprocessing, and subsequent to refabrication prior to reinsertion in the core,
- The numerical procedure used to achieve the "infinite recycle" isotopic ratios in the discharge fuel – so as to be uncontaminated by the startup transient;

3. Recycle stream isotopic mass fractions
Isotope mass in recycle stream/total TRU mass in recycle stream for each actinide isotope;

4.

$$\text{Blending ratio} = \frac{\text{TRU mass from external feed}}{\text{TRU mass from recycle}} \quad \text{used to fabricate fresh fuel}$$

5. Fresh fuel assembly isotopic mass fractions.

BOEC to EOEC transition and mass flow reporting

1. $\text{Burnup swing} = (k_{EOEC} - k_{BOEC})/k_{BOEC} \; k_{EOEC} \Big|_{\text{constant position rod}}$

2.

$$\text{TRU breeding ratio} = \frac{\text{EOEC TRU mass inventory}}{\text{BOEC TRU mass inventory}}$$ [3]

3. Mass increments by heavy metal isotope

- Isotopic mass drawn from the LWR TRU for fabrication of the fresh fuel assemblies for each TRU isotope,
- Sum over entire model of isotopic mass at BOEC for each TRU isotope,
- Sum over entire model of change in mass, δ(mass), due to burnup for each TRU isotope,
- Sum over TRU isotopes of the previous item divided by the energy extraction (MWth days) delivered during the burn cycle.

[3] Note that this definition excludes U-235.

4. Safety parameters reporting

- βeff, in unit of Δk/k,
- Fuel Doppler coefficient
  i.e., of heavy metal isotopes (with a narrative synopsis of how the calculation is made and what isotopes are accounted for),
- Sodium void worth
  - *of core, i.e., excluding blankets and reflectors,*
  - *of core plus blanket/reflector regions above core,*
- Burnup swing of the cycle – defined above under BOEC to EOEC transition
  – with rods at constant position
- Decay heat level for decay times of 1 hour, 1 month, 1 year, 10 years, 100 years, 1000 years, 10 000 years
  - *Total,*
  - *Heavy metal component,*
  - *Fission product component*

5. Radioactivity and decay
   - Provide a narrative synopsis of the radioactivity chain representation used for long-term out-of-core physics representations
     - *Isotopes treated,*
     - *Detailed chain representation specifically for the actinides showing all transitions and the values of all decay constants, branching ratios, etc.*
   - Describe the numerical solution approach for the equations,
   - Describe how the decay heat is computed

6. Curie increments at the times of 1, 10, 100, 1000, 10 000, 100 000, 1 000 000 years from the time of BOEC
   - Isotopic mass •λ for the TRU masses drawn from the LWR TRU for fabrication of the fresh fuel assemblies for each TRU isotope (expressed in Curies),
   - Sum over entire model of BOEC mass •λ for each TRU isotope,
   - Sum over entire model of δ(mass) •λ by isotope for each TRU isotope,
   - Sum over isotopes of previous item divided by the energy extraction (MWth days) delivered during the burn cycle

7. Toxicity hazard increments at the times of 1, 10, 100, 1000, 10 000, 100 000, 1 000 000 years from the time of BOEC
   - {(mass • (λ) • (toxicity index)} by isotope for each TRU isotope drawn from the LWR spent fuel for fabrication of the fresh fuel assemblies, expressed in long-term cancer deaths via oral intake,
   - Sum over the entire model of BOEC {(mass • (λ) • (toxicity index)} by isotope for each TRU isotope expressed, in long-term cancer deaths via oral intake, at the times 1, 10, 100, 1000, 10 000, 100 000, 1 000 000 years from the BOEC,
   - Sum over entire model of {δ (mass • (λ) • (toxicity index)} by isotope for each TRU isotope,
   - Sum over isotopes of previous item divided by the energy extraction (MWth days) delivered during the burn cycle.

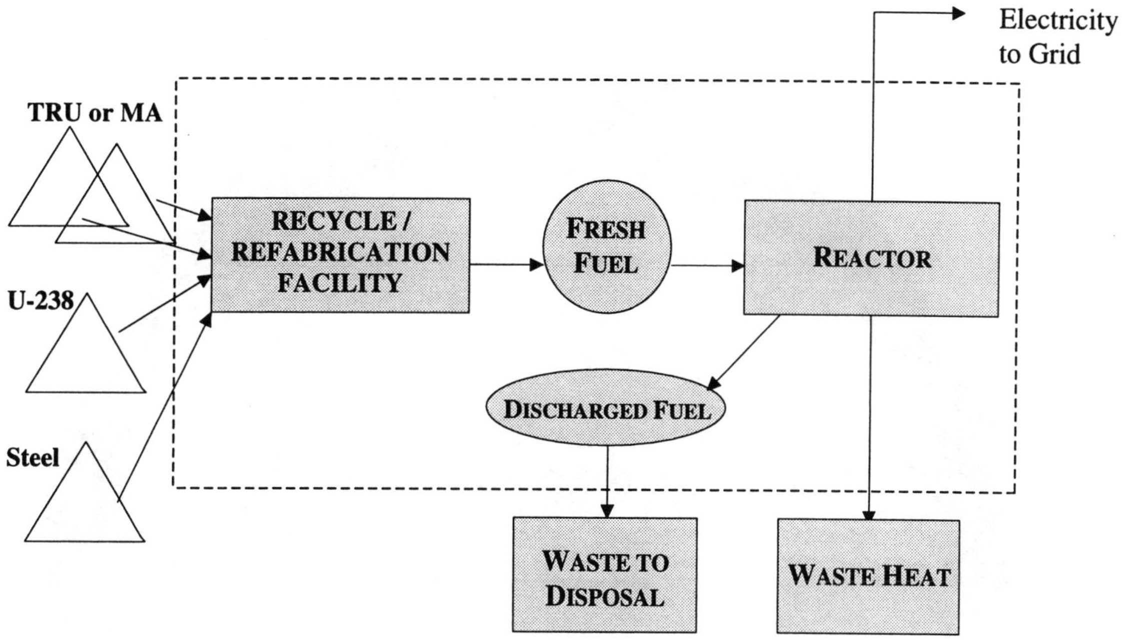

Figure B.1 ***LMR multiple recycle burner core benchmark***

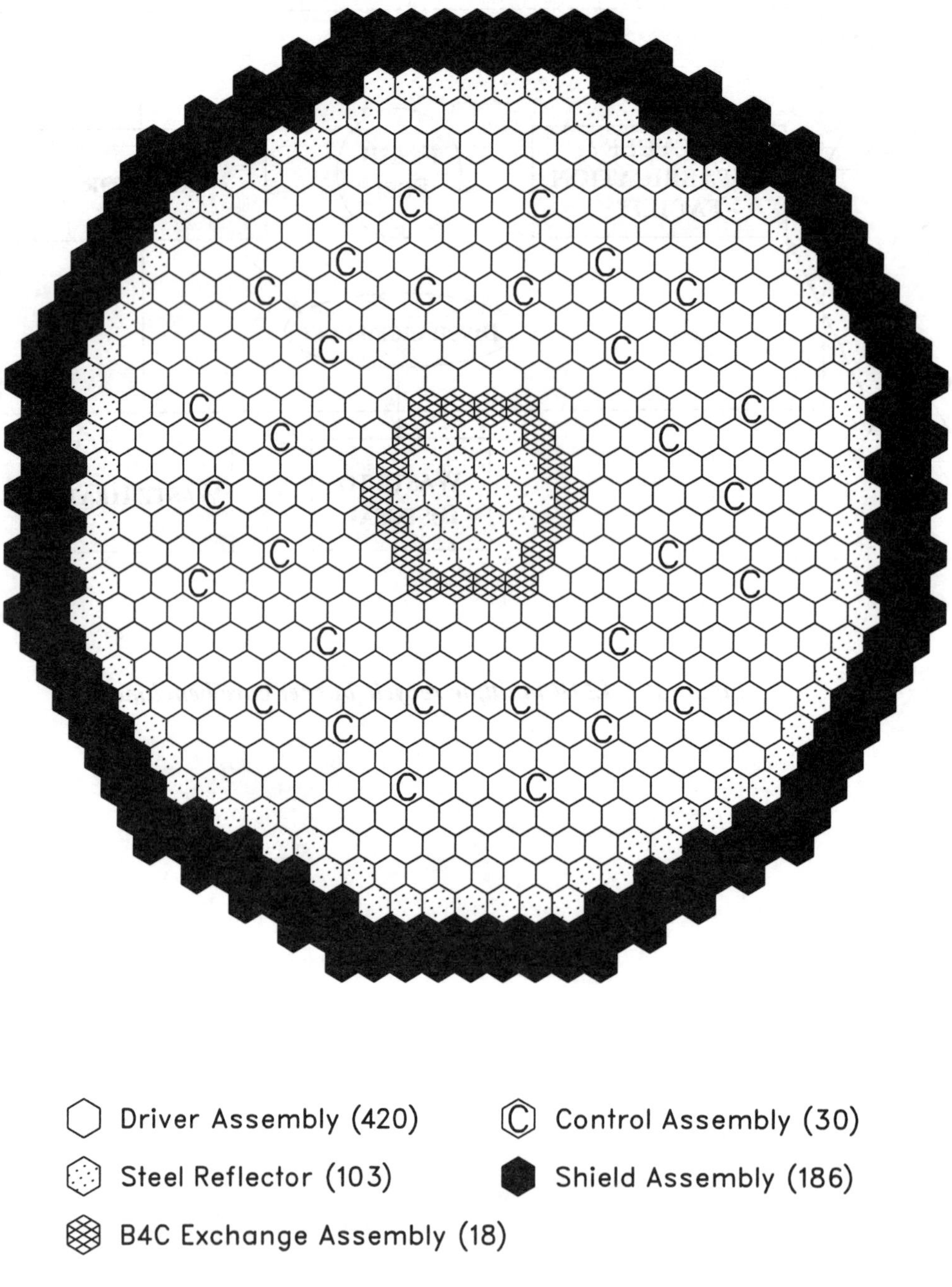

Figure B.2.1 ***Geometry of breeding ratio ≈ 0.5 core*** [B.1]

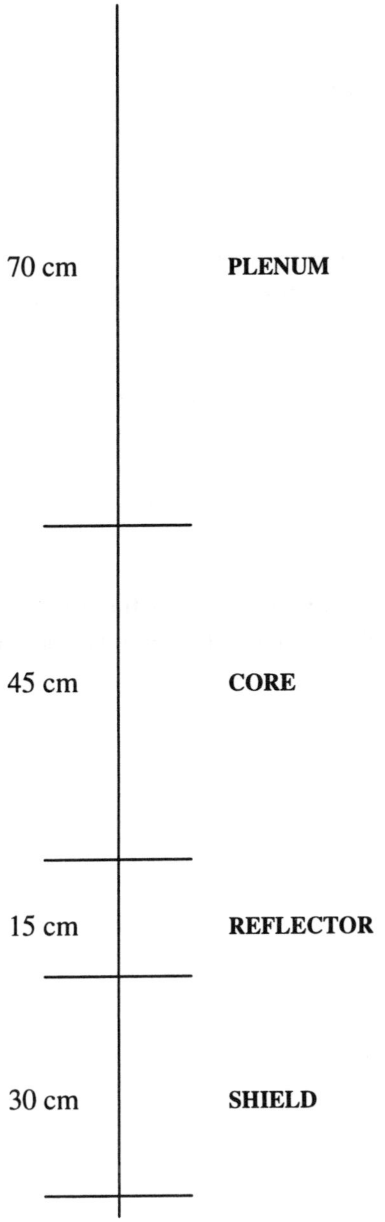

All assemblies have an axial height of 160 cm with a 15.617 lattice pitch and are arranged in a configuration with 1/6 core symmetry, as shown on previous page. Only nine distinct material zones are specified. In the driver assemblies, a 30-cm thick lower axial shield is below a 15-cm thick lower reflector zone which is adjacent to the 45-cm tall active core; there is a 70-cm plenum region above the active core. The absorber regions of the control assemblies are parked above the active core. All other assemblies have uniform axial compositions. The isotopic number densities of each non-driver, non-blanket assembly region are specified in Table B.1 (see next). Table B.1 contains the driver and blanket compositions for the first benchmark only.

Figure B.2.1 (cont.) ***Geometry of breeding ratio ≈ 0.5 core*** [B.1]

Figure B.2.II ***Geometry of breeding ratio ≈ 1.0 core*** [B.2]

Geometry is same as Figure B.2.I, except:

1. A lower axial blanket of 45-cm thickness is added under all driver assemblies (but no upper axial blanket is utilised). The composition of this lower axial blanket in fresh reload assemblies is given in Table B.2;

2. A single row of radial blankets displaces the row of steel reflectors one row outward and eliminates the innermost row of radial shield assemblies of Figure B.2.I;

3. Nineteen radial blankets displace the three rows (19 assemblies) of steel assemblies in the central island;

4. All blanket assemblies are "fuelled" in the elevations from the top of the driver fuel to the bottom of the lower axial blanket region of the drivers and their fresh composition is specified in Table B.2.

Figure B.2.III ***Geometry of breeding ratio ≈ 0.75 core***

Geometry is same as Figure B.2.I [4], except:

1. The lower axial blanket on the fuel assemblies is only 15-cm below the bottom of the driver fuel zone.

[4] Central island and radial blanket assemblies are absent in CR = 0.75 option, e.g., external breeding **only** in 15-cm lower axial blanket.

Table B.1
Material composition specifications
(Number Densities in atoms/barn-cm)

ISOTOPE	DRIVER				CONTROL		EXCHANGE	REFLECTOR	SHIELD
	SHIELD	*REFLECTOR*	*CORE*	*PLENUM*	*IN*	*OUT*			
Na-23	7.447-3	7.447-3	7.637-3	1.678-2	8.865-3	2.080-2	6.075-3	3.546-3	3.546-3
Fe	1.179-2	4.821-2	1.790-2	1.790-2	1.538-2	4.865-3	1.405-2	6.088-2	1.516-2
Cr	1.761-3	7.201-3	2.674-3	2.674-3	2.297-3	7.265-4	2.100-3	9.092-3	2.263-3
Mo	7.952-5	3.252-4	1.207-4	1.207-4	1.038-4	3.281-5	9.485-5	4.106-4	1.022-4
Ni	6.499-5	2.658-4	9.868-5	9.868-5	8.479-5	2.681-5	7.751-5	3.356-4	8.354-5
Mn-55	2.777-5	1.136-4	4.217-5	4.217-5	3.624-5	1.146-5	3.313-5	1.434-4	3.570-5
B-10	9.278-3				2.783-2		8.017-3		9.495-3
B-11	3.758-2				3.092-3		3.247-2		3.845-2
C-12	1.171-2				7.731-3		1.012-2		1.199-2
Zr			3.189-3						
U-235			1.632-5						
U-238			8.144-3						
Np-237			1.521-4						
Pu-236			3.155-10						
Pu-238			2.845-5						
Pu-239			1.431-3						
Pu-240			5.606-4						
Pu-241			3.775-4						
Pu-242			1.093-4						
Am-241			7.071-5						
Am-242m			3.127-7						
Am-243			6.987-5						
Cm-242			2.741-8						
Cm-243			2.214-7						
Cm-244			1.555-5						
Cm-245			1.431-6						
Cm-246			1.778-7						

Table B.2 **Blanket material composition specifications** [5]

	NUMBER DENSITIES IN ATOM/BARNS CM	
	Axial blankets	*Radial and Central island blankets*
U-238	1.244-2	1.604-2
U-235	2.493-5	3.215-5
Na-23	7.637-3	5.655-3
Fe	1.790-2	1.574-2
Cr	2.674-3	2.350-3
Mo	1.207-4	1.061-4
Ni	9.868-5	8.674-5
Mn-55	4.217-5	3.707-5
Zr	3.614-3	4.661-3

[5] Apply to reload assemblies.

Table B.3
LWR transuranic isotopics

Isotopic values are the weight fraction of the individual isotope in the total transuranic mass

LWR

ISOTOPE	AT 3.17 YEARS COOLING
Np-237	5.40-2
Pu-236	1.12-7
Pu-238	1.01-2
Pu-239	0.508
Pu-240	0.199
Pu-241	0.134
Pu-242	3.88-2
Am-241	2.51-2
Am-242m	1.11-4
Am-243	2.48-2
Cm-242	9.73-6
Cm-243	7.86-5
Cm-244	5.52-3
Cm-245	5.08-4
Cm-246	6.31-5
MA/fiss. Pu	0.172
MA/Pu	0.124
Np-237/MA	0.490
Am-241/MA	0.228
Am-243/MA	0.225
Np-chain	0.213

MA = sum of minor actinides;
fiss. Pu = Pu-239 + Pu-241;
Np-chain = Np-237 + Am-241 + Pu-241.

Table B.4
Fuel cycle assumptions

REACTOR SEGMENT OF CYCLE		
Cycle Length	365 days	
Capacity Factor	85%	
Power Rating	1575 MWth	
Core Driver Refuelling	$^1/_3$ per cycle	
Blanket Refuelling	$^1/_4$ per cycle	
RECYCLE SEGMENT OF CYCLE		
Cooling Interval	365 days	
Chemical Separation	done on day 1 of second year	
Blending & Fabrication	done on day 184 of second year	
Re-insertion into reactor	done on day 1 of third year	
CHEMICAL PARTITIONING FACTORS	*% to Product*	*% to Waste*
All TRU isotopes	99.9%	0.1%
Rare Earth Fission Products* (excluding Y, Sm, and Eu)	5%	95%
All Other Fission Products*	0%	100%

* Recommend for Benchmark purposes, recycle zero fission products and send all to waste.
ANL solutions are provided for recommended and for fission product recycle cases in the benchmark volume.

References

[B.1] R. N. Hill, "Calculational Benchmark Comparisons for a Low Sodium Void Worth Actinide Burner Core Design", Proceedings of ANS Topical Meeting on Advances in Reactor Physics, Charleston, SC., U.S.A., March 1992.

[B.2] R. N. Hill, "LMR Design Concepts for Transuranic Management in Low Sodium Void Worth Cores", Proceedings of International Conference on Fast Reactor and Its Fuel Cycle, Kyoto, Japan, October 1991.

List of symbols and abbreviations

barns	nuclear physics' unit for measurement of cross-section $= 10^{-28}$ m^2
β_{eff}	effective delayed neutron fraction
BOEC	*B*eginning *O*f *E*quilibrium *C*ycle
BOL	*B*eginning *O*f *L*ife
BR	*B*reeding *R*atio
CD	cancer dose CD/Ci CD per Curie CD/g CD per gram
Ci	curie, radiation activity unit 1 Ci = 3.7×10^{10} Bq (becquerel)
CR	conversion ratio
EOEC	*E*nd *O*f *E*quilibrium *C*ycle
EOL	*E*nd *O*f *L*ife
Δ, δ	variation
FP	fission product
fiss. Pu	fissile plutonium
g/MWd	gram per megawattday
HM	heavy metal
IFR	*I*ntegral *F*ast *R*eactor
k	neutron multiplication factor
K	degrees Kelvin
kg/y	kilogram per year
λ	radioactive decay constant
LMFBR	*L*iquid-*M*etal *F*ast *B*reeder *R*eactor
LMR	*L*iquid-*M*etal *R*eactor
LWR	*L*ight-*W*ater *R*eactor

MA	minor actinide
MeV	megaelectronvolt
MOX	*M*ixed *OX*ide (uranium and plutonium)
MW	megawatt
MWd	megawattday
MWe	megawatt electric
MWth	megawatt thermal
ν	neutrons per fission
OECD/NEA	OECD Nuclear Energy Agency
Pu	plutonium
PUREX	*P*lutonium *U*ranium *EX*traction
PYRØ	pyrometallurgical-based reprocessing
$	unit of reactivity
T	temperature
Tdk/dT	Doppler coefficient
TRU	Transuranium elements
TRUEX	*TRU EX*traction American method for reprocessing spent fuel
U	uranium
WPPR	Working Party on Physics of Plutonium Recycling
y	year

MAIN SALES OUTLETS OF OECD PUBLICATIONS
PRINCIPAUX POINTS DE VENTE DES PUBLICATIONS DE L'OCDE

ARGENTINA – ARGENTINE
Carlos Hirsch S.R.L.
Galería Güemes, Florida 165, 4° Piso
1333 Buenos Aires Tel. (1) 331.1787 y 331.2391
Telefax: (1) 331.1787

AUSTRALIA – AUSTRALIE
D.A. Information Services
648 Whitehorse Road, P.O.B 163
Mitcham, Victoria 3132 Tel. (03) 9873.4411
Telefax: (03) 9873.5679

AUSTRIA – AUTRICHE
Gerold & Co.
Graben 31
Wien I Tel. (0222) 533.50.14
Telefax: (0222) 512.47.31.29

BELGIUM – BELGIQUE
Jean De Lannoy
Avenue du Roi 202 Koningslaan
B-1060 Bruxelles Tel. (02) 538.51.69/538.08.41
Telefax: (02) 538.08.41

CANADA
Renouf Publishing Company Ltd.
1294 Algoma Road
Ottawa, ON K1B 3W8 Tel. (613) 741.4333
Telefax: (613) 741.5439

Stores:
61 Sparks Street
Ottawa, ON K1P 5R1 Tel. (613) 238.8985
211 Yonge Street
Toronto, ON M5B 1M4 Tel. (416) 363.3171
Telefax: (416)363.59.63

Les Éditions La Liberté Inc.
3020 Chemin Sainte-Foy
Sainte-Foy, PQ G1X 3V6 Tel. (418) 658.3763
Telefax: (418) 658.3763

Federal Publications Inc.
165 University Avenue, Suite 701
Toronto, ON M5H 3B8 Tel. (416) 860.1611
Telefax: (416) 860.1608

Les Publications Fédérales
1185 Université
Montréal, QC H3B 3A7 Tel. (514) 954.1633
Telefax: (514) 954.1635

CHINA – CHINE
China National Publications Import
Export Corporation (CNPIEC)
16 Gongti E. Road, Chaoyang District
P.O. Box 88 or 50
Beijing 100704 PR Tel. (01) 506.6688
Telefax: (01) 506.3101

CHINESE TAIPEI – TAIPEI CHINOIS
Good Faith Worldwide Int'l. Co. Ltd.
9th Floor, No. 118, Sec. 2
Chung Hsiao E. Road
Taipei Tel. (02) 391.7396/391.7397
Telefax: (02) 394.9176

CZECH REPUBLIC – RÉPUBLIQUE TCHEQUE
Artia Pegas Press Ltd.
Narodni Trida 25
POB 825
111 21 Praha 1 Tel. (2) 2 46 04
Telefax: (2) 2 78 72

DENMARK – DANEMARK
Munksgaard Book and Subscription Service
35, Nørre Søgade, P.O. Box 2148
DK-1016 København K Tel. (33) 12.85.70
Telefax: (33) 12.93.87

EGYPT – ÉGYPTE
Middle East Observer
41 Sherif Street
Cairo Tel. 392.6919
Telefax: 360-6804

FINLAND – FINLANDE
Akateeminen Kirjakauppa
Keskuskatu 1, P.O. Box 128
00100 Helsinki

Subscription Services/Agence d'abonnements :
P.O. Box 23
00371 Helsinki Tel. (358 0) 121 4416
Telefax: (358 0) 121.4450

FRANCE
OECD/OCDE
Mail Orders/Commandes par correspondance:
2, rue André-Pascal
75775 Paris Cedex 16 Tel. (33-1) 45.24.82.00
Telefax: (33-1) 49.10.42.76
Telex: 640048 OCDE
Internet: Compte.PUBSINQ @ oecd.org

Orders via Minitel, France only/
Commandes par Minitel, France exclusivement :
36 15 OCDE

OECD Bookshop/Librairie de l'OCDE :
33, rue Octave-Feuillet
75016 Paris Tel. (33-1) 45.24.81.81
(33-1) 45.24.81.67

Dawson
B.P. 40
91121 Palaiseau Cedex Tel. 69.10.47.00
Telefax : 64.54.83.26

Documentation Française
29, quai Voltaire
75007 Paris Tel. 40.15.70.00

Economica
49 rue Héricart
75015 Paris Tel. 45.78.12.92
Telefax : 40.58.15.70

Gibert Jeune (Droit-Économie)
6, place Saint-Michel
75006 Paris Tel. 43.25.91.19

Librairie du Commerce International
10, avenue d'Iéna
75016 Paris Tel. 40.73.34.60

Librairie Dunod
Université Paris-Dauphine
Place du Maréchal de Lattre de Tassigny
75016 Paris Tel. 44.05.40.13

Librairie Lavoisier
11, rue Lavoisier
75008 Paris Tel. 42.65.39.95

Librairie des Sciences Politiques
30, rue Saint-Guillaume
75007 Paris Tel. 45.48.36.02

P.U.F.
49, boulevard Saint-Michel
75005 Paris Tel. 43.25.83.40

Librairie de l'Université
12a, rue Nazareth
13100 Aix-en-Provence Tel. (16) 42.26.18.08

Documentation Française
165, rue Garibaldi
69003 Lyon Tel. (16) 78.63.32.23

Librairie Decitre
29, place Bellecour
69002 Lyon Tel. (16) 72.40.54.54

Librairie Sauramps
Le Triangle
34967 Montpellier Cedex 2 Tel. (16) 67.58.85.15
Tekefax: (16) 67.58.27.36

A la Sorbonne Actual
23 rue de l'Hôtel des Postes
06000 Nice Tel. (16) 93.13.77.75
Telefax: (16) 93.80.75.69

GERMANY – ALLEMAGNE
OECD Publications and Information Centre
August-Bebel-Allee 6
D-53175 Bonn Tel. (0228) 959.120
Telefax: (0228) 959.12.17

GREECE – GRÈCE
Librairie Kauffmann
Mavrokordatou 9
106 78 Athens Tel. (01) 32.55.321
Telefax: (01) 32.30.320

HONG-KONG
Swindon Book Co. Ltd.
Astoria Bldg. 3F
34 Ashley Road, Tsimshatsui
Kowloon, Hong Kong Tel. 2376.2062
Telefax: 2376.0685

HUNGARY – HONGRIE
Euro Info Service
Margitsziget, Európa Ház
1138 Budapest Tel. (1) 111.62.16
Telefax: (1) 111.60.61

ICELAND – ISLANDE
Mál Mog Menning
Laugavegi 18, Pósthólf 392
121 Reykjavik Tel. (1) 552.4240
Telefax: (1) 562.3523

INDIA – INDE
Oxford Book and Stationery Co.
Scindia House
New Delhi 110001 Tel. (11) 331.5896/5308
Telefax: (11) 332.5993
17 Park Street
Calcutta 700016 Tel. 240832

INDONESIA – INDONÉSIE
Pdii-Lipi
P.O. Box 4298
Jakarta 12042 Tel. (21) 573.34.67
Telefax: (21) 573.34.67

IRELAND – IRLANDE
Government Supplies Agency
Publications Section
4/5 Harcourt Road
Dublin 2 Tel. 661.31.11
Telefax: 475.27.60

ISRAEL
Praedicta
5 Shatner Street
P.O. Box 34030
Jerusalem 91430 Tel. (2) 52.84.90/1/2
Telefax: (2) 52.84.93

R.O.Y. International
P.O. Box 13056
Tel Aviv 61130 Tel. (3) 546 1423
Telefax: (3) 546 1442

Palestinian Authority/Middle East:
INDEX Information Services
P.O.B. 19502
Jerusalem Tel. (2) 27.12.19
Telefax: (2) 27.16.34

ITALY – ITALIE
Libreria Commissionaria Sansoni
Via Duca di Calabria 1/1
50125 Firenze Tel. (055) 64.54.15
Telefax: (055) 64.12.57

Via Bartolini 29
20155 Milano Tel. (02) 36.50.83

Editrice e Libreria Herder
Piazza Montecitorio 120
00186 Roma Tel. 679.46.28
Telefax: 678.47.51

Libreria Hoepli
Via Hoepli 5
20121 Milano Tel. (02) 86.54.46
Telefax: (02) 805.28.86

Libreria Scientifica
Dott. Lucio de Biasio 'Aeiou'
Via Coronelli, 6
20146 Milano Tel. (02) 48.95.45.52
Telefax: (02) 48.95.45.48

JAPAN – JAPON
OECD Publications and Information Centre
Landic Akasaka Building
2-3-4 Akasaka, Minato-ku
Tokyo 107 Tel. (81.3) 3586.2016
Telefax: (81.3) 3584.7929

KOREA – CORÉE
Kyobo Book Centre Co. Ltd.
P.O. Box 1658, Kwang Hwa Moon
Seoul Tel. 730.78.91
Telefax: 735.00.30

MALAYSIA – MALAISIE
University of Malaya Bookshop
University of Malaya
P.O. Box 1127, Jalan Pantai Baru
59700 Kuala Lumpur
Malaysia Tel. 756.5000/756.5425
Telefax: 756.3246

MEXICO – MEXIQUE
OECD Publications and Information Centre
Edificio INFOTEC
Av. San Fernando no. 37
Col. Toriello Guerra
Tlalpan C.P. 14050
Mexico D.F.
Tel. (525) 606 00 11 Extension 100
Fax : (525) 606 13 07

Revistas y Periodicos Internacionales S.A. de C.V.
Florencia 57 - 1004
Mexico, D.F. 06600 Tel. 207.81.00
Telefax: 208.39.79

NETHERLANDS – PAYS-BAS
SDU Uitgeverij Plantijnstraat
Externe Fondsen
Postbus 20014
2500 EA's-Gravenhage Tel. (070) 37.89.880
Voor bestellingen: Telefax: (070) 34.75.778

**NEW ZEALAND
NOUVELLE-ZÉLANDE**
GPLegislation Services
P.O. Box 12418
Thorndon, Wellington Tel. (04) 496.5655
Telefax: (04) 496.5698

NORWAY – NORVÈGE
Narvesen Info Center – NIC
Bertrand Narvesens vei 2
P.O. Box 6125 Etterstad
0602 Oslo 6 Tel. (022) 57.33.00
Telefax: (022) 68.19.01

PAKISTAN
Mirza Book Agency
65 Shahrah Quaid-E-Azam
Lahore 54000 Tel. (42) 353.601
Telefax: (42) 231.730

PHILIPPINE – PHILIPPINES
International Booksource Center Inc.
Rm 179/920 Cityland 10 Condo Tower 2
HV dela Costa Ext cor Valero St.
Makati Metro Manila Tel. (632) 817 9676
Telefax : (632) 817 1741

POLAND – POLOGNE
Ars Polona
00-950 Warszawa
Krakowskie Przedmieácie 7 Tel. (22) 264760
Telefax : (22) 268673

PORTUGAL
Livraria Portugal
Rua do Carmo 70-74
Apart. 2681
1200 Lisboa Tel. (01) 347.49.82/5
Telefax: (01) 347.02.64

SINGAPORE – SINGAPOUR
Gower Asia Pacific Pte Ltd.
Golden Wheel Building
41, Kallang Pudding Road, No. 04-03
Singapore 1334 Tel. 741.5166
Telefax: 742.9356

SPAIN – ESPAGNE
Mundi-Prensa Libros S.A.
Castelló 37, Apartado 1223
Madrid 28001 Tel. (91) 431.33.99
Telefax: (91) 575.39.98

Mundi-Prensa Barcelona
Consell de Cent No. 391
08009 – Barcelona Tel. (93) 488.34.92
Telefax: (93) 487.76.59

Llibreria de la Generalitat
Palau Moja
Rambla dels Estudis, 118
08002 – Barcelona
(Subscripcions) Tel. (93) 318.80.12
(Publicacions) Tel. (93) 302.67.23
Telefax: (93) 412.18.54

SRI LANKA
Centre for Policy Research
c/o Colombo Agencies Ltd.
No. 300-304, Galle Road
Colombo 3 Tel. (1) 574240, 573551-2
Telefax: (1) 575394, 510711

SWEDEN – SUÈDE
CE Fritzes AB
S–106 47 Stockholm Tel. (08) 690.90.90
Telefax: (08) 20.50.21

Subscription Agency/Agence d'abonnements :
Wennergren-Williams Info AB
P.O. Box 1305
171 25 Solna Tel. (08) 705.97.50
Telefax: (08) 27.00.71

SWITZERLAND – SUISSE
Maditec S.A. (Books and Periodicals - Livres et périodiques)
Chemin des Palettes 4
Case postale 266
1020 Renens VD 1 Tel. (021) 635.08.65
Telefax: (021) 635.07.80

Librairie Payot S.A.
4, place Pépinet
CP 3212
1002 Lausanne Tel. (021) 320.25.11
Telefax: (021) 320.25.14

Librairie Unilivres
6, rue de Candolle
1205 Genève Tel. (022) 320.26.23
Telefax: (022) 329.73.18

Subscription Agency/Agence d'abonnements :
Dynapresse Marketing S.A.
38 avenue Vibert
1227 Carouge Tel. (022) 308.07.89
Telefax: (022) 308.07.99

See also – Voir aussi :
OECD Publications and Information Centre
August-Bebel-Allee 6
D-53175 Bonn (Germany) Tel. (0228) 959.120
Telefax: (0228) 959.12.17

THAILAND – THAÏLANDE
Suksit Siam Co. Ltd.
113, 115 Fuang Nakhon Rd.
Opp. Wat Rajbopith
Bangkok 10200 Tel. (662) 225.9531/2
Telefax: (662) 222.5188

TURKEY – TURQUIE
Kültür Yayinlari Is-Türk Ltd. Sti.
Atatürk Bulvari No. 191/Kat 13
Kavaklidere/Ankara Tel. 428.11.40 Ext. 2458
Dolmabahce Cad. No. 29
Besiktas/Istanbul Tel. (312) 260 7188
Telex: (312) 418 29 46

UNITED KINGDOM – ROYAUME-UNI
HMSO
Gen. enquiries Tel. (171) 873 8496
Postal orders only:
P.O. Box 276, London SW8 5DT
Personal Callers HMSO Bookshop
49 High Holborn, London WC1V 6HB
Telefax: (171) 873 8416
Branches at: Belfast, Birmingham, Bristol, Edinburgh, Manchester

UNITED STATES – ÉTATS-UNIS
OECD Publications and Information Center
2001 L Street N.W., Suite 650
Washington, D.C. 20036-4910 Tel. (202) 785.6323
Telefax: (202) 785.0350

VENEZUELA
Libreria del Este
Avda F. Miranda 52, Aptdo. 60337
Edificio Galipán
Caracas 106 Tel. 951.1705/951.2307/951.1297
Telegram: Libreste Caracas

Subscriptions to OECD periodicals may also be placed through main subscription agencies.

Les abonnements aux publications périodiques de l'OCDE peuvent être souscrits auprès des principales agences d'abonnement.

Orders and inquiries from countries where Distributors have not yet been appointed should be sent to: OECD Publications Service, 2 rue André-Pascal, 75775 Paris Cedex 16, France.

Les commandes provenant de pays où l'OCDE n'a pas encore désigné de distributeur peuvent être adressées à : OCDE, Service des Publications, 2, rue André-Pascal, 75775 Paris Cedex 16, France.

10-1995

OECD PUBLICATIONS, 2 rue André-Pascal, 75775 PARIS CEDEX 16
PRINTED IN FRANCE
(66 96 02 1) ISBN 92-64-14704-7 - No. 48469 1996